2nd International Conference on Advanced Materials Research and Manufacturing Technologies (AMRMT 2017)

AA005970

IOP Conference Series: Materials Science and Engineering
Volume 229

Phuket, Thailand
2 – 5 August 2017

ISBN: 978-1-5108-4868-9
ISSN: 1757-8981

Printed from e-media with permission by:

Curran Associates, Inc.
57 Morehouse Lane
Red Hook, NY 12571

Some format issues inherent in the e-media version may also appear in this print version.

Copyright© (2017) by the Institute of Physics
All rights reserved. The material featured in this book is subject to
IOP copyright protection, unless otherwise indicated.

Printed by Curran Associates, Inc. (2017)

For permission requests, please contact the Institute of Physics
at the address below.

Institute of Physics
Dirac House, Temple Back
Bristol BS1 6BE UK

Phone: 44 1 17 929 7481
Fax: 44 1 17 920 0979

techtracking@iop.org

Additional copies of this publication are available from:

Curran Associates, Inc.
57 Morehouse Lane
Red Hook, NY 12571 USA
Phone: 845-758-0400
Fax: 845-758-2633
Email: curran@proceedings.com
Web: www.proceedings.com

Table of contents

Volume 229

2017 2nd International Conference on Advanced Materials Research and Manufacturing Technologies (AMRMT 2017)

2–5 August 2017, Phuket, Thailand

Accepted papers received: 5 September 2017
Published online: 22 September 2017

Preface

Metals and processing Technology

Nanomaterials

Advanced Manufacturing

PREFACE

AMRMT 2017 was held in Phuket, Thailand during August 2-5, 2017. AMRMT 2017 was organized by Hong Kong Society of Mechanical Engineers(HKSME), supported by Swinburne University of Technology and other organizations. The conference provides a useful and wide platform both for display the latest research and for exchange of research results and thoughts in broad Advanced Materials Research and Manufacturing Technologies. The participants of the conference were from almost every part of the world, with background of either academia or industry, even well-known enterprise. The success and prosperity of the conference is reflected high level of the papers received.

The proceedings are a compilation of the accepted papers and represent an interesting outcome of the conference. This book covers 4 chapters: Metals and processing Technology; Nanomaterials; Advanced Manufacturing and Mechanical Proeprties, Wearing and Tribology.

We would like to acknowledge all of those who supported AMRMT 2017. Each individual and institutional help were very important for the success of this conference. Especially we would like to thank the organizing committee for their valuable advices in the organization and helpful peer review of the papers.

We sincerely hope that AMRMT 2017 will be a forum for excellent discussions that will put forward new ideas and promote collaborative researches. We are sure that the proceedings will serve as an important research source of references and the knowledge, which will lead to not only scientific and engineering progress but also other new products and processes.

Prof. Alan Kin Tak Lau, Swinburne University of Technology, Australia

Content from this work may be used under the terms of the Creative Commons Attribution 3.0 licence. Any further distribution of this work must maintain attribution to the author(s) and the title of the work, journal citation and DOI.
Published under licence by IOP Publishing Ltd

IOP Conf. Series: Materials Science and Engineering **229** (2017) 011002 doi:10.1088/1757-899X/229/1/011002

COMMITTEES

Conference Co-Chair

Prof. Alan Kin-Tak Lau, Swinburne University of Technology, Australia

Ms. Chan Yuen Shan, Lee Kee Holdings Limited, Hong Kong

Program Committee Chair

Prof. Gong Hao, National University of Singapore, Singapore

Advisory Committee

Prof. Rossi Setchi, Cardiff University, UK

Prof. Jennifer Loy, Griffith University, Australia

Prof. Dong Ruan, Swinburne University of Technology, Australia

Prof. Min Wang, The University of Hong Kong, Hong Kong SAR

Prof. Haitao Huang, Hong Kong Polytechnic University, Hong Kong SAR

Prof. Jun Wang, The University of New South Wales, Australia

Prof. Zhengwei You, Donghua University, China

Prof. Jae-Won Choi, The University of Akron, USA

Prof. Pei Li, The Hong Kong Polytechnic University, Hong Kong SAR

Dr. Yu Dong, Curtin University, Australia

Prof. Shaoyun Fu, Chongqing University, China

Prof. Debes Bhattacharyya, University of Auckland, New Zealand

Prof. Joong Hee Lee, Chonbuk National university, Korea

Prof. Ada Pui Ying Hung, Swinburne University of Technology, Australia

Content from this work may be used under the terms of the Creative Commons Attribution 3.0 licence. Any further distribution of this work must maintain attribution to the author(s) and the title of the work, journal citation and DOI.
Published under licence by IOP Publishing Ltd

IOP Conf. Series: Materials Science and Engineering **229** (2017) 011002 doi:10.1088/1757-899X/229/1/011002

Prof. Xiao Feng Li, Shanghai Maritime University, China

Clara Chan, Lee Kee Group, Hong Kong SAR

Sam TS Chin, Hong Kong Aircraft Engineering Company Limited, Hong Kong SAR

Prof. Jinsong Leng, Harbin Institute of Technology, China

Prof. Thanchanok Ratvijitvech, Mahidol University, Thailand

Prof. Nat Vorayos, Chiang Mai University, Thailand

Prof. Watchanida Chinpa, Prince of Songkla University, Thailand

Prof. Chatchai Putson, Prince of Songkla University, Thailand

Prof. Varunee Premanond, King Mongkut's University of Technology Thonburi, Thailand

Dr. Hing-Ho Tsang, Swinburne University of Technology, Australia

International Technical Committee

Prof. Teik-Thye Lim, Nanyang Technological University, Singapore

Prof. Christopher Cheeseman, Imperial College London, UK

Prof. Zhongwen Yao, Queen's University, Canada

Dr. Mihai Caleap, University of Bristol, UK

Dr. Grzegorz Lisak, Nanyang Technological University, Singapore

Dr. Prof. Qian Shunzhi, Nanyang Technological University, Singapore

Dr. Prabhudev M.S, Governement Polytechnic Kalgi, India

IOP Conf. Series: Materials Science and Engineering **229** (2017) 011003 doi:10.1088/1757-899X/229/1/011003

Peer review statement

All papers published in this volume of *Journal of Physics: Conference Series* have been peer reviewed through processes administered by the proceedings Editors. Reviews were conducted by expert referees to the professional and scientific standards expected of a proceedings journal published by IOP Publishing.

Content from this work may be used under the terms of the Creative Commons Attribution 3.0 licence. Any further distribution of this work must maintain attribution to the author(s) and the title of the work, journal citation and DOI.
Published under licence by IOP Publishing Ltd

IOP Conf. Series: Materials Science and Engineering **229** (2017) 012001 doi:10.1088/1757-899X/229/1/012001

Study on the Property Change of Rhizoma Coptidis and Its Ginger Juice Processed Products Based on 5-Ht Level and Brain Tissues Morphology of Rats

Lingyun Zhong[1,*], Hengli Tong[1], Mu Lv[1] and Yufen Deng[1]

[1]School of Pharmacy, Jiangxi University of Traditional Chinese Medicine, Nanchang 330004, Jiangxi Province, China

*zhongly163@126.com

Abstract. According to the theory of traditional Chinese Medicine (TCM), all Chinese materia medica need to be processed using Pao zhi which is a processing technology before being used in clinic. Ginger juice, made from dried or fresh ginger, is one of the main TCM processing accessories and always used to help change some Chinese materia medica's properties for its warm or hot nature. The purpose of this paper is to discuss the influence of ginger juice on Rhizoma Coptidis (RC) by determining 5–hydroxytryptamine (5-HT) content and observing morphological changes in the harns tissue of rats. Raw Rhizoma Coptidis (RRC), fresh ginger juice processed Rhizoma Coptidis (FGJPRC), dried juice processed Rhizoma Coptidis (DGJPRC), dried ginger juice (DGJ) and fresh ginger juice (FGJ) were prepared using appropriate methods. Immunohistochemical staining was used to observe the distribution of 5-HT and fluorescence spectrophotometry was applied to determine 5-hydroxytryptamine content in the brain tissue of rats. 5 - HT in brain tissue of the rats of RRC group was distributed most densely, with the highest content. Compared to the blank group, RRC and different ginger processed RC groups could lead to increasing content of 5-HT in rat encephalon, and significant differences in RRC. Compared with the RRC, the 5-HT content in rat encephalon in DGJPRC, FGJPRC, FGJ and DGJ groups reduced, and DGJPRC, FGJPRC groups showed significant difference, FGJ and DGJ groups showed extreme significant differences. The research showed that processing with hot, warm accessories would moderate the cold nature of RC. The cold and hot nature of Traditional Chinese Materia Medica could be expressed by the difference of 5-HT contents and morphological changes of rats' brain tissue. Simultaneously, the research showed the different excipient of ginger juice would have different effects on the processing of RC.

1. Introduction

The medicinal properties of TCM are known as drug biased. Drugs through processing can change biased. RC is the dried rhizome which comes from *Ranunculaceae Coptis chinensis Franch., Coptis teeta Wall., Coptisdeltoidea C.Y.Cheng* et Hsiao., often called *"Coptis chinensis*"," *Coptis teeta*", and "*Radix Coptidis Deltoideae*". It is mainly distributed in Sichuan, Hubei and Yunnan province, etc. The nature of RC is bitter and cold. Its medical effects are always considered as on the heart, spleen, liver, stomach, gallbladder and large intestine meridian. It has the function of clearing away heat , eliminating damp evil and removing toxic material [1]. 5-HT is an significant monoamine neurotransmitter within the cerebrospinal axis. 5-HT neuron cell bodies located mainly in the Raphe

Content from this work may be used under the terms of the Creative Commons Attribution 3.0 licence. Any further distribution of this work must maintain attribution to the author(s) and the title of the work, journal citation and DOI.
Published under licence by IOP Publishing Ltd

nuclei of the brainstem and its adjacent areas, their projection fibers are widely distributed in the cerebral cortex, subcortical region, brainstem, cerebellum and spinal cord, the brain region. 5-HT transmitters with different receptor subtypes activate second messengers and ion channels which have effects on the cardiovascular and respiratory systems, body temperature, apetite, sexual behavior, sleep, pain and emotional regulation of a variety of complex physiological and pathological process [2-5]. In addition, 5-HT transmitter also play an important role in learning and memory [6]. Research shows that Chinese herbal compound has different effects in central monoamine neurotransmitter [7], such as 5-HT, prompted by the central monoamine neurotransmitter and cold-heat nature exist correlation. In this article, through different dosages on groups of rats, the effect on brain morphological and distribution and the change of the content of 5-HT research is discussed the influence of different ginger systems of RC properties.

2. Material and equipment

2.1. Reagents

Serotonin standard was purchased from NICPBP and its NICPBP was 111656-200401.The n-Butanol, n-Heptane and Ether were obtained from Tianjin Hengxing chemical reagents company. L–cysteine and o-Phthalaldehyde were obtained from the Shanghai Kaiyang biotechnology company. SP-Histostain TM-Plus kits, the condensed reagents kits together with the serotonin antibody were obtained from Wuhan doctor de sheng bioengineering company and the Hcl, paraformaldehyde, glutaraldehyde were from Sinopharm Group chemical reagents company. Sucrose was prepared too.

2.2. Apparatus

The Cary Eclipse Fluoscence was obtained from the American varian Technology Corporation, China. Desktop low temperature and high speed refrigerated centrifuge purchased from Sartorius (Germany), an Electro-Thermostatic Water Bath from Shanghai Instrument Company was used along with pipettes from Shanghai tuo mo si scientific instrument company. A manual homogenizer from Beijing Younikang biotechnology scientific company, Rayngor ST infrared thermometer from American Raytek, herb grinder from Changsha Xulang grinding mechanical company, RE52CS rotary evaporator from Shanghai Yarong bio-chemical instrument company, SHZ-D (III) rotary water vacuum pump from Gongyi Yuhua instrument company, HH-1508R III computer rapid cold-tome from Huahai scientific and teaching instrument company, Jinhua, Zhejiang, E220LED biologic microscope from Xiamen MOTIC company was used for the experimental analysis of samples.

3. Experimental methods

3.1. Preparation of Rhizoma Coptidis and its processed product

Rhizoma Coptidis was purchased from the Tianqi Hall medicine plants and produced from Sichuan. The preparation of FGIPRC contained mixing 100g RC Tablets with 10% FGJ moistening and stir frying 15 minutes at 120 °C in stir fry- container, sieving and allowing debris to cool. DGJPRC was prepared with a same method, where 100 g of RRC and 10% DGJ were mixed with each other, moistening and stir frying 15 minutes at 120 °C in stir fry- container, sieving and allowing debris to cool.

3.2. Sample preparation

RRC, FGJPRC and DGJPRC powders (150 g each) were sifted through a 5 mm mesh screen, added 10 times the bulk of water and allowed to soak for 60 minutes and decanted after that. The solutions filtered after boiled for 30 minutes. Residues were decanted again by putting in 8 times the bulk of water, filtering after boiling for 30 minutes. Both percolates were gathered and concentrated to 1g•ml-1. RRC, FGJPRC and DGJPRC were extracted separately by 150 ml, the concentration decantation was kept at 4 °C before use.

4. Observation of the morphology of cerebrum tissues

4.1. Grouping and medicine administration

There had 72 female SD rats weighing 180–220 g offered by Experimental Animals Co., Ltd of Cavins in Changzhou (Certificate No.SCXK2011-0003) were grouped into six groups randomly, and each group including 12 rats as follows: blank control group, RRC, FGJPRC, DGJPRC, FGJ and DGJ groups. The groups were treated with RRC, FGJPRC, DGJPRC, FGJ and DGJ at dosage of 3 ml (about containing crude drug 3g) once a day for 3 weeks. In each group, eight rats were put into death based on medicine administration to take cerebrum tissues for the mensuration of 5-HT content. And the rest four rats were used in morphological researches.

4.2. Preparation of section

Cerebrums were harvested (foramen magnum with scissors transection, carefully in the foramen magnum inclined insert the scissors to cut open the parietal bone, hemostatic forceps disc parietal bone broken on both sides, the olfactory bulb was shangdi parietal carefully removed, one side with scissors to cut the optic nerve and agent to the base of skull, the entire block of brain tissue is removed) 1h after the RRC, FGJPR, DGJPRC, FGJ, DGJ, blank groups rats in 22 days for 1 hour were sacrificed by cardiac perfusion-fixation under chloral hydrate anesthesia (fixed liquid: 10% paraformaldehyde, 0. 1% glutaraldehyde, and 0. 1 M phosphate buffer into 250 ml), and were cleaned with ice normal saline. After washing, brain tissues were placed in post-fixation fluid (to remove the glutaraldehyde from the fixing solution) for 3-4 hours; the liquor was substituted by phosphate buffer containing 30% sugar and kept at 4 °C past a night. After being organization fully at the bottom of the sink solution, from the optic decussation to the cortex, a vertical dissection was executed to excise the prosencephalon. Another vertical dissection was executed to cut off a section of the cortex from the hinder boundary of the hypothalamus to the cortex. After the spinal cord and epencephalon were cleared away, the remainder was the stem of brain (including the quadrigeminal body, pons, and medulla oblongata) was removed. The stem of brain was cut into coronal sections and a series of cryostat sections 35 μm thick were used for immunohistochemical staining.

4.3. Operating method of Rat Cardiac perfusion fixation

Two syringes separately filled with normal saline and fixative were adjusted to make the tube free of bubbles. The rats relaxed their front and hind legs after anesthetized, and were laid on the autopsy table with their legs fixed. Abdominal skin was raised with tweezers and a small incision in the right hand side of the abdomen made with scissors. The skin was cut from the incision to the jaw along midabdominal centre line and xiphoid centre line for separation of subcutaneous tissue, and then the skin turned to two sides. The sternum was cut upwards along the midabdominal centre line and xiphoid centre line again, and cut along the two sides of the diaphragm muscle using a hemostat to clamp tightly the skin of sternum and chest, which was rolled outwards to expose the heart fully. The pericardium was opened with the tweezers saline solution used to keep the heart moist. The aorta was isolated the aorta prepared for ligation and infusion using silk thread. A small incision was cut in the left ventricle with ophthalic scissors, then an infusion needle inserted into the left ventricle with a silk thread ligation whilst infusing with the physiological saline at a rate around 20 ml/min. At the same time, the right atrium was cut to discharge the blood. The liver was observed until it gradually became white. After the fixed liquid went into the blood vessels of rats, the rats, limbs appeared to twitch gradually, confirming that the perfusion fluid had gone into the brain

4.4. Immunohistochemical Staining

Staining and color development were performed by ASP kit and concentrated DAB kit, Each tissue parts were put on poly-L-lysine-Prep slides and dried with fan at temperature 58 °C~ 60 °C for 1 hour, fixed with 4% paraformaldehyde for 30 min and incubated in 3% (v/v) H_2O_2 deionizer water for 5–10 minutes to dispel endogenous peroxidase activity. Then, Samples were washed 1 ~ 2 times with water

IOP Conf. Series: Materials Science and Engineering **229** (2017) 012001 doi:10.1088/1757-899X/229/1/012001

and 5% BSA sealing fluid, rinsed with tap water for 20 minutes at room temperature, washed in distilled water for 2 min, followed by soaking in 75% ethanol, 95% ethanol for 3 minutes then anhydrous ethanol soak for 3 min, This was followed by washing in xylene 1 for 5 min, xylene 2 soak for 5 min.

4.5 .Microscopic observation of section

A multiple lensed microscope adjusted to times 40 magnification was used. The brain sections obtained from the RRC, FGJMRC, DGJMRC, FGJ, DGJ and blank groups were observed after color development and several micrographs were taken from each group at biological microscope.

5 Effection of 5-HT concentraction in rats cerebrum tissues

5.1. Configuration of solutions for processing of cerebrum tissue needed

Prepared Acidic n-butanol by mixing 0.85 ml concentrated HCl with 1L n-butanol. Cysteine solution (0.25%) was newly prepared with 0.1 M HCL. 0.002%o-phthalaldehyde-10 M HCl solution was newly prepared before use.

5.2. Preparation of 5-HT standard solution

5-HT standard solutions were prepared by putting 2 mg concentrated 0.01 M HCl in 200ml and mixed. Take solution 0.5 ml, 1.0ml, 1.5 ml, 2.0 ml, 2.5 ml, 3.0 ml, respectively with 0.01 mol/l HCL constant volume to 10 ml, at concentrations of 0.5, 1.0, 1.5, 2.0, 2.5 and 3.0μg/ml and set aside.

5.3. Brain tissue processing and 5 - HT content determination

The rat brains of RRC, FGJPRC, ZGJPRC, FGJ, DGJ and blank groups were fast obtained after putting rats to death. The cerebrum tissues were washed with saline of ice-cold tempreture to clear away epencephalon, the olfactory bulb and blood components. Weighed Half of the cerebrum tissue after filter paper absorbed excess water. Put ten times bulk of the pre-cooled acidic n-butanol into whole cerebrum tissues. Centrifuged samples for 5 minutes and 2.5 ml of supernatant were sub-packed and put in 1 ml 0.1 M HCl. Afterwards vibrated samples for 5 minutes, centrifuged for 5 minutes and probably 0.4 ml of the bottom phase was received. To this liquor, 0.1 ml of newly prepared 0.25% cysteine and 3 ml 0.002% o-phthalaldehyde-10 M HCl was added, mixed and boiled for 10 minutes. This mixed liquor was used to mensurate the concentration of 5-HT after chilled down. The fluorescence at 360 nm was mensurated in a Cary Eclipse fluorescence spectrophotometer (Varian Inc., USA) by using an excitation wavelength of 480 nm [8]. 0.5ml 5-HT standard solutions with different concentration were disposed with ten times of acidic n-butanol and mixed as standard. The remaining steps taken were similar as those used in the samples. Another tube applied as blank was full of 0.5ml 0.01 M HCl, with 5 ml acidic n-butanol put in and mixed each other. The rest steps were used for the samples as a blank pipe were used for zero.

6. Results

6.1. Brain tissue morphology

Comparing with blank control group, the 5-HT-immunopositive neurons in the brain stem raphe nuclei showed denser distribution in RRC, FGJPRC and DGJPRC groups, but compared with RRC group, FGJPRC and DGJPRC stained more sparsely and showed lighter coloring. Compared with the blank group, 5-HT neurons in cell showed lighter coloring and only found in the brainstem raphe nucleus in FGJ and DGJ groups, other areas appeared to be circular and blank. Due to less immune response, and dyeing, DGJ group was less stained than that FGJ group. The results was shown in Figure 1.

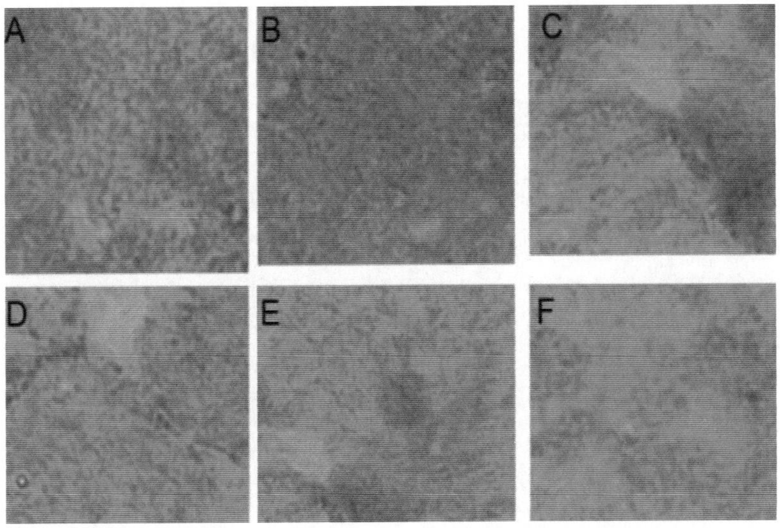

Figure.1 Brain tissue sections treated by six products. A: RRC group; B: Blank group; C: FGJPRC group; D: DGJPRC group; E: FGJ group; F: DGJ group.

6.2. The change of brain 5 - HT content

Using the blank tube to set the spectrophotometer to zero and then fluorescent intensity of the known concentration of 5-HT standard solutions was determined.

The fluorescent intensity of each group measured value were taken into 5-HT standard linear regression equation and used to calculate the content of 5-HT in different samples.

The results showed that, compared to the blank group, RRC and different ginger processed RC groups could lead to increasing content of 5-HT in rat harns, and significant differences in RRC. DGJ and FGJ groups could decrease the content of 5-HT in the harns of rats, without significant difference. Compared with the RRC, the content of 5-HT in rat brain in DGJPRC, FGJPRC, FGJ and DGJ groups reduced, and DGJPRC, FGJPRC groups showed significant difference, FGJ and DGJ groups showed extreme significant differences. FGJ compared with DGJ, DGJPRC compared with FGJPRC, although there were no significant difference, but there was a trend of reduce in the content of 5-HT. The results were shown in table 1, table 2 and figure 2

Table 1. The concentration and fluorescence intensity 5 - HT standard solution

Label	Solution[μg/ml]	Fluorescent intensity
1	0.5	7.421
2	1.0	13.572
3	1.5	20.135
4	2.0	28.98
5	2.5	39.341
6	3.0	43.245

Table 2. 5-HT contents in different samples.

Group	5-HT content[μg/g][$\bar{x}$±s]
Blank	19.05±1.15
RRC	21.11±1.78[a]
DGJPRC	20.02±1.15[b]

FGJPRC	20.11±1.10[b]
FGJ	18.68±1.50[c,d,e]
DGJ	18.38±1.98[c,d,e]

[a] compared with blank group P < 0.05.
[b] compared with RRC group P < 0.05.
[c] compared with RRC group P<0.01.
[d] compared with FGJPRC group P<0.05.
[e] compared with DGJPRC group P<0.05.

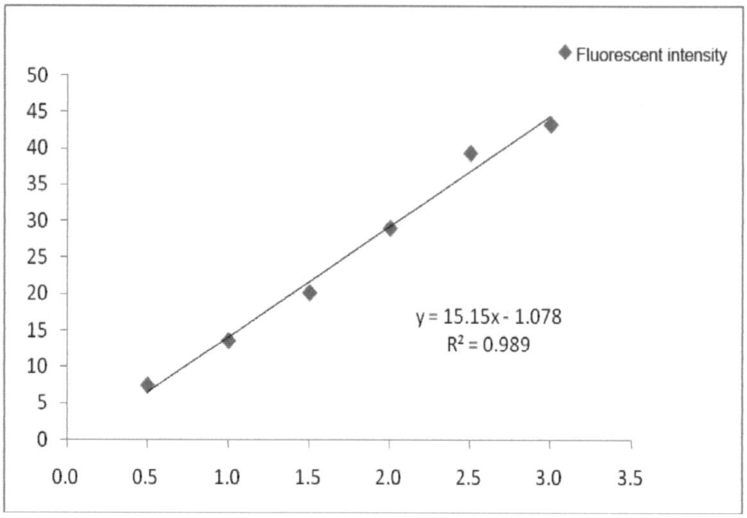

Figure. 2 5-HT standard curve

7. Discussion

As neurotransmitters, 5-HT is mainly distributed in the pineal gland and the hypothalamus, and functions an inhibitory neurotransmitter. The study of Huang etc [9] found that Cold syndrome in brains of rats, NE and DA content decrease, and 5-HT content increased. Wu et al. [10] and Huang et al [11] had demonstrated that, warm medicine could make the brain NE, DA content increase with decreasing 5 - HT content and enhanced excitability. This paper investigated that the FGJ, DGJ, FGJPRC and DGJPRC affect the distribution of brain tissue. Its content was determined to prove the difference of different ginger juices and as well as the processing accessories under the influence of drugs.

The results showed that, 5-HT in brain tissue of the rats of RRC group was distributed most densely, with the highest content. But after processing with different ginger juices, distribution and 5-HT content had the varying degree changes. These results showed that cold coptidis after warm medicinal properties of the ginger processing were eased with the cold eased; among them DGJPRC was less than FGJPRC on 5-HT distribution and content was also reduced. Schematic diagram is shown in Figure. 1,Figure. 2, Table.1 and Table.2

8. Conclusions

RRC by DGJ and FGJ of the two different hot degree as accessories was processing, medicinal properties of two processing varieties were changed obviously, after 5-HT content and changes of brain morphosis of rats to reflect the degree of the two changes, The results "ginger broil coptis by hot resisting cold", "dried ginger is hot in nature" , and "fresh ginger is cold in nature " are consistent with the traditional theory.

Acknowledgements

This study was supported by the National Natural Science Foundation of China (No. 81260643; 81460605)

Conflict of interest
None.

References

[1] Teng H, Choi YH: Optimization of ultrasonic-assisted extraction of bioactive alkaloid compounds from rhizoma coptidis (Coptis chinensis Franch.) using response surface methodology. *Food Chem* **142**:299-305, 2014.

[2] Kaumann AJ: Surprises from a cardiac 5-HT_4TG mouse: spontaneous atrial arrhythmias by endogenous 5-HT of atrial origin Different mechanism of arrhythmias through 5-HT_4 receptors and β-adrenoceptors. *Naunyn Schmiedebergs Arch Pharmacol* **386**:353-6, 2013.

[3] Navailles S, Lagière M, Le Moine C, De Deurwaerdère P: Role of 5-HT2C receptors in the enhancement of c-Fos expression induced by a 5-HT2B/2C inverse agonist and 5-HT 2 agonists in the rat basal ganglia. *Exp Brain Res* **230**:525-35, 2013.

[4] Aronsen D, Bukholt N1, Schenk S: Repeated administration of the 5-HT1B/1A agonist, RU 24969, facilitates the acquisition of MDMA self-administration: role of 5-HT1A and 5-HT1B receptor mechanisms. *Psychopharmacology (Berl)* **233**:1339-47, 2016.

[5] Seyedabadi M, Fakhfouri G, Ramezani V, Mehr SE, Rahimian R: The role of serotonin in memory: interactions with neurotransmitters and downstream signaling. *Exp Brain Res* **232**:723-38, 2014.

[6] Bombardi C, Di Giovanni G: Functional anatomy of 5-HT2A receptors in the amygdala and hippocampal complex: relevance to memory functions. *Exp Brain Res* **230**:427-39, 2013.

[7] Szyndler J, Maciejak P, Turzyńska D, Sobolewska A, Bidziński A, Płaźnik A: Time course of changes in the concentrations of monoamines in the brain structures of pentylenetetrazole-kindled rats. *J Neural Transm (Vienna)* **117**:707-18, 2010.

[8] Möller M, Du Preez JL, Viljoen FP, Berk M, Harvey BH: N-Acetyl cysteine reverses social isolation rearing induced changes in cortico-striatal monoamines in rats. *Metab Brain Dis* **28**:687-96, 2013.

[9] Huang YJ, Qin HZ, Li SY, Shi B, Liang YJ, Liu JH, Liu L, Zheng ZW: 3 kinds of Cloves extracts on the effect of neurotransmitter, cAMP, cGMP in brain of rats with cold. *Zhongguo Shiyanfangjixue* **16**:153-155, 2010.

[10] Wu Y, Pan X, Xu Y, Lu X, He S, He R, Gong M: Optimization of combinations of ginsenoside-Rg1, ginsenoside-Rb1, evodiamine and rutaecarpine for effective therapy of mouse migraine. *J Nat Med* **70**:207-16, 2016.

[11] Huang YQ, Qing HZ,Li SY, Shi B, Liang YJ, Liu JH, Liu L, Zheng ZW: The effect of 3 kinds of clove extract on cold syndrome rat brain neurotransmitter and cyclic adenosine monophosphate, Cgmp. *Zhongguo Shiyan Fangjixue Zazhi* **16**:153-155, 2010.

IOP Conf. Series: Materials Science and Engineering **229** (2017) 012002 doi:10.1088/1757-899X/229/1/012002

Rate Controlling Step in the Reduction of Iron Oxides; Kinetics and Mechanism of Wüstite-Iron Step in H_2, CO and H_2/CO Gas Mixtures

Abdel-Hady A. El-Geassy[1,*]

[1]Central Metallurgical Research and Development Institute (CMRDI), Cairo, Egypt

* elgeassy@yahoo.com

Abstract. Wüstite (W1 and W2) micropellets (150-50 μm) were prepared from the reduction of pure Fe_2O_3 and 2.1% SiO_2-doped Fe_2O_3 in 40%CO/CO_2 gas mixture at 1000°C which were then isothermally reduced in H_2, CO and H_2/CO gas mixtures at 900-1100°C. The reduction reactions was followed by Thermogravimetric Analysis (TG) technique. The effect of gas composition, gas pressure and temperature on the rate of reduction was investigated. The different phases formed during the reduction were chemically and physically characterized. In SiO_2-doped wüstite, fayalite (Fe_2SiO_3) was identified. At the initial reduction stages, the highest rate was obtained in H_2 and the lowest was in CO gas. In H_2/CO gas mixtures, the measured rate did not follow a simple additive equation. The addition of 5% H_2 to CO led to a measurable increase in the rate of reduction compared with that in pure CO. Incubation periods were observed at the early reduction stages of W1 in CO at lower gas pressure (<0.25 atm). In SiO_2-doped wüstite, reaction rate minimum was detected in H2 and H2-rich gas mixtures at 925–950°C. The influence of addition of H_2 to CO or CO to H_2 on the reduction reactions, nucleation and grain growth of iron was intensively studied. Unlike in pure wüstite, the presence of fayalite enhances the reduction reactions with CO and CO-rich gas mixtures. The chemical reaction equations of pure wüstite with CO are given showing the formation of carbonyl-like compound $[Fe_m(CO_2)_n]^*$. The apparent activation energy values, at the initial stages, ranged from 53.75 to 133.97 kJ/mole indicating different reaction mechanism although the reduction was designed to proceed by the interfacial chemical reaction.

1. Introduction

About 90% of iron and steel are produced by Blast Furnace–Basic Oxygen Furnace (BF-BOF) route and the balance is by Direct Reduction (DR) and Smelting Reduction (SR) routes. The BF-BOF route nowadays suffers from the high energy consumption and the CO_2 emission [1]. Great interest is now being devoted to increasing the iron production by Direct Reduction–Electric Arc Furnace (DR-EAF) process. New generation of DR process is recently developed to use coal, coal off -gases and coke oven gases as a source of heat and reducing agents [2-3]. The gaseous reduction of iron oxides is of practical importance, and more in-depth studies were carried out to understand the reaction kinetics and mechanism. The reduction of hematite (Fe_2O_3) to iron is a complicated reaction process due to the formation of lower solid oxides (magnetite and/or wüstite) prior to metallic iron. The experimental conditions have a paramount importance on the kinetic data which are influenced by the gases surrounding the reacting particles [4-5]. Most of previous studies were carried out with H_2 or CO [5-10] and scant attentions were paid to H_2/CO mixtures [7]. Below 570°C, hematite (Fe_2O_3) is reduced to

Content from this work may be used under the terms of the Creative Commons Attribution 3.0 licence. Any further distribution of this work must maintain attribution to the author(s) and the title of the work, journal citation and DOI.
Published under licence by IOP Publishing Ltd

magnetite (Fe_3O_4) then directly to iron and above which, non-stoichiometric wüstite ($Fe_{1-y}O$) is developed prior to iron. The composition of wüstite varies with temperature and the oxygen partial pressure in the gas phase. Although extensive works were done in the reduction of iron oxides directly to iron ($Fe_2O_3 \rightarrow Fe$), the intermediate reaction steps ($Fe_2O_3 \rightarrow Fe_3O_4$), ($Fe_3O_4 \rightarrow Fe_{1-y}O$) and ($Fe_{1-y}O \rightarrow Fe$) did not take such attention. The wüstite–iron transformation step is the most important step in the reduction process and determines the overall reaction rate and the degree of utilization of the reducing gases. In heterogeneous gas-solid reactions, the structure of the intermediate solid products plays significant roles on the reaction kinetics [4-5]. The rate of removal of O_2 from solids in the reduction reactions is a function of partial pressure of gases in the surrounding atmosphere and its activity of the oxide surface [8-10]. The magnitude of the dependence of the rate on the partial pressure is important to determine the order of reaction in the absence of inter-granular gas diffusion in porous product layer [4-11]. In previous studies by the author, Fe_2O_3, CaO-Fe_2O_3 and MgO-Fe_2O_3 compacts were reduced stepwisely at 900-1100°C to either; magnetite in CO/CO_2 then directly to iron in CO [12], or directly to wüstite in CO/CO_2 then to iron in CO [13], or to magnetite then to wüstite in CO/CO_2 and then to iron in CO [14]. The reduction mechanism from one step to another was predicted from correlations between the kinetics, grain structures and gas-solid mathematical model.

The present investigation aimed at studying the isothermal reduction behaviour of pure and SiO_2-doped wüstites at 900-1100°C in H_2, CO and their mixtures using micropellets to overcome the gaseous diffusion effects. The different factors affecting the reduction were intensively studied. The pattern of nucleation and the grain growth of metallic iron on the wüstite surface were used for better understanding the role of each gas in the mixture on the reduction reaction mechanism.

2. Materials and experimental technique

2.1 Materials

Analytical grades of Fe_2O_3 (99.5%) and 2.1% SiO_2-doped Fe_2O_3 compacts were made and sintered at 1250°C for 3hrs, then crushed to produce uniform size of micropellets (150-50µm). Pure Fe_2O_3 and SiO_2-doped Fe_2O_3 micropellets were isothermally reduced to wüstite in 40%CO/CO_2 gas mixture at 1000°C. The produced wüstites named herein after as W1 and W2 respectively. The reducible O_2-content in W1 and W2 was measured from non-isothermal reduction tests of wüstites to iron in H_2 at 1100°C from which the oxygen content can be calculated. The chemical composition of W1 and W2 are $Fe_{0.941}O$ and $Fe_{0.947}O$ respectively. Wüstite compositions were also confirmed by XRF elemental analysis. Fig.1 (a) and (b) shows the morphologies and the grain structures of W1 and W2 respectively. Unlike in W1 where dense grains with faceted structure are formed, porous grains with parallel plates are formed in W2. The surface area of W1 and W2 are 0.0520 and 0.1218 m^2/g respectively.

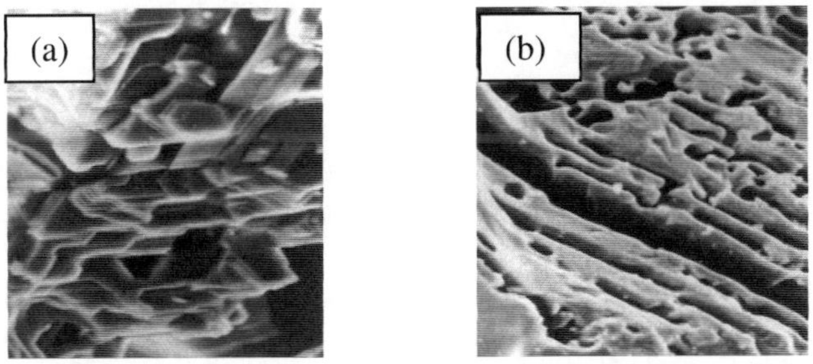

Figure 1. SEM image for wüstite micropellet(a) W1 (b) W2 [5000X]

X-ray diffraction shows the presence of well crystalline structure in W1 and the presence of fayalite and wüstite in W2. Fayalite is formed by the solid-state reaction between the freshly reduced wüstite and SiO_2; ($2FeO + SiO_2 \rightarrow 2FeO.SiO_2$).

2.2 Reduction technique

The reduction of W1 and W2 micropellets in H_2, CO and H_2/CO mixtures was carried out isothermally at 900-1100°C. TG (STA-504, USA, accuracy = ± I μg) was applied for the continuous recording of the O_2-weight loss resulted from the reduction reactions. The schematic diagram showing gas purification system, reduction assembly and the gas pressure controller were given elsewhere [9]. At a given temperature, the extent of reduction at time (t) was calculated from;

$$\%R_t = [\ (W_0 - W_t\ ) / W_{O2}\ ]\ 100 \qquad (1)$$

Where; R_t is the reduction extent at a given time (t), W_0 is the weight of wüstite micropellets, W_t is the weight of reaction product and the W_{O2} is the total weight of the reducible O_2 in wüstite. In some tests, reduction was carried out at 0.1–1.0 atm. Partially and completely reduced products were examined by XRD, XRF, SEM, and RLM. The carbon content in the reduced products was also analysed.

3. Results and discussion

3.1 Reduction of pure wüstite (W1)

Figs. 2 (a) and (b) shows the reduction behaviour of W1 at 1000 and 1100°C respectively. At ≤ 950°C, the reduction in pure CO and in CO-rich gas mixtures (≥ 50% CO) was not completed and stopped at different extents. This is resulted from the thermal decomposition of CO, (reaction 2) and the rate of carbon deposition is higher than the rate of O_2-removal giving false recording in the weight-loss and was corrected from carbon analysis of the reduced samples. The highest rate of reduction was obtained in pure H_2 and the lowest rate was in pure CO. In CO/H_2 gas mixtures, the rate increases with rise with the increase in the H_2 content in the gas mixture. The reduction behaviour is generally characterized by almost constant rate up to certain extent depending on the applied temperature, followed by a gradual decrease in the rate at later stages. At ≥ 95% extents, the reduction in H_2 and H_2 rich gas mixture shows slowing down in the rate due to the sintering of metallic iron which hinder gas diffusion at the FeO/Fe interface and thus, the solid-state diffusion of Fe^{2+} and O^{2-} contributes in the reduction process. The higher the temperature, the higher sintering effect and the lower the reduction rate obtained. On the other hand, in CO and CO-rich gas mixtures; the slowing down in the rate at latter stages was eliminated due to the carbon deposition the subsequent reactions as follows;

$$2\,(CO)_g \rightarrow 2\,(CO)_{ads} \rightarrow (CO_2)_g + C_{gr} \qquad (2)$$

$$3\,Fe + C_{gr} \rightarrow Fe_3C \qquad (3)$$

$$2\,FeO + C_{gr} \rightarrow 2\,Fe + (CO_2)_g \qquad (4)$$

$$3\,FeO + 5\,(CO)_{ads} \rightarrow Fe_3C + 4\,(CO_2)_g \qquad (5)$$

$$Fe_3C + 2\,FeO \rightarrow 5\,Fe + (CO_2)_g \qquad (6)$$

The CO_2 formed from reactions (4-6) may be partially entrapped and build up pressure at the FeO/Fe interfaces. With progress in reduction, the pressure of the entrapped gases increased up to certain level (4-5 atm) above which the particles are disintegrated giving large numbers of smaller micro-grains [15]. The newly formed FeO micrograins could further reacts with CO or with soot carbon to metallic iron and generate CO_2. The disintegration at the Fe/FeO interface retarded the sintering between metallic iron grains in the micropellets. Fig.3 (a,b and c) shows the structure of 75% reduced wüstite with H_2, 50%H_2/CO and CO gases respectively. It shows the formation of relatively larger and denser grains of iron in H_2 than that in H_2/CO mixtures. In pure CO, large numbers of small particles of iron are produced from the disintegrations at the FeO/Fe interfaces. The pattern of nucleation and grain growth of metallic iron on the wüstite surface at the early reduction stages was examined. Figs. 4 (a, b and c) shows SEM image in 25% reduced samples with H_2, 50%H_2/CO and pure CO respectively. In H_2, the wüstite surface is almost nucleated with iron showing small areas of the original wüstite surfaces (Fig.4 a). In 50%H_2/CO, relatively large areas of unreduced wüstite surfaces are also observed (Fig.4 b). In CO, iron is nucleated on small areas of wüstite surfaces showing higher grain growth (Fig.4, c). This indicated that gas composition has a great influence on the mode of nucleation of iron on the wüstite surface and their grain growth. The presence of H_2 in the gas mixture enhanced the nucleation of iron on the entire surface of wüstite and the presence of CO accelerates the grain growth of the formed iron nuclei.

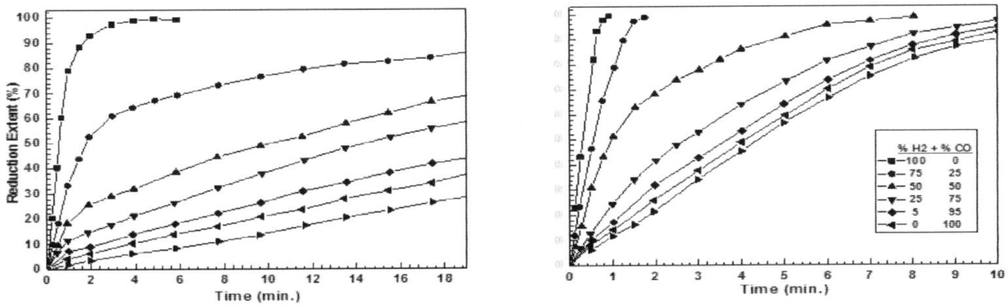

(a) (b)

Figure 2. Reduction path of W1 with H_2, CO and H_2/CO mixtures at; (a) 1000°C; (b) 1100°C

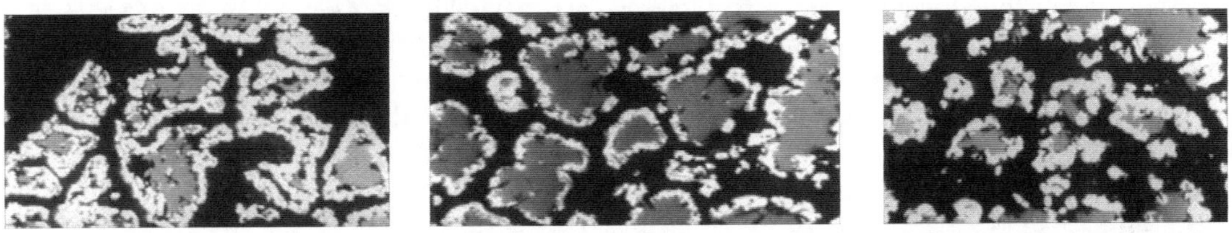

(a) (b) (c)

Figure 3. LM structure of W1, 75% extent at 1000oC in; (a) H_2 (b) 50% H_2/CO, (c) CO [320 X]

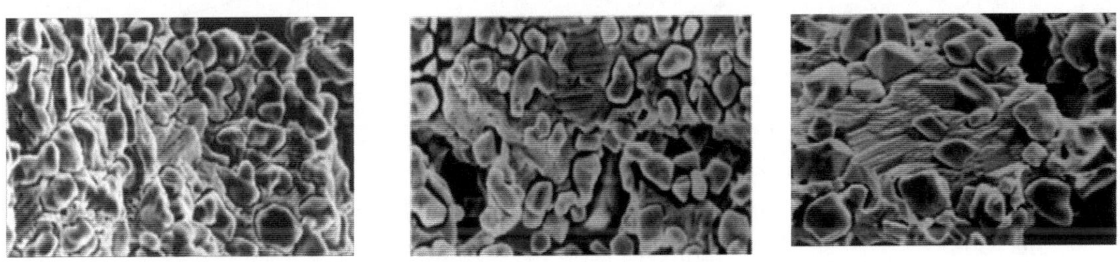

(a) (b) (c)

Figure 4. SEM image of W1, reduced up to 25% at 1000°C in: (a) H2; (b) 50% H2/CO; (c) CO [5000X]

To investigate the effect of H_2 and CO contents in the gas mixture on the reduction, W1 was reduced at 1000°C at different total gas pressures. The total pressure of reducing gas was adjusted at 0.1 up to atmospheric pressure. Fig. 5 (a, b and c) shows the reduction path of wüstite micropellets in H_2, 50%H_2/CO and CO at different gas pressures respectively. It can be seen that, for a given gas composition, the rate of reduction increases with the increase in the total gas pressure. In CO, the reduction was delayed at the initial stages showing the presence of incubation periods at$\leq$ 0.25 atm, (where no weight-loss was recorded). The measured incubation period increases with the decrease in P_{CO} as shown in Fig. 5 (c). Also, it was found that the reduction in pure CO is very low as compared with that in H_2 or even in H_2/CO gas mixture due to the difficulties of the formation of iron nuclei indicating that the reduction reactions of wüstites in H_2 are differed than that in CO.

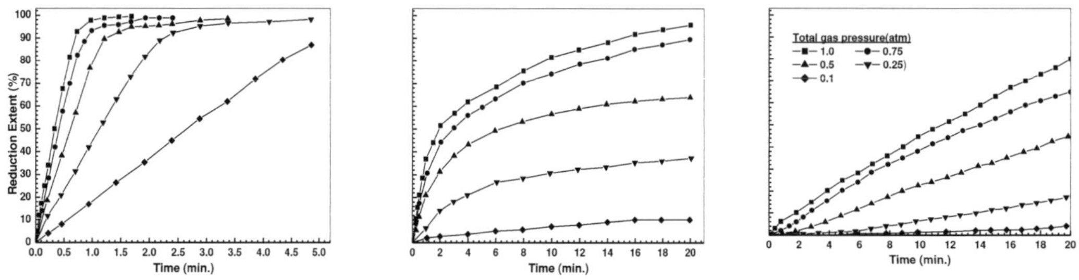

(a) (b) (c)

Figure 5. Influence of total gas pressure on the reduction of W1 micropellets at 1000°C in:
(a) H_2; (b) 50% H_2/CO; (c) CO

Fig.6 shows the variation between the rate of reduction of W1 calculated at different total gas pressure, while Fig.7 depicts the variation between the rates calculated at 900-1100°C and the gas composition. It was found that the rates are gradually decreasing with the increase in the CO content in the gas mixture up to 5-10% H_2 containing gas mixture, below which an abrupt decrease in the rate was observed in pure CO. This indicated that the small addition of H_2 to CO in the mixture greatly promotes the nucleation of iron on the wüstite surface and consequently increases the rate of reduction.

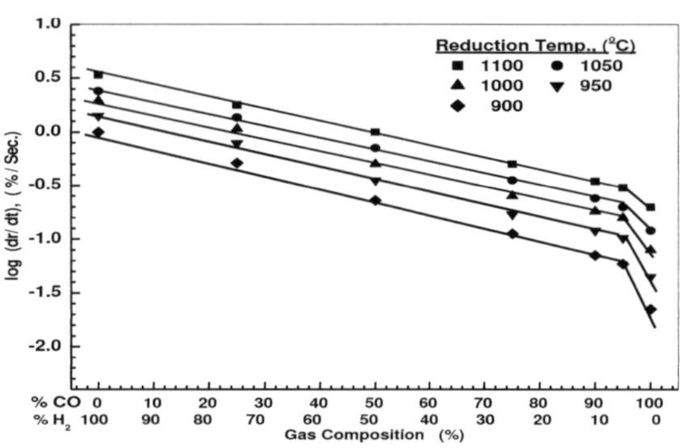

Figure 6 Influence of gas composition on the reduction rate of W1 at 900-1100°C

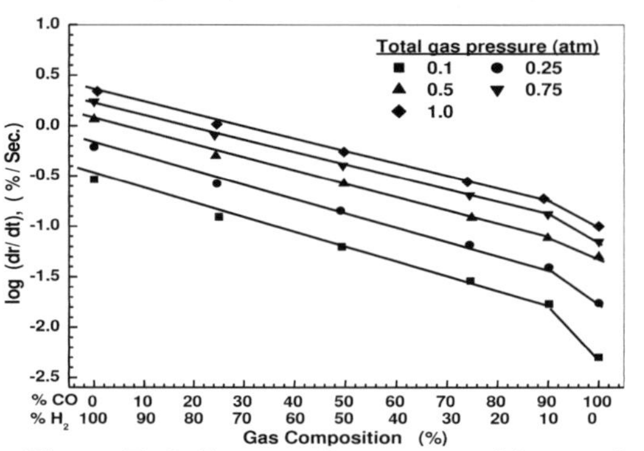

Figure 7. Influence of gas composition on the reduction rate of W1 at 0.1 -1.0 atm

In order to follow up the effect of the H_2 or CO in the gas mixtures on the reduction rate of wüstite, the following tests were adopted at 1000°C;

a- *Role of H_2 in H_2/CO gas mixture*; W1 was reduced in H_2 up to 25% extent, then H_2 was replaced by N_2. After a while, the reduction was continued to iron in pure CO. The reduction behavior of this test is given in Fig.8 and compared with that in pure CO from the beginning. It can be observed that the reduction rate is greatly increased by nucleating the wüstite surface with iron in H_2. The nucleation of iron, at 25% extent, decreased the time of reduction from 40 min. in pure CO to 23 min. in H_2 followed by CO. This indicated that the presence of H_2 in gas mixture greatly accelerated the reduction of wüstite due to the easier formation of iron nuclei on the surface of wüstite which is subsequently grew up in the presence of CO gas.

b- *Role of CO in H_2/CO mixture*; this was carried out by the reduction of wüstite in H_2, CO, H_2/CO and H_2/Ar gas mixtures. In gas mixtures, H_2 partial pressure was kept constant (P_{H2}= 0.25 atm) and CO or Ar are flowing at different partial pressure to adjust the total gas pressure as seen in Table 1. Fig.9 shows the variation between the rate of reduction (dr/dt) at the initial stages in H_2, H_2/CO, CO and H_2/Ar and the gas composition respectively. The result obtained showed that, in pure H_2, H_2/Ar mixtures and CO, the rate decreases gradually in a linear relationship with the increase in P_{H2} or P_{CO}. On the other hand, the rate in H_2/CO gas mixtures showed non-linear relationship between (dr/dt) and gas composition indicating that the rate of reduction in H_2/CO gas mixtures did not obeyed the simple additive formulation in the form;

$$\text{Rmixture} = \text{RCO} . \text{XCO} + \text{RH2} .\text{XH2} \quad (7)$$

where; R_{CO}, R_{H2} are the rates in pure CO and pure H_2 at the pressure corresponding to the pressure of gas mixture, X_{CO}, X_{H2} are the mole fraction of CO and H_2 in the gas mixture. It is interesting to note that the dr/dt in pure H_2 and in H_2/Ar gas mixture are closer to each other indicating that the wüstite surface is saturated by H_2 molecules and Ar has no effect. The rates in H_2/CO mixtures are much lower than that in H_2 and H_2/Ar and the effect increased with the increase in CO mole% giving a higher effect in a gas mixture containing 25%CO and below which the effect is gradually decreased. This can be resulted from poisoning of the wüstite surface by CO molecules that prevent H_2 molecules to react with oxygen from wüstite.

Table 1. Variation between the P_{CO} in the mixture and the total gas pressure (P_{H2}=0.25atm)

P_{H2} + PCO	(0.250 + 0.00)	(0.250 + 0.025)	0.250 + 0.083)	(0.250 + 0.250)	(0.250 + 0.750)
Pressure (atm)	0.250	0.275	0.333	0.500	1.000

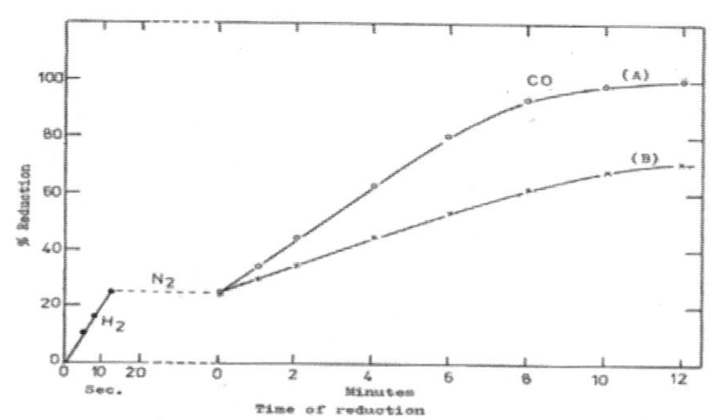

Figure8. Reduction of W1 at 1000°C in; (A) H_2 followed by CO; (B) pure CO

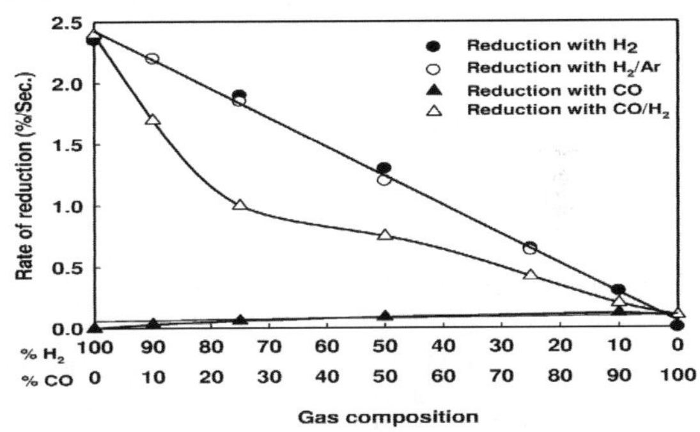

Figure 9. Influence of gas composition, P_{H2}, P_{CO} on the initial rate of reduction at 1000°C.

3.1.1 Chemical reaction of pure wüstite micropellets. Based on the reduction behaviour, pattern of nucleation and grain growth of iron at various gas pressure and the presence of incubation periods in CO at lower gas pressure, the chemical reaction wüstite in H_2 and CO can be presented as follows;

a) Reaction of wüstite with H_2;

$$(H_2)_g = (H_2)_{ads} \rightarrow 2H^+ + 2e^- \quad (8)$$
$$FeO \rightarrow Fe^{2+} + O^{2-} \quad (9)$$
$$2H^+ + O^{2-} \rightarrow (H_2O)_{ads} + (H_2O)_g \quad (10)$$
$$Fe^{2+} + 2e \rightarrow (Fe)_{metal} \quad (11)$$

b) Reaction of wüstite with CO;

$$(CO)_g = (CO)_{ads} \quad (12)$$
$$FeO \rightarrow Fe^{2+} + O^{2-} \quad (13)$$
$$(CO)_{ads} + (O)^{2-} \rightarrow (CO_2)^{2-}{}_{ads} \quad (14)$$
$$n\,(CO_2)^{2-}{}_{ads} + m\,Fe^{2+} \rightarrow [\,Fe_m(CO_2)_n]^* \rightarrow m\,Fe + n\,(CO_2)_{ads} \rightarrow n(CO_2)_{g+} \quad (15)$$

Where; n is the number of $(CO_2)^{2-}$ molecules reacted with m molecules of $(Fe)^{2+}$ to produce the intermediate carbonyl-like compound which is subsequently dissociated to Fe_{metal} and $(CO_2)_g$.

3.2 Reduction of SiO_2-doped wüstite (W2)

Micropellets of W2 were reduced with H_2, CO and H_2/CO mixtures at 900-1100°C. Fig 10 (a) and (b) shows the reduction path obtained at 1000 and 1100°C respectively. It can be seen that the rate of reduction, up to certain extent, increased with the increase in the H_2 content in the gas mixtures. The reduction with H_2 is characterized by an initial constant rate up to a certain extent followed by slowing down till the end of reduction.

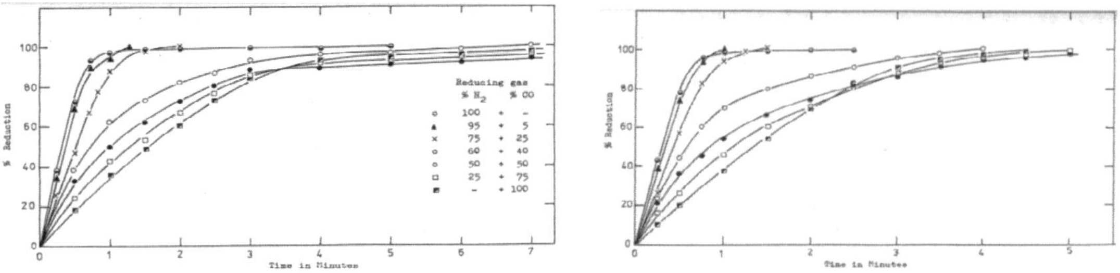

(a) (b)

Figure 10. Influence of gas composition on the reduction of W2 at; (a) 1000°C (b) 1100°C.

The reduction in pure CO shows a continuous decrease in the rate with progress in reduction. At ≤ 1000°C, the reduction was stopped before completion due the higher rate of carbon deposition than the rate of oxygen removal from the reduction. In H_2/CO mixtures, the rate increases with the increase in H_2 content in H_2/CO mixtures. The SEM image of 25% and 85% reduced samples in H_2, 50% H_2/CO and CO at 1000°C are given in Figs. 11 and 12 (a), (b) and (c) respectively. The structure formed at 25% extent in H_2 and H_2/CO mixture is characterized by the formation of numbers of hexagonal iron on the wüstite surface with well defined-triangular pores. While in pure CO (Fig.11 c), iron whiskers were formed, even at early stages of reduction. Iron whiskers were formed by the formation of smaller iron nuclei growing up in a direction perpendicular to the wüstite surface.

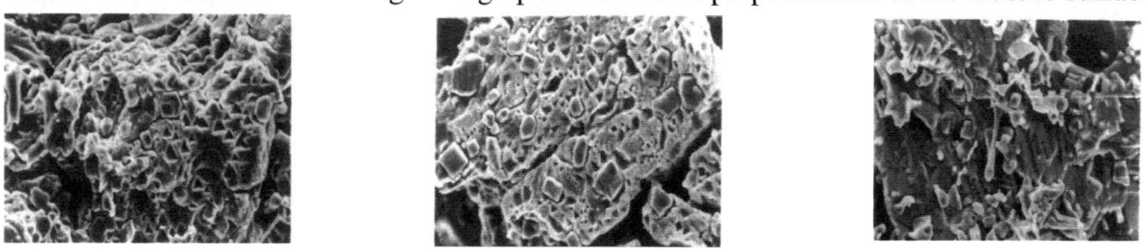

(a) (b) (c)

Figure 11. SEM image of 25% reduced W2 in; (a) H_2; (b) 50%H_2/CO; (c) CO [5000 X]

The structure developed at higher reduction extents in H_2 and H_2/CO mixtures shows the formation of iron layer partially surrounding the wüstites grains as shown in Fig.12 (a, b). In CO, the reduction is characterized by the formation of dense iron layer surrounding wüstite surface forming entrapped wüstite relics as shown in (Fig.12 c). In this case, wüstite relics cannot be reacted directly with gases (H_2 or CO) and the solid-state diffusion are contributed in the reduction process.

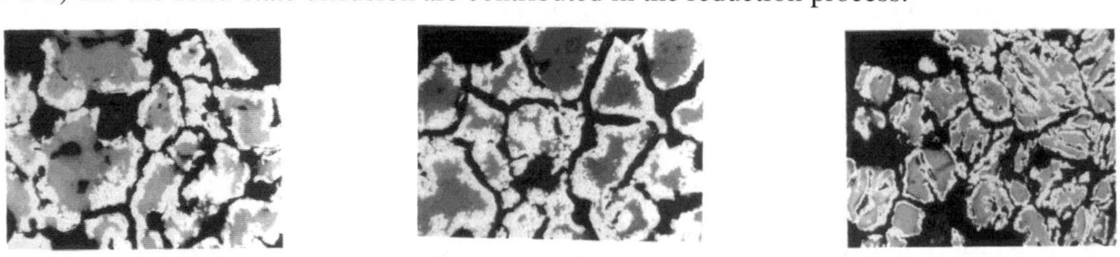

(a) (b) (c)

Figure 12. LM microstructures of 85% reduced W2 in; (a) H_2 (b) 50%H_2/CO (c) CO [320 X]

3.3 Reduction kinetics and mechanism

The nucleation and grain growth of iron on the surface of W1 and W2 in H_2, H_2/CO and CO differs and have a measurable influence on the reduction mechanisms. Moreover, the presence of even smaller molar ratio of H_2 in H_2/CO mixtures promotes the reduction of pure wüstite. The role of either H_2 or CO in the gas mixtures was intensively studied and the morphology and the grain structure showed that W1 is dense than W2 due to the presence of fayalite ($2FeO.SiO_2$). From the reduction bath of W1 and W2, the reaction rate constants (*K*) were calculated from the relationship;

$$R = K A P^{n} \qquad (16)$$

Where; R = rate of reduction (% sec^{-1}), A = surface area of wüstites ($m^2 g^{-1}$), P = reducing gas pressure (atm) and n = order of reaction. The variation between K values and the gas composition of W1 and W2 are depicted in Fig. 13, from which the followings can be noticed;

(a) In pure H_2, the rate constant for W.II is approximately two times lower than that for W.I. This due to the presence of hardly reducible fayalite (7.8 %) in H_2 gas[16]. It is interesting to note that there are significant differences in the pattern of nucleation and growth of iron on the surface of W1 and W2. In W1, the entire surface is almost nucleated with iron whereas the surface of W2 grains is partially nucleated with a small numbers of iron grains since hydrogen is unable to attack the regions where fayalite is present. Thus the reduction is hindered giving lower rate of reduction in W2 than in W1.

(b) The reduction with CO shows that the rate constant (*K*) in W2 is about 3 times higher than that in W1. This could be due to the presence of fayalite phase which is easily reduced in CO rather than in H_2 [16]. Accordingly the reduction in CO would proceed on the regions of the grains surfaces where fayalite is present.

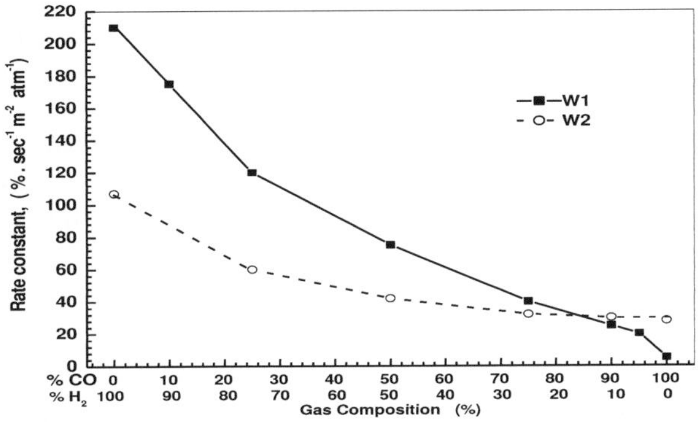

Figure 13. Variation between reaction rate constant (K) and the gas composition at 1000°C.

The apparent activation energy (*Ea*) for the reduction of W1 and W2 at 900-1100°C were calculated from Arrhenius equation [8] and the computed values are given in Table 2

Table 2. Apparent activation energy values (Ea) computed from Arrhenius plots at 900-1100°C

Gas Composition		Ea values (kJ/mole)		Gas Composition		Ea values (kJ/mole)	
% H_2 +	%CO	W1	W2	% H_2 +	%CO	W1	W2
100	0.0	53.75	58.39	25	75	75.65	35.53
75	25	60.61	48.07	10	90	80.84	
50	50	64.79	41.8	5	95	89.24	
				0.0	100	133.97	28.34

For W1, as the CO content in the reducing gas mixture increases the *Ea* values gradually increases up to 95 % CO beyond which an abrupt increase in values were obtained. This indicates that the reduction of W1 is much easier in H_2 than that in CO. For W2, on the other hand, the *Ea* values gradually decrease with the increase in the H_2 content in the gas mixture indicating the easier reduction of W2 in CO than in H_2. This is resulted from the nucleation and grain growth of metallic iron on the wüstite surfaces as shown in Figs 3 and 11 for W1 and W2 at 25% reduction extents (Figs 4 and 11). The values of *Ea* obtained for W1 and W2 indicated several different rate controlling processes ranging from gas diffusion (≤30 kJ/mol) to interfacial chemical reaction (30-50 kJ/mol) and to solid-state diffusion (≥80 kJ/mol). The measured overall reduction rate is not equivalent to the specific rate of chemical reaction but it also depends on the active surface area of the oxide and the diffusion coefficient of the reducing gas. Thus, the activation energy (Ea) obtained does not necessarily corresponding to the specific rate of chemical reaction but more reasonably to whatever process is controlling. In the present the followings conditions were considered to justify the rate controlling step in the reduction of wüstite to iron;

a) Micropellets (150 - 50μm) were used to minimize the resistance to gas diffusion in the inter-granular pores. Accordingly, the gaseous diffusion effect can be minimized.

b) The reduction of wüstites was carried out directly to iron (one step reduction) to eliminate the complication arises from the formation of intermediate lower oxides.

c) At the early stages of reduction, the reduced metallic iron formed around the individual wüstite grains is very thin and consequently, the resistance to the solid state diffusion of Fe^{2+} and O^{2-} can also be neglected.

Under these conditions, the interfacial chemical reaction is most probably controlling the reduction of wüstite. The differences in the apparent activation energy values obtained could be related to the nucleation and the grain growth of iron and thus the *Ea* values cannot be taken as decisive evidence to predict the reduction mechanism but can be monitored to an indicative for the process.

4. Conclusions

The results obtained concluded the followings;

a) The rate of reduction increased with rise in temperature and with the increase in the H_2 content in the gas mixture. Small addition of H_2 to CO greatly accelerated the reduction rate, while the addition of CO to H_2 retarded the reduction process.

b) The nucleation of iron on the wü stitesurface was dependant on the wü stitescomposition and its morphology. In H_2 and H_2-rich mixtures, iron is nucleated on the pure wü stitesurface, while in CO, iron whiskers were formed in W2 due to the presence of fayalite ($2FeO.SiO_2$).

c) The reduction of W1 in H_2, H_2/Ar increased linearly with the increase in the H_2 content in the mixture. In H_2/CO gas mixtures, non-linear relation was obtained showing that the measured rate (dr/dt) is not additive.

d) The apparent activation energy values calculated for W1 and W2 showed two different trends. For W1 the values gradually increased with the increase in CO content in the gas mixture up to 95% followed by an abrupt decrease in pure CO. For W2, the Ea values were gradually decreased with the increase in CO content due to the presence of fayalite which promote the reduction in CO.

e) The reduction of W1 and W2 at the initial stages is most probably controlled by interfacial chemical reaction mechanism although the discrepancy in the values of Ea obtained. This is arises from the use of micropellets and the formation of thin film of iron on the wü stite surface at 5% reduction extent. The magnitudes of the apparent activation energy values could not be taken as a decisive conclusion but could be taken as an indicative approach for the rate controlling mechanism.

References

[1] Geerdes M, Toxopeus H and Vliet C 2009 *"Modern Blast Furnace Ironmaking"*, 2nd Edn 164, Delft . IOS Press BV.
[2] Chalttetjee A 2010 *"Sponge Iron Production by Direct Reduction of Iron Oxide"*, Textbook, PHI
Learnibg Private Limited, New Delhi, P 353.
[3] Mousa EA, Senk D and Babich A 2010 *Steel Res. Int.* **81** (9), 706
[4] Bradshaw AV 1970 *Trans IMM*, **79**, C281.
[5] Unal A and Bradshaw AV 1983 *Met. Trans.*, **14B**, 743.
[6] Shehata KA and. Ezz SY 1973 *Trans. Inst. Min. Met., Sec.* **C 82**, 38.
[7] El-Geassy AA, Shehata KA and Ezz SY 1977 *Trans ISIJ*, **17**, 629.
[8] El-Geassy AA 1986 *J. of Mater Sci* , **12**, 3889.

[9] El-Geassy, Rajakumar V 1985 *Trans ISIJ,* **25**, 449
[10] El-Geassy AA 1985 *Trans ISIJ,* **25**, 1037
[11] Mousa EA, Omar AA, Nasr MI and El-Geassy 2007, *Steel Res.Int.*, **78** (8), 579.
[12] El-Geassy AA 1997 *J. IS I J Int,* **37**, 844-853.
[13] El-Geassy AA 1998 *Scand, J. of Metallurgy,* **27** (.5), 205-213.
[14] El-Geassy AA 1999 *J. Ironmaking and Steelmaking,* **26,** (1), 41-52,
[15] Bohnenkamp K Riecke E and Engell HJ 1967 *Arch Eisenhüttenw.,* **38**, 240
[16] Mazanek E and Jasieneka S 1966 *JISI,* **204**, 344.

IOP Conf. Series: Materials Science and Engineering **229** (2017) 012003 doi:10.1088/1757-899X/229/1/012003

Experimental Investigations on the effect of Additive on the Tensile Properties of Fiber Glass Fabric Lamina

Nava Sai Divya A[1,*], Dr Raghu Kumar B[2], and Dr Lakshmi Narayana G[3]

[1]Research Scholar, KL University, Vaddeswaram, Guntur, Andhra Pradesh, India
[2]Dept. of Mech. Engg., PVP Siddhartha Inst. of Tech., Kanuru, Andhra Pradesh, India
[3]Faculty of Mechanical Division, School of Engg., Bahrain Polytechnic, Kingdom of Bahrain

[*]ryali_divya@yahoo.com

Abstract. The main objective of this work is to investigate the effect of additives on tensile behaviour of fiber glass fabric at lamina level to explore an alternative skin material for the outer body of aerospace applications and machines. This experimental work investigates the effect of silica concentration in epoxy resin lapox L-12 on the tensile properties of glass fabric lamina of 4H-satin weave having 3.6 mm thickness. The lamina was prepared by using hand lay-up method and tests were conducted on it. Various tensile properties values obtained from experimentation were compared for four glass fiber lamina composites fabricated by adding the silica powder to resin bath. The effect of variations in silica concentration (0% SiO_2, 5% SiO_2, 10% SiO_2 and 15% SiO_2) on the tensile properties of prepared material revealed that maximum stiffness was obtained at 15% and yield strength at 10% SiO_2 concentration in glass fiber lamina. Increasing the silica concentration beyond 10% had led to deterioration in the material properties. The experimentation that was carried out on test specimen was reasonably successful as the effect of silica powder as an additive in glass fiber lamina enhanced the mechanical properties up to certain limit. The underpinning microscopic behaviour at the source of these observations will be investigated in a follow up work.

1. Introduction

Glass fiber/epoxy composites find extensive use in many applications such as aerospace, ship building, wind turbine blades, automotive sectors, thermal and electrical insulations, manufacturing sectors etc. These are used instead of metallic composites because of their inherent properties like low manufacturing cost, high strength, high stiffness, high chemical resistance, and good insulating properties [1].

Various studies have shown that glass fiber composites are excellent at handling tensile stress but are, however, weak in terms of enduring compression due to its brittle property and have some drawbacks like low elastic modulus, low fatigue strength and sensitivity to abrasion [2]. The silica content in glass fiber retains its strength at high temperature and has higher fatigue strength for that reason S-glass having high silica content is used extensively for aerospace applications [3].

The objective of the work is to investigate the effect of silica concentration in epoxy resin on the tensile properties of glass fabric lamina of 4H-satin weave having 3.6 mm thickness. Some advantages of the modified composites were anticipated such as improvements in tensile properties. The paper presents results associated with the influence of silica additive to glass fiber/Epoxy composite at lamina level on tensile behaviour.

Content from this work may be used under the terms of the Creative Commons Attribution 3.0 licence. Any further distribution of this work must maintain attribution to the author(s) and the title of the work, journal citation and DOI.
Published under licence by IOP Publishing Ltd

IOP Conf. Series: Materials Science and Engineering **229** (2017) 012003 doi:10.1088/1757-899X/229/1/012003

2. Materials and Experimental Procedure

This study involved epoxy glass fabric as reinforcements whose specifications are shown in Table 1, epoxy resin lapox L-12 and hardener K6 as shown in Table2 and Table3 respectively.

Table 1. Glass Fibre Properties.

Description	Specification
Variety	13ml Fibre Glass Fabric (epoxy)
Type of weave	4H Satin
Width/Thickness	40″/3.6 mm
Construction	WARP 48 threads/inch; WEFT 36 threads/inch
Weight per sqm.	457.100 grams
Breaking strength per 50mm	WARP 366.000 grams; WEFT 275.000 grams

Table 2. Epoxy Resin Lapox L-12 Specifications.

Specification	ResultUnit	Requirementrange
Color on Gardnr Scale	GS	NMT 1
Viscosity @25°C by Brookfield	mPas	10000 to 12000
Epoxy value	Eq/kg	5.25 to 5.45
Hydrolisable Chlorine-PO	%	NMT 0.1
Marten's value	25°C	145 to 200
Volatile content	% w/w	NMT 0.55

Table 3. Epoxy Hardner Lapox K-6 Specifications.

Specification	ResultUnit	Requirementrange
Color on Gardner Scale	GS	NMT 3
Refractive Index	NA	1.494 to 1.5
Water by KF	%	NMT 1
Shear Strength	Kg/mm^2	1.4 to 100

The epoxy glass fabric and epoxy resin were supplied by Harsh-Deep Industries and Atul Ltd, Gujarat. Silica powder was used as additive with varing concentration from 0% to 15% and its influence on the tensile properties of the epoxy glass fiber composite at lamina level of 3.6 mm thickness (single layer) was investigated by experimental method. The lamina was prepared by using hand lay-up method. The mould metal plate was cleaned with 80 grade fine emery paper in order to remove rust and unwanted particles, greased with acetone with cotton waste and applied mould release agent (wax polish) and allowed it to dry for few minutes. Laminate was fabricated by pouring mixture of epoxy resin of 66 grams and hardener of 9.9grams by weight in a glass fiber of dimensions 380 mm×380 mm×3.6 mm.The resin that was pored on the fiber should be distributed uniformly and rolling device was used for required bonding[4]. The samples were removed from mould and four test specimen were cut from glass fiber lamina (sample) as per ASTM-D882 standards as shown in Figure1 for various silica concentrations (5%, 10%, & 15% by weight) after a curing period of 24 hours at room temperature.To avoid degradation of material properties, a product called AGM-9 has been used as a stabilizer for fiber composite properties and prevent the fiber strength to drop due to the effect of moisture absorption from the air. Clear dimensions of test specimen for each silica concentration have been shown in Figure 2. Total 16 specimens were cut by using diamond tipped blade saw for good surface finish[5].

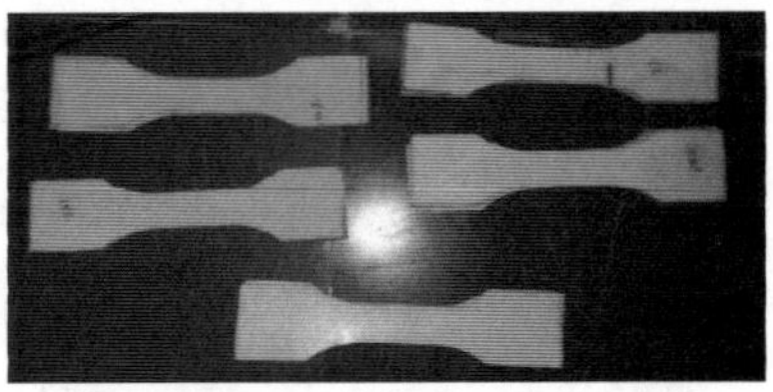

Figure 1. Specimen prepared for experimentation

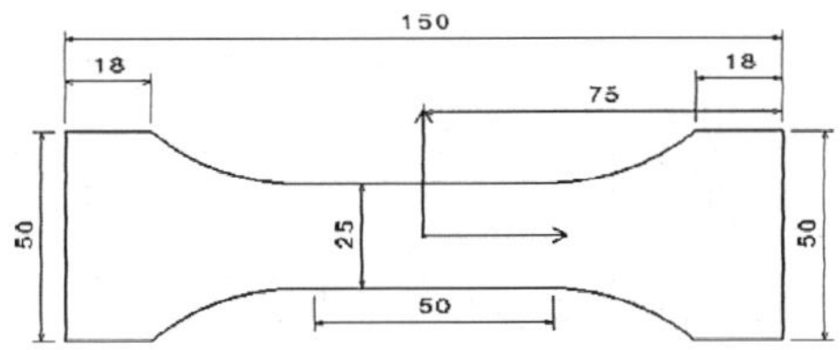

Figure 2. Dimensios of test specimen

The obtained specimens were provided high quality surface finish by using emery paper. The tensile specimens that were cut from the lamina were subjected to uni-axial tension using Universal Testing Machine having 20kN capacity as shown in Figure3 with the surrounding room temperature of 32°C.

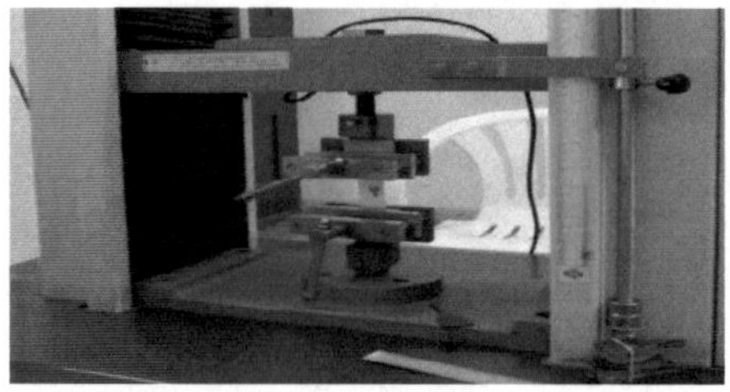

Figure 3. Universal Testing Machine with 20kN capacity

The load was applied till fracture with a grip displacement rate was maintained at 1mm/min. Tests were conducted 4 times for each silica concentration and the average values of the tensile properties of composite laminas with varying silica concentration have been calculated and tabulated as shown in Table4.

Table 4. Average tensile properties obtained from tests conducted on specimen with variaions in silica concentrations.

Silica Concentration	Max. elongation (mm)	Tensile strength @yield point(MPa)	Tensile strength @break point(MPa)	Elastic modulus (MPa)
0%	2.23	123.72	125.37	3728.13
5%	2.1	159.69	160.07	4004.13
10%	1.78	191.04	192.03	4432.56
15%	1.61	145.41	146.92	4872.22

3. Results and Discussions

The infinite number of combinations of fiber and matrix, enormou variety of spatial arrangements of the fiber and their volume fraction make mechanical characterization a complex scenario for composite materials [6]. Mechanical properties such as tensile, compresssive, flexural, hardness, fatigue etc. have been investigated on various composite materials analytically and experimentally [7] to predict their behavior.The four types of laminas with varying additive concentration (0% SiO_2, 5% SiO_2, 10% SiO_2 and 15% SiO_2) have been tested by applying uniaxial tensile load. Table5 summarises the average values of various tensile properties of four test specimen per each silica concentration varying from 0% wt to 15% wt of resin bath.The results show that there is significant change in tensile properties with the change in additive concentration.

Table 5. Average values of various tensile properties of four test specimen.

MechanicalProperty	SilicaConcentration	Specimen 1	Specimen 2	Specimen 3	Specimen 4
Tensile strength at yield (MPa)	0%	127.63	124.68	112.78	129.80
	5%	152.35	164.23	155.66	166.52
	10%	194.46	200.38	177.70	191.61
	15%	152.48	138.00	142.00	149.13
Tensile strength at break (MPa)	0%	130.57	125.53	114.57	130.77
	5%	152.35	164.23	157.07	166.62
	10%	195.47	200.38	179.67	192.57
	15%	153.62	139.09	145.84	149.13
Modulus of elasticity (MPa)	0%	3778.75	3633.88	3711.24	3788.64
	5%	3994.97	3880.16	4028.45	4112.91
	10%	4346.87	4283.45	4526.54	4496.02
	15%	4793.19	4909.73	4804.00	4981.94

Maximum elongation (mm)	0%	2.38	2.44	2.03	2.06
	5%	2.32	2.06	2.01	2.01
	10%	1.48	1.64	2.06	1.95
	15%	1.85	1.53	1.43	1.64

For analysis purpose, the graphs were drawn to show the influence of silica additive on tensile properties. The changes in yield strength with respect to silica concentration have been shown in Figure 4.The yield strength was found to be maximum at 10% silica concentration with 200.389 MPa. Though there was drop in yield strength for 15% silica concentration, it was higher than 0% silica concentration (lamina without additive) by 17.52%.

The changes in tensile strength at breaking point with respect to silica concentration have been shown in Figure 5. As can be seen, this graph shows the same behavior as yield strength. The tensile strength at breaking was found to be maximum at 10% silica concentration with a value of 200.38MPa. Though there was a drop in tensile strength at breaking for 15% silica concentration when compared with 5% and 10% silica concentration, it was higher than 0% silica concentration (lamina without additive) by 17.20%. It was observed that yielding strength and breaking strength values were found to be same for some specimen. This is due to the fact that brittle property of the composite. Further analysis has to be carried out to observe the yielding point for such specimens.

From the tests that were conducted, it was clearly found from Figure 6 that there was significant increase in the stiffness of the composite due to the fact that increase in silica concentration. The effect of variations in silica concentration (0% SiO_2, 5% SiO_2, 10% SiO_2 and 15% SiO_2) revealed that maximum stiffness is obtained at 15% SiO_2 concentration in glass fiber lamina.

The variation of maximum elongtion of the test specimen with the variations in silica concentrations have clarly shown in Figure 7. It was observed that maximum elongation ws found to be occurred at 0% silica concentration and gradually decreases as silica concentration increases. Though there was a raise in maximum elongation for 15% silica concentration, it was lower than 0% silica concentration (lamina without additive) by 27.8%.

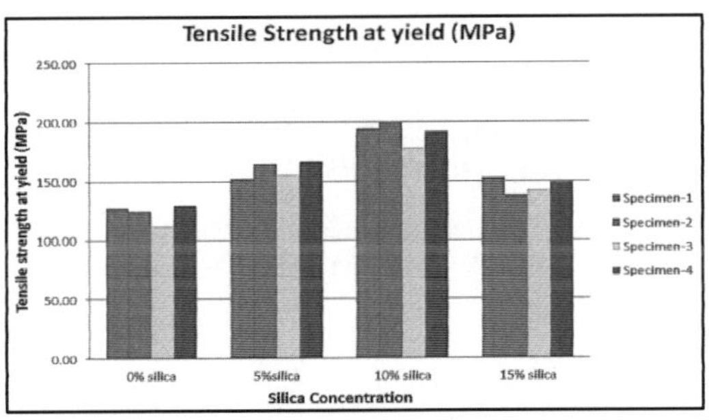

Figure 4. Variation of Tensle strength at yield with variations in silicon concentration

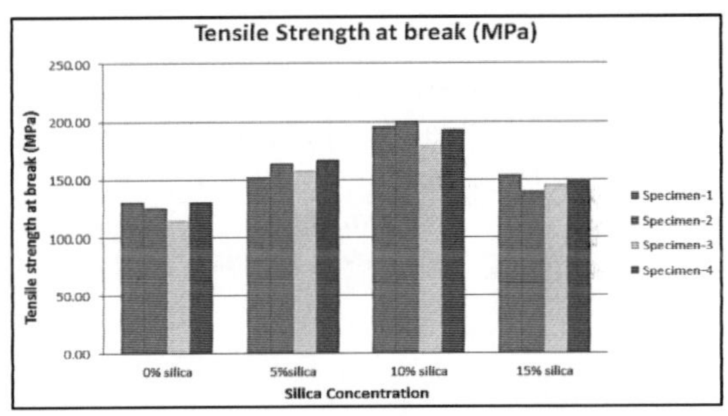

Figure 5. Variation of Tensle strength at break with variations in silicon concentration

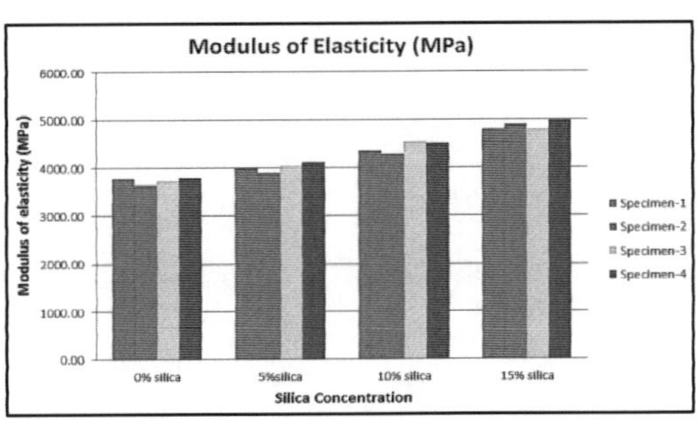

Figure 6. Variation of modulus of elsticity with variations in silicon concentration

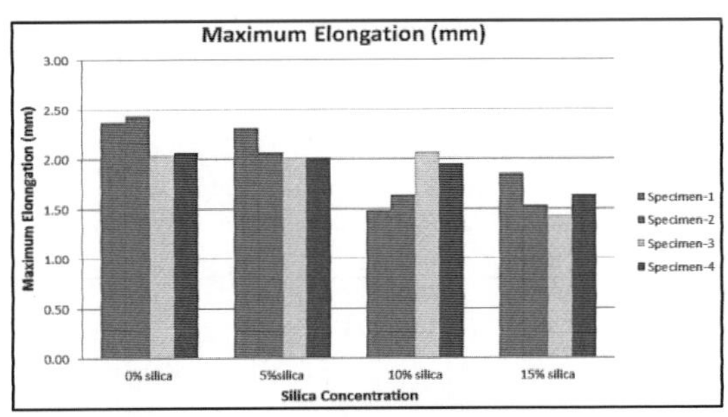

Figure 7. Variation of Maximum elongation with variations in silicon concentration

4. Conclusions

Based on the experimental results on tensile properties of glass/epoxy composite lamina with different additive concentration, the following conclusions were drawn. Intensity in increasing yield strength was observed with the increase in silica concentration up to 10%. Furthermore, there was a drop in yield strength for 15% silica, yet the yield strength obtained at 15% silica concentration was more than that of the composite without additive. The tensile strength at breaking was found to be maximum at 10% silica concentration. There were no significant changes in elognation percentage with the influence of additive. The elastic modulus was found to be increased with increase in additive concentration, stiffness of the composite was enchanced. The epoxy glass fiber lamina had shown good results for yield strength with 10% silica concentration and 15% for stiffness.

Acknowledgments

The authors would like to acknowledge SphoorthyEngg. College (Hyderabad, India) for providing all necessary facilities and also for giving us permission to use 20 kN capacity of Universal Testing Machine for our experimentation work. Authors would also thank Dr. Subramanian C.Pillai, Faculty of Mechanical Engineering, Bahrain Polytechnic, Kingdom of Bahrain for his inspirational and valuable inputs and continuous moral support given to us in completing this paper successfully.

References

[1] Tomita Y, Morioka K, Iwasa M. “Bending fatigue of long carbon fiber-reinforced epoxy composites”,*Mater SciEng A* 2001:319-321:679-82.

[2] Erchard, 2006 Gunter,“Designing with plastics”, *Trans. Martin Thompson. Munich: Hanser publications.*

[3] PD Mangalgiri,“Composite materials for aerospace applications”,*Bulletin of Materials Science*, Volume 22, Issue 3, pp 657-664.

[4] Nestor Perez, “Fracture Mechanics”, *Kluwer Academic publishers NewYork, Bostan, Dordrecht, London, Moscow.*

[5] J.M.Ferreira, J.T.B.Pires, J.D. Costa, Z.Y.Zhang, “Fatigue damage analysis of aluminized glass fiber composites”, *Materials Science and EngineeringA407*(2005)1-6,Elsevier(2005).

[6] K. Harries, “Fatigue behavior of bonded FRP used for flexural retrofit”, *Proceedings of the International Symposium on Bond Behavior of FRP structures (BBFS 2005)*

[7] Ascione.F, Feo.L, Maceri F,“An experimental investigation on the bearing failure load of glass fiber/epoxy laminates”, *Composites Part B: Engineering,*2009, pp 88-97.

IOP Conf. Series: Materials Science and Engineering **229** (2017) 012004 doi:10.1088/1757-899X/229/1/012004

Fundamental Characteristics of Grooving Aiming at Reduction of Kerf Loss Using an Ultrafine Wire Tool

Satoshi Sakamoto[1,*], Keitoku Hayashi[1], Masaya Gemma[1], Yasuo Kondo[2], Kenji Yamaguchi[3], Takao Yakou[1] and Susumu Arakawa[4]

[1]Yokohama National University, 79-2 Tokiwadai, Hodogaya-ku, Yokohama, Japan
[2]Yamagata University, 4-3-16 Jonan, Yonezawa, Japan
[3]Yonago National College of Technology, 4448 Hikona-cho, Yonago, Japan
[4]Kurashiki Boring Kiko Co., Ltd., 2-4-20 Matsue, Kurashiki, Japan

*sakamoto-satoshi-tv@ynu.ac.jp

Abstract. Thinning of silicon wafers and reduction of kerf loss can minimize the manufacturing costs of semiconductor products. Currently, the volume of kerf loss is about the same as the volume of the wafer itself. Therefore, we study slicing techniques for silicon wafers that result in reduced kerf loss by using an ultrafine wire tool and fine abrasive grains. As a first step, grooving characteristics using an ultrafine tungsten wire tool and fine abrasive grains are investigated in this paper. A borosilicate glass is used as the work material. The main conclusions are as follows: Precision machining using ultrafine wire tool is possible and the kerf loss decreases because the groove width decreases. However, a larger diameter of the wire tool results in a deeper groove. A faster relative speed produces a shorter wire tool lifetime, but a deeper groove. To supply enough abrasive grains to the machined portion, it is necessary to use abrasive grains having a suitable particle size for the specific diameter of the ultrafine wire tool.

1. Introduction

Multi-wire sawing is an excellent slicing method for hard and brittle materials, such as silicon ingots, magnetic materials, ceramics, and sapphire [1]. Multi-wire sawing has become the mainstream slicing system for large-diameter ingots in recently years.

Presently, the volume of kerf loss is about the same as the volume of the wafers when several thin wafers are sliced from a silicon ingot using a multi-wire saw [2]. Thinner wafers and a decrease in kerf loss can minimize manufacturing costs [3]. Additionally, from the viewpoint of resource savings, drastic reduction of kerf loss is desired. The diameter of the wire tool and particle size of the abrasive grain that are used for slicing greatly affect the volume of the kerf loss. In other words, when the wire tool is thin and the abrasive particles are small, the kerf loss can be drastically reduced. However, there are few research reports of slicing performance using the fine wire tool because of frequent breaking of the wire tool and deterioration in slicing performance. On the other hands, the slicing method of silicon wafers using electric discharge machining or chemical processing, such as etching to reduce kerf loss, have been proposed [4, 5]. However, these methods still have problems because it is difficult to slice a high resistance material or use a high processing speed. Therefore, these methods are still not practical for regular use.

Content from this work may be used under the terms of the Creative Commons Attribution 3.0 licence. Any further distribution of this work must maintain attribution to the author(s) and the title of the work, journal citation and DOI.
Published under licence by IOP Publishing Ltd

IOP Conf. Series: Materials Science and Engineering **229** (2017) 012004 doi:10.1088/1757-899X/229/1/012004

The main purpose of this study is to reduce kerf loss using multi-wire sawing. Therefore, the possibility of using an ultrafine wire tool and fine abrasive grains as the slicing tool for multi-wire sawing are examined. Both slicing and grooving methods are applied using the relative motion of the wire tool and the work material. In other words, both machining mechanisms are the same mechanism. Therefore, as a first step to realize slicing with less kerf loss, we discuss the fundamental grooving characteristics using ultrafine wire tools and fine abrasives in this paper.

2. Experimental procedure

Figure 1 showed the schematic of the slicing experimental setup and the experimental conditions and materials used for the experiments were summarized in Table 1. A jig that held the wire tool under constant tension was installed on the table of a computerized numerical control (CNC) milling machine. The work material was attached to the work holder, which was installed along the Z-axis of the CNC milling machine. Prior to the experiments, the work material was pushed 15 mm toward the wire tool and grooving experiments were conducted using rotational motion of the spindle, as shown in Figure 2. A borosilicate glass round bar with a diameter of 10 mm was used as the work material, and the rotational speed of the work material was set to 28 m/min and 228 m/min. Previous studies demonstrated when the contact length of the wire tool and the work material (the length of the wire tool contributing to grooving) were short, the wire tool was damaged in a short amount of time due to wear [6]. In this study, to increase the length of the wire tool contributing to grooving, the X-Y table reciprocated at a low speed (278 mm/min) in the X-axis direction. The X-Y table movement distance was 90 mm. Therefore, the length of the wire tool contributing to the grooving was 90 mm. In addition, a free software (Mach 3, Artsoft USA) was used as the numerical control (NC) software.

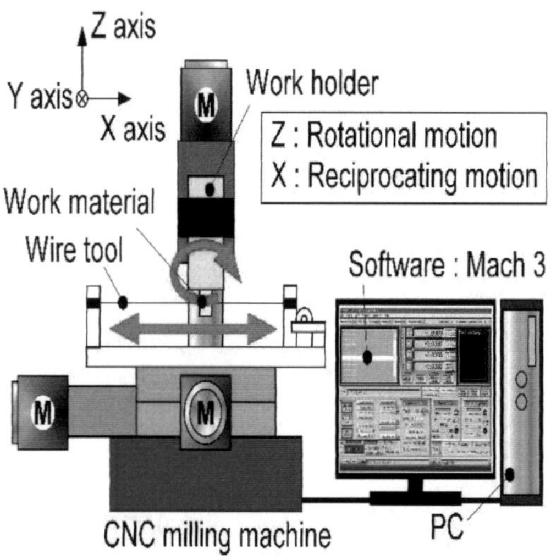

Figure 1. Schematic of the grooving experimental setup.

Table 1. Materials used and experimental conditions.

Wire tool	
Diameter	50–100 [μm]
Material	Tungsten
Applied tension	3.6 [N]
Work material	Borosilicate glass (Pyrex)
Diameter	10 [mm]
Rotational speed	28–228 [m/min]
Slurry	
Base oil	Glycol-based water-soluble coolant
Abrasives	Green silicon carbide (GC)
Particle size	1.2–6.7 [μm]
Concentration	40 [wt%]
Supply amount	50 [mL/min]
X-axis	
Table speed	278 [mm/min]
Moving distance	90 [mm]

The scanning electron microscope (SEM) image of the unused wire tools is shown in Figure 3. The diameter of each wire tool is 50 μm, 80 μm, 100 μm, and 140 μm. A wire tool with a diameter of 140 μm is a commonly used steel wire tool. Experiments are conducted using tungsten wire tools with diameters of 50–100 μm. The tension applied to the wire tools is 3.6 N. Green silicon carbide (GC) with three different particle sizes is used as the abrasive and the mean diameters of the abrasives are 1.2 μm, 3.0 μm, and 6.7 μm. The abrasives are mixed with a water-soluble coolant and is based on polyethylene glycol, which is used as the water-soluble slurry. The supplied slurry includes 40 wt% of the abrasive. The feed amount is 50 mL/min.

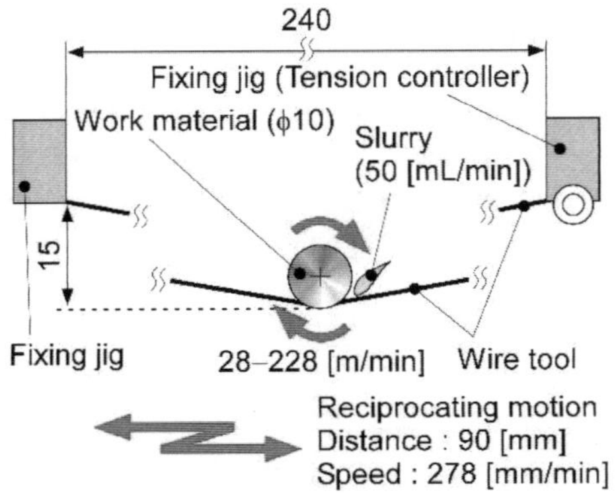

Figure 2. Schematic diagram of the grooving experiments.

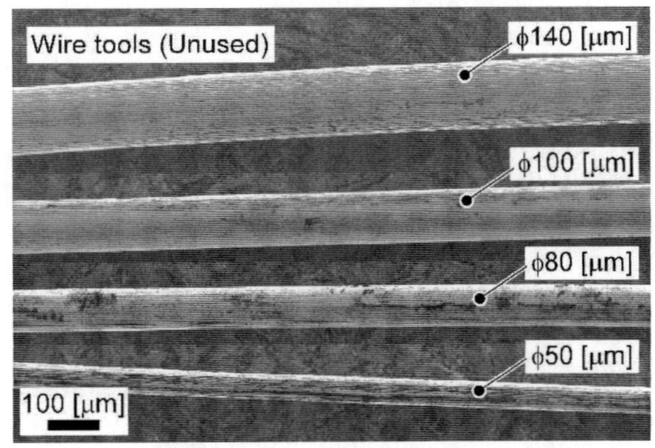

Figure 3. Appearance of the ultrafine wire tools.

3. Experimental results and discussion

3.1. Possibility of grooving using ultrafine wire tool

First, it is investigated whether grooving can be performed with ultrafine wire tools when borosilicate glass is used as the work material. Figure 4 shows an example of the work material after grooving. The diameter of the ultrafine wire tool is 50 μm and the particle size of the abrasives is 6.7 μm. The rotational speed of the work material is 28 m/min. From Figure 4, it is sufficiently possible to groove the material when using ultrafine wire tools.

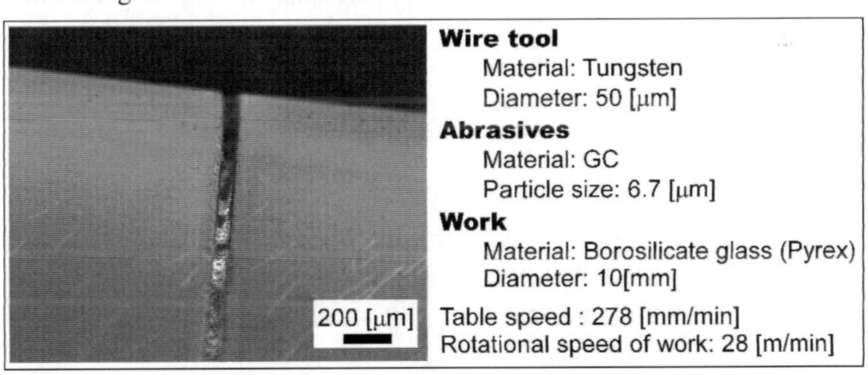

Figure 4. Example of work material after grooving.

3.2. Influence of diameter of ultrafine wire tool

The influence of the diameter of the ultrafine wire tool on the grooving characteristics is investigated while maintaining a constant particle size of the abrasives at 6.7 μm. Figure 5 shows the processing time necessary to break the ultrafine wire tool. It is clear from this figure that increasing the diameter of the wire tool, the processing time up to the breaking is increased. In addition, as the relative speed increases, the processing time that produces a broken tool is shortened. However, when the relative speed was fast, the influence of the wire diameter appears slightly. This is because the wire tool breaks in a very short amount of time when the relative speed is high.

Figure 6 shows the influence of the diameter of the ultrafine wire tool on the groove depth. When the diameter of the wire tool increases, the groove depth tends to deepen. When the diameter of the wire tool is larger, the processing time to break the wire tool increases. However, as the relative speed increases, the groove depth tends to deepen although the processing time to break the wire tool decreases. This is due to the improvement of the cutting performance of individual abrasives when there is an increase in relative velocity.

Figure 7 shows the influence of the diameter of the ultrafine wire tool on the groove width (kerf width). Naturally, the width of the groove is directly related to the volume of kerf loss. The empirical formula for the groove width is known [7]. Therefore, the calculated groove width is given by Equation 1.

$$W = 3G \tag{1}$$

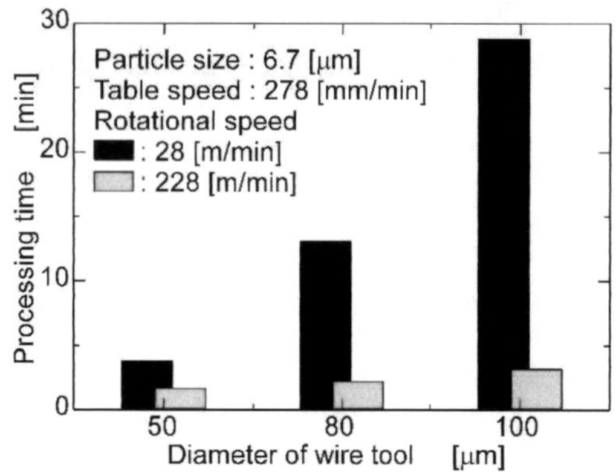

Figure 5. Influence of wire diameter on processing time, until the wire tool breaks.

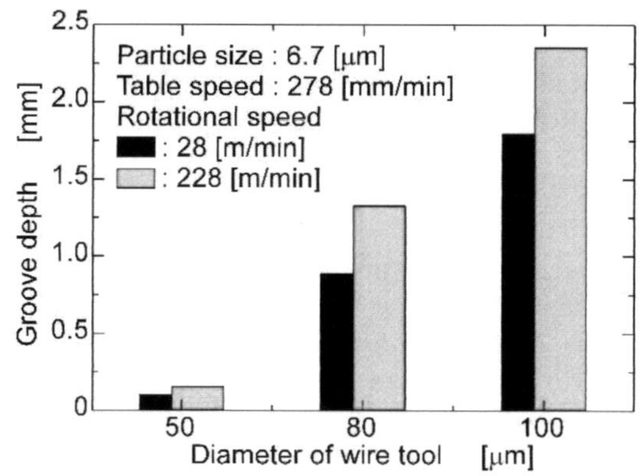

Figure 6. Influence of wire diameter on groove depth.

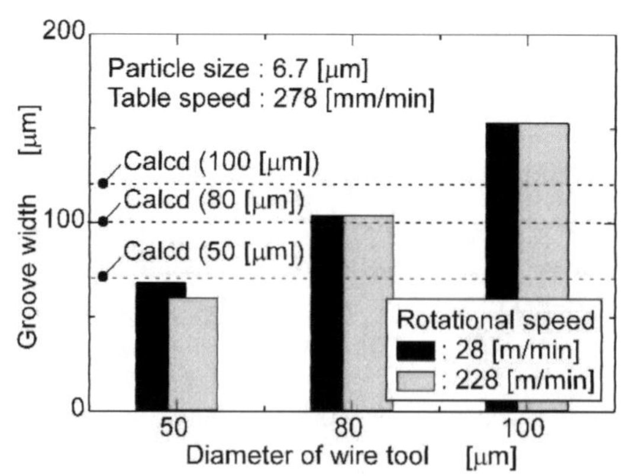

Figure 7. Influence of wire diameter on the groove width.

Here, *W* is the calculated groove width, *T* is the diameter of the wire tool, and *G* is the particle size of abrasives. *W* is called a calculated value (calcd) in this study and it is indicated by a dashed line in the figure. Generally, the groove width in the experiments tends to be larger than the calcd because it is influenced by the accuracy of the experimental apparatus. It is clear from this figure the groove width does not depend on the relative speed, but depends on the diameter of ultrafine wire tool. In addition, a larger diameter of the wire tool creates a greater difference between the experimental value and the calcd. This is the result of the particle size of the abrasives remaining constant irrespective of changes in the diameter of the wire tool. Therefore, when the diameter of the wire tool is small, a sufficient number of abrasives cannot be conveyed to the processing part and the experimental value becomes smaller than the calcd.

3.3. Influence of particle size of abrasives

The influence of particle size of the abrasives on grooving characteristics is investigated while the diameter of the ultrafine wire tool is kept constant at 50 μm. Figure 8 shows the influence of particle size of abrasives on processing time, running until the wire tool breaks. A larger particle size produces a shorter time necessary for the wire tool to break. This is due to increased damage to the wire tool by the abrasives. Additionally, the faster the relative speed, the time necessary to break the wire tool is shortened.

Figure 9 shows the influence of the particle size of the abrasives on the groove depth. As the particle size of the abrasives increases, the groove depth tends to deepen. Additionally, a faster relative speed creates a deeper groove depth.

Figure 10 shows the influence of the particle size on the groove width. No clear relationship is found between the different particle sizes of the abrasives, the difference in relative speed, and the groove width. Additionally, when the particle size of the abrasives is 6.7 μm, the experimental value is smaller than the calcd. This phenomenon is particularly noticeable when the relative speed is fast. The particle size is too large relative to the diameter of the ultrafine wire tool (50 μm), and it is not able to carry enough abrasives to the processing part. A larger particle size for the abrasives makes it easier to fall off from the ultrafine wire tool. Therefore, it is believed the amount of slurry adhering to the ultrafine wire tool decreases. In other words, it becomes clear the balance between the wire diameter and the particle diameter greatly influences the grooving characteristics. Therefore, to supply enough abrasive grains to the machined portion, it is necessary to use abrasive grains having a size suitable for the specific diameter of ultrafine wire.

It is clear from Figure 10 that groove width can be reduced to about 60 μm by using an ultrafine wire tool and fine abrasives.

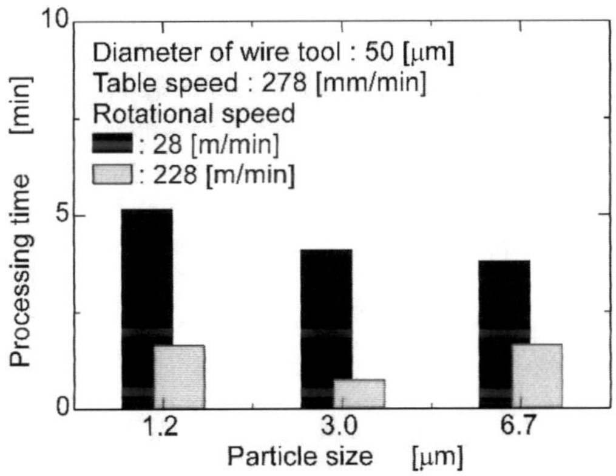

Figure 8. Influence of particle size on processing time, until the wire tool breaks.

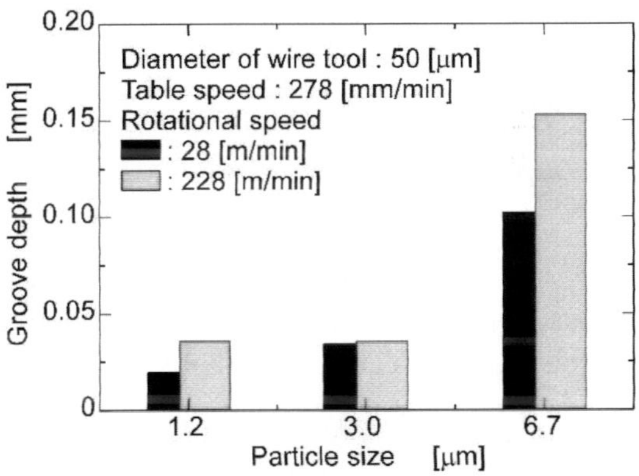

Figure 9. Influence of particle size on groove depth.

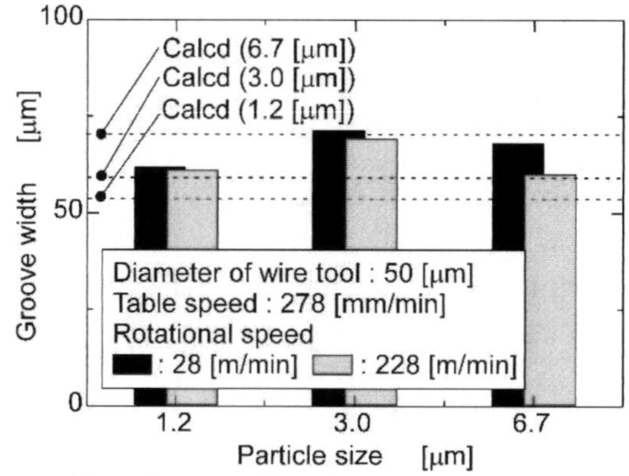

Figure 10. Influence of particle size on groove width.

4. Conclusions

We investigated grooving characteristics using an ultrafine wire tool. The main conclusions obtained from this study showed the larger the diameter of the wire tool, the deeper the groove depth became. Additionally, a faster relative speed resulted in a shorter lifetime of the wire tool, but a greater groove depth. The groove width did not depend on the relative speed, but depended on the diameter of ultrafine wire tool used. To supply sufficient abrasives to the machined portion, it was necessary to use abrasives that had a particle size suitable for the diameter of the wire tool used. In this study, we could reduce groove width (kerf width) to about 60 μm by using a fine wire tool and precision abrasive.

Acknowledgments

We thank Mr. Takeshi Kosugi (Synthesis Marketing Bureau Co., LTD.) and Instrumental Analysis Center of Yokohama National University for their support and cooperation. Additionally, part of this work was supported by JSPS KAKENHI Grant Number JP12345678.

References

[1] H Wu 2016 *Precision Engineering* **43** 1–9
[2] Hoshiyama T 2009 *J. of the Japan Society for Abrasive Technology* **53** No. 11 655–8
[3] NEDO 2009 *Roadmap (PV2030+)* (Tokyo: New Energy and Industrial Technology Development Organization)
[4] Uno Y, Okada A, Okamoto Y, Suzuki K, Hirota S, Hirano T, Takata S and Ishikawa K 2005 *Proc. of 3rd Int. Conf. on Leading Edge Manufacturing in 21st century (Nagoya)* (Tokyo: The Japan Society of Mechanical Engineers) pp 783–6
[5] Murata J, Tsuchida T, Tani Y and Zhang Y 2014 *Trans. of the JSME* **80** No. 815 SMM0183
[6] Sakamoto S, Gemma M, Hayashi K, Kondo Y, Yamaguchi K, Yamaguchi M and Fujita T 2016 *Key Engineering Materials* **703** pp 17–21
[7] Kinutani K and Kanamichi 1996 *Electronic Materials and Parts* **35** No. 7 (Tokyo: Kogyo Chosakai Publishing Co., Ltd.) pp 29–33

Browsing Through Closed Books: Evaluation of Pre-processing Methods for Page Extraction of a 3-D CT Book Volume

D Stromer[1,a], V Christlein[1,b], T Schön[2,d], W Holub[2,e] and A Maier[1,c]

[1] Pattern Recognition Lab, Friedrich-Alexander-University Erlangen-Nuremberg, Erlangen, Martensstraße 3, Germany
[2] Fraunhofer Development Center for X-ray Technology, Flugplatzstraße 75, Fuerth, Germany

[a]daniel.stromer@fau.de, [b]vincent.christlein@fau.de, [c]andreas.maier@fau.de, [d]tobias.schoen@iis.fraunhofer.de, [e]wolfgang.holub@iis.fraunhofer.de

Abstract. It is often the case that a document can not be opened, page-turned or touched anymore due to damages caused by aging processes, moisture or fire. To counter this, special imaging systems can be used. One of our earlier work revealed that a common 3-D X-ray micro-CT scanner is well suited for imaging and reconstructing historical documents written with iron gall ink – an ink consisting of metallic particles. We acquired a volume of a self-made book without opening or page-turning with a single 3-D scan. However, when investigating the reconstructed volume, we faced the problem of a proper automatic extraction of single pages within the volume in an acceptable time without losing information of the writings. Within this work, we evaluate different appropriate pre-processing methods with respect to computation time and accuracy which are decisive for a proper extraction of book pages from the reconstructed X-ray volume and the subsequent ink identification. The different methods were tested for an extreme case with low resolution, noisy input data and wavy pages. Finally, we present results of the page extraction after applying the evaluated methods.

1. Introduction

The standard procedure of using a book scanner is not applicable for imaging many historical documents. Although the page-turning process is done by an air flow, the documents may have a moisture, fire or aging damage that prohibits turning or opening. However, some methods exist for imaging and reconstructing such documents.

For our work, we assumed that writings are made with iron gall ink, an ink that has been widely used since the 5th century. Galilei, Mozart, da Vinci and even the Romans used this ink [1] for writing or drawing their work. This kind of ink mainly consists of three materials – gum arabic, tannic acid and iron(II) sulfate. The latter element has a higher X-ray energy absorption allowing to differentiate it from the paper. Mocella et. al used X-ray phase contrast imaging (Synchrotron) for recovering papyri rolls from Herculaneum [2] having the drawback of being a highly complex, expensive, and not feasible approach for everyday applications due to the immobility of the scanning system. Another approach is based on Terahertz radiation [3] which has the drawback of a limited depth resolution meaning that only a few pages can be imaged at once and has not been proofed for historical inks.

Content from this work may be used under the terms of the Creative Commons Attribution 3.0 licence. Any further distribution of this work must maintain attribution to the author(s) and the title of the work, journal citation and DOI.
Published under licence by IOP Publishing Ltd

Our approach is using a common micro-CT X-ray system, used for medical imaging or material testing. We showed that such systems are also capable of imaging such documents [4,5]. We encountered the problem, that performing 3-D X-ray scans on books differ a lot from scrolls which is due to the shape of the objects. While the scroll has a uniform diameter, a book's diameter differs with regard to the position of the gantry. This varying X-ray penetration length produces artifacts that do not occur with scrolls. Another major challenge when visualizing such documents by using 3-D imaging techniques is the extraction of single pages to inspect the writings on the paper. With higher resolutions and reduced X-ray energies, the acquisition and reconstruction times get higher. Also the extraction of rather thin pages out of the volume is a problem. They are stacked together in a very small area (sometimes more than 100 pages per cm) and squeezed at many points, but they are also wavy and the writings have a higher absorption than the paper. Without extraction, it is hardly possible to inspect single pages in the output volume of a reconstructed X-ray 3-D scan. Figure 1 shows the center slice of our generated volume. The green lines denote the transitions between two neighboring pages. To counter this problem, Seals et al. showed that using the Structure Tensor [6] performs good. However, the approach has been developed for scrolls and not for books, where many pages are squeezed together in the volume and the pages are not rolled up. A drawback of the structure tensor is that it has a poor corner response and edges are not a proper model for pages. The Vesselness filter defined [7] and normally used in medical imaging eliminates the drawbacks of the Structure Tensor and models the page as a curved surface.

In this work, we concentrate on the pre-processing of the acquired X-ray images. We show different approaches to create a binarized volume that sets the pages to '1' and the air to '0'. Therefore, we use global and adaptive thresholding methods as well as the Structure Tensor and the Vesselness filter. Afterwards, a selected page is extracted by a straight-forward segmentation algorithm. After applying all methods on a small snippet of the volume and comparing it to a manually segmented version with respect to computation time and exactness, we apply the methods on a larger part of the page.

2. Materials and Methods

2.1. The Book Model

We built a book model with ten pages. The paper is handmade and has a thickness of about 200 μm and the writings were made with iron gall ink and a quill. Letters of different font sizes, line thicknesses and paper saturations have been written on the pages. The height of the letters range from 1 cm to 2.5 cm. The pages were put in a pressboard holder during the X-ray scan to allow an upright book scan. The book consists of three different materials. Iron gall ink is assumed to be iron(II) sulfate heptahydrate ($FeSO_4 \cdot 7H_2O$) with a density 1.89 $g \cdot cm^{-3}$ having the highest intensities in the output. Pages were expected to consist of cellulose with the chemical formula $C_6\,H_{10}\,O_5$ and a specific density 1.5 $g \cdot cm^{-3}$ resulting in medium range intensities. Gaps between pages are assumed to be air and appear dark in the images.

2.2. CT Scan Trajectory and 3-D Reconstruction

For the scan, we used a micro-CT system built for non-destructive material testing and performed a 360° circular scan with 2400 projections using cone-beam geometry. This geometry is commonly used for 3-D X-ray CT scans and derives its name from the cone-shaped field-of-view as shown in Figure 2. We used a quadratic $2864 \times 2864\ px$ flat panel detector with a pixel shift of 80 μm and set up an energy of 70 kVp which is sufficient to see larger objects written with the ink. The book model was positioned upright on the turntable and y denotes the axis of rotation.

For the 3-D reconstruction the commonly used FDK algorithm [8] was used. The algorithm consists of three steps – cosine weighting, ramp filtering and back projection. The chosen voxel size was 68 μm^3 such that approximately 4 px cover one page. In total, the scan time was kept relatively low (about 1.5 hours). All calculations were performed with the CONRAD framework for cone beam geometries [9].

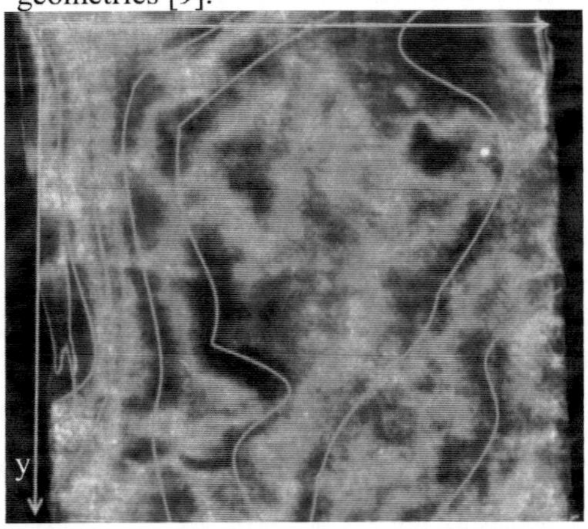

Figure 1. Center slice of the generated 3-D volume with the green lines denoting the transition between two pages.

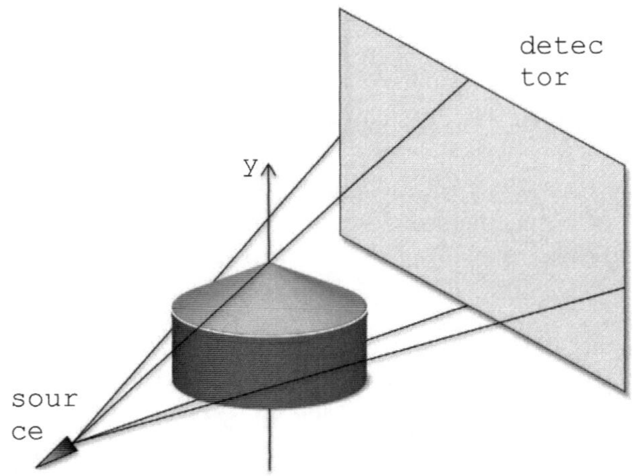

Figure 2. X-ray CT scan trajectory where the object is placed on a turntable and rotated with small angles. At every angle a projection image is taken.

2.3. Page Extraction

As we want to investigate single pages, we need to extract single pages out of the volume. We use an Figure 3 shows a xz-layer slice of the 3-D X-ray volume showing the ten book pages. We select two start points $p_1$and p_2 on the page-of-interest (green points), one close to the left and one close to the right. Afterwards, the algorithm walks along the page, segments the page (orange line) and crops the rest of the book. The key for the distinction between page and air is the use of two thresholds since one intensity threshold is not sufficient enough due to the squeezed pages. The red box shows such an area where the lower page is clearly differentiable from the other two but the upper two pages' overlap such that a simple threshold method would run into problems. Other problems are: ink that is brighter than the paper, as well as X-ray artifacts and noise disrupting the image.

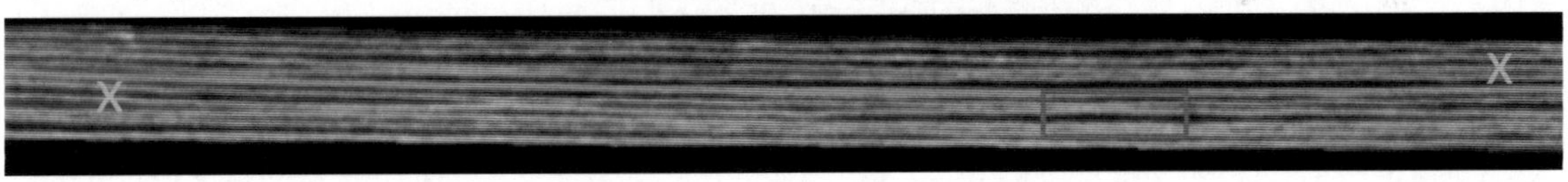

Figure 3. Center xz-layer slice of the 3-D X-ray volume showing the ten book pages that were stacked together in the wooden holder.

Figure 4 shows the complete processing pipeline. In the following, we will show and evaluate different approaches to perform the binarization process where the page is set to '1' and air is set to '0' such that the presented algorithm extracts (Page Extraction) only those parts of the page that really belong to it. After the binarization, we multiply the binary image with the original volume and receive the final page.

Figure 4. Processing pipeline for the extraction and 2-D mapping of a single page out of the volume.

2.4. 3-D Volume Pre-processing

Different approaches for binarization exist, where we picked out six common approaches to compare them to each other. Before applying those algorithms to the volume, we initially smoothed the single slices with a Gaussian Blur filter ($\sigma = 2$). This will not affect the output, because we afterwards will multiply the generated binary volume with the original volume, but we get rid of the noise and smooth the writings. Changes in the kernel size or σ had only small effects on the end result.

2.4.1. Otsu's Thresholding. The first method of choice is Otsu's thresholding [10], a global thresholding method trying to minimize the intra-class variance between two classes (in our case: air and page) by

$$\sigma_w^2(t) = w_1(t)\sigma_1^2(t) + w_2(t)\,\sigma_2^2(t), \tag{1}$$

where $w_{1,2}$ are the class probabilities, $\sigma_{1,2}^2$ are the class variances and t is the threshold. $w_{1,2}(t)$ is calculated by iterating through the histogram where the goal is to minimize the intra-class variance while maximizing the inter-class variance.

2.4.2. Adaptive Thresholding Approaches. In comparison to the global thresholding methods, local thresholding methods work on small patches that are slid over the whole image [11]. We picked out two commonly used approaches for our tests. The first method is the Adaptive Mean Threshold putting the same weight on every pixel in the given neighborhood $\mathrm{N_n}$ with n denoting the kernel size. This leads to the filter kernel $\mathrm{k_m(x,y)}$ of Equation (2) where $\mathrm{i(x,y)}$ denotes the specific intensity value.

$$k_{m,n\times n}\,(x,y) = \frac{1}{n\cdot n}\sum_{k,l\in N_n} i(k,l) \tag{2}$$

Secondly, the Adaptive Median Threshold function is tested. It uses the same filter kernel as the Mean filter, but sorts the pixel intensities in ascending order and considers the middle value as threshold. The third approach is the Adaptive Gaussian Threshold with the filter kernel $k_g(x,y)$ shown in Equation (3) where the standard deviation σ defines the range of the filter function.

$$k_g\,(x,y) = \frac{1}{2\pi\sigma} exp\left(-\frac{x^2+y^2}{2\sigma^2}\right) \tag{3}$$

The entire volume is filtered with the specific filter kernel function and afterwards, the original pixel value is compared to the filter output value. If the value is smaller than the output, it is set to '0', else it is set to'1'. As all filters are separable in each direction, the filtered image can be computed within a very short time.

2.4.3. Structure Tensor. Another approach that is part of this evaluation is the Structure Tensor. Seales et al. used this method for unwrapping the En-Gedi scroll successfully [6]. However, it has to be investigated that this method also fits on a book whose structure differs from the scroll in the way that the pages are compressed. The Structure Tensor is defined by the spatial averaged tensor product of the gradients in x- and y-direction as shown in Equation (4), with $\nabla \mathrm{f}_{\sigma_2} = (\nabla \mathrm{k}_{\sigma_2}) * \mathrm{f}$.

$$\boldsymbol{J}_{\sigma_1,\sigma_2}(x,y) = k_{\sigma_1}\,(x,y) * (\nabla \mathrm{f}_{\sigma_2} \otimes \nabla \mathrm{f}_{\sigma_2}), \tag{4}$$

Afterwards, the eigenvalues λ_1 and λ_2 (with $\lambda_1 \geq \lambda_2$) are computed which describe the structure of the image by the following three relations with ϵ denoting a small threshold value. $\lambda_1 \approx \lambda_2 \leq \epsilon$ describes a flat area, $\lambda_1 \gg \epsilon > \lambda_2$ describes an edge and $\lambda_1 \geq \lambda_2 > \epsilon$ describes a corner. The edge image is set as the binarization image, but it has to be mentioned that an edge is not a good model for a page and the corner response of the Structure Tensor is poor such that is has to be proofed if this method is appropriate for the given task.

IOP Conf. Series: Materials Science and Engineering **229** (2017) 012005 doi:10.1088/1757-899X/229/1/012005

2.4.4. Vesselness filter. The Vesselness filter eliminates the drawbacks of the Structure Tensor by modeling the page as a curved surface and highlighting tube-similar structures. It is widely used in medical imaging for segmenting vessels with differing diameters. Instead of the tensor product, the image's hessian matrix is computed by Equation (5) and (6). The second derivative of the Gaussian kernel at scale σ is used to measure the contrast between the regions inside and outside the range $(-\sigma, +\sigma)$ in the derivatives' directions.

$$\boldsymbol{H}_\sigma = \begin{bmatrix} \frac{\delta^2}{\delta x^2} I(x,\sigma) & \frac{\delta}{\delta x \delta y} I(x,\sigma) \\ \frac{\delta}{\delta x \delta y} I(x,\sigma) & \frac{\delta^2}{\delta y^2} I(x,\sigma) \end{bmatrix}, \qquad \begin{aligned} I(x,\sigma) &= I(x) * k_g(x,\sigma) \\ &= I(x) * \frac{1}{\sqrt{2\pi\sigma^2}} exp\left(-\frac{\|x\|^2}{2\sigma^2}\right) \end{aligned} \tag{5,6}$$

As we know that the paper has a constant diameter of about 4 px over the entire length and all slices, we can simplify the method and use only a single σ, which also reduces the computation time. Afterwards, the vessel parameters are calculated by Equation (7) and (8) and the total 'Vesselness' is computed by Equation (9) where β and c are control parameters ($\beta = 0.5, c$ depends on the scale). λ_1 and λ_2 are the eigenvalues of $\boldsymbol{H}_\sigma$ (with $\lambda_1 \geq \lambda_2$). The output $\boldsymbol{V}(x,\sigma)$ is the finally generated probability for the vessel, or in our case the page. To binarize this output, we set the threshold value to 95 % meaning that only the most-likely part is considered to be part of the page.

$$R_B = \frac{\lambda_2}{\lambda_1}, \qquad S = \sqrt{\lambda_1^2 + \lambda_2^2}, \qquad \boldsymbol{V}(x,\sigma) = \begin{cases} 0 & ,\lambda_1 \approx 0 \\ exp\left(-\frac{R_B^2}{2\beta^2}\right)\left(1 - exp\left(-\frac{S^2}{2c^2}\right)\right) & ,\lambda_1 > 0 \end{cases} \tag{7,8,9}$$

2.5. Measurements

The abovementioned methods are compared with regard to the total computation time (in seconds) and common segmentation measures ($RMSE$, Accuracy (acc) and f1-score ($f1$)) [12]. Finally, the resulting 2-D images will also be inspected with the naked eye. For the $RMSE$, a smaller value denotes a better match. For the remaining four accuracy measurements, the result gets more exact the closer the value is to '1'.

3. Results

First, a small snippet with a total size of 380 × 180 × 85 $voxel$ was cropped and manually segmented. For the Vesselness filter, the Structure Tensor, and the Adaptive Gaussian Thresholding, $\sigma = 2$ was the best solution ($tested: \sigma = [0.25, 0.5, 1, 2, 3, 5, 10]$). For Mean and Median filtering, a kernel size of (5,2) was configured. Table 1 shows that the Vesselness approach is the most exact approach while the Structure Tensor performs worst. For the adaptive thresholding methods, the Gaussian performs best with respect to the $RMSE$, accuracy and $f1$-score. The Vesselness filter had the longest computation time while the Mean and Gaussian thresholding methods performing about eight times faster.

Table 1. Results for the comparison of the manually segmented volume with the proposed methods with regard to accuracy and computation time.

	Otsu	Mean	Median	Gaussian	Structure	Vesselness
$RMSE$	0.509	0.397	0.390	0.387	0.606	**0.374**
acc	0.740	0.843	0.848	0.850	0.633	**0.860**
$f1$	0.723	0.785	0.773	0.790	0.522	**0.797**
$time\ [s]$	**0.081**	0.415	1.101	0.502	3.554	4.129

IOP Conf. Series: Materials Science and Engineering **229** (2017) 012005 doi:10.1088/1757-899X/229/1/012005

Next, we performed the calculations on the complete volume. Table 2 shows the results. The computation time for the Vesselness filter was $38.542\ s$ for processing a volume of $1430 \times 500 \times 85\ voxel$. Ignoring Otsu's thresholding, the Gaussian filter performed best again with being approximately 17 times faster than the Vesselness filter.

Table 2. Computation time of complete volume with regard to computation time.

	Otsu	Mean	Median	Gaussian	Structure	Vesselness
$time\ [s]$	**0.300**	3.891	11.484	2.292	33.626	38.542

Figure 5(a) shows the original part of the page that should be extracted out of the volume. The extraction algorithm of the binarized volume mentioned in Section 2.3. took about 2.5 minutes with identical starting points. Figures 5(b-f) show the outputs starting with the Vesselness filter as object of comparison at the top left position, and the outputs sorted by decreasing $f1$-score starting at the top left cell (Vesselness) ending at the bottom right position (Otsu's Thresholding). The result of the Structure Tensor was not considered. The examples show that the result gets slightly worse, visible through the increasing black holes, which confirms the measurements. The $RMSE$ of the Vesselness 2-D image (Figure 5(b)) and the Gaussian 2-D image (Figure 5(c)) is 0.075. Additionally, it should be mentioned, that applying the 3-D versions of the Structure Tensor and the Vesselness filter did not yield to improved results but led to higher computation times.

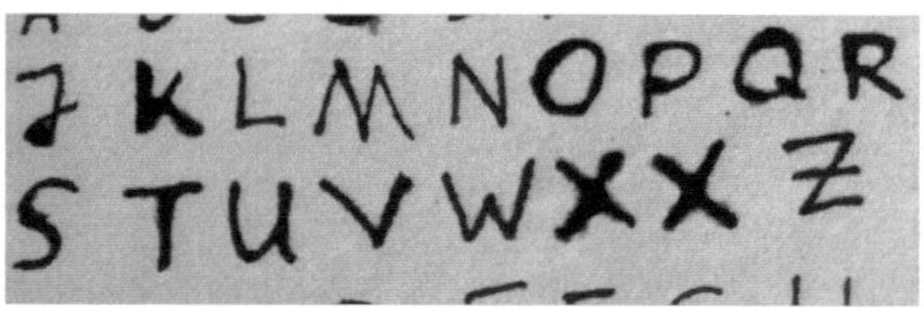

Figure 5(a). Original part of segmented page.

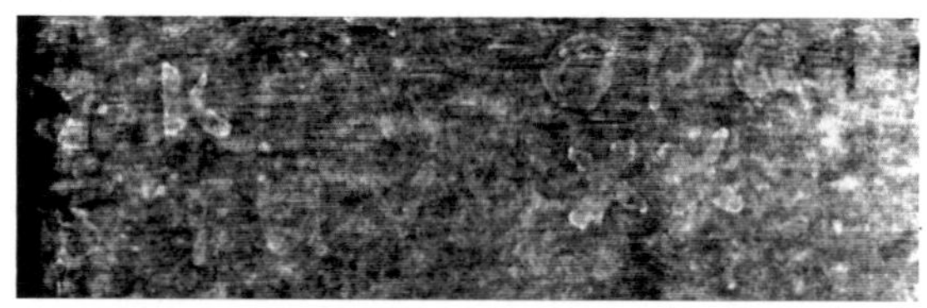

Figure 5(b). Vesselness filter result.

Figure 5(c). Adaptive Gaussian filter result.

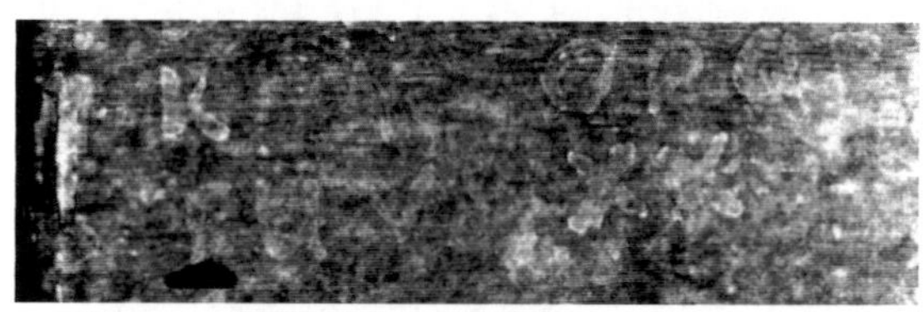

Figure 5(d). Adaptive Mean filter result.

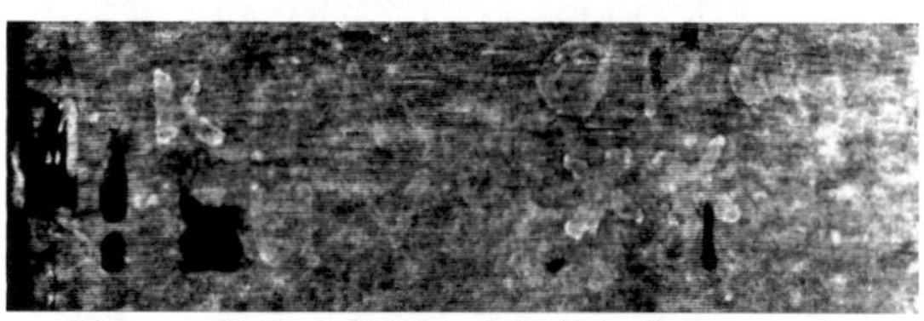

Figure 5(e). Adaptive Median filter result.

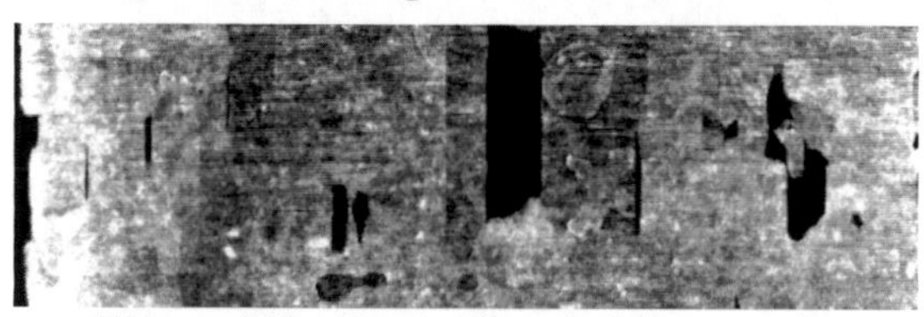

Figure 5(f). Otsu's Thresholding result.

4. Discussion and Outlook

In this work, we evaluated different approaches for the binarization of a book volume acquired with a micro-CT X-ray system which is an essential part for the investigation of single page's writings. Our pipeline consists of an initial volume smoothing and denoising followed by one of the binarization approaches. Finally, the page-of-interest is getting extracted and 2-D mapped. The total computation time takes about $3-4$ minutes (volume size $= 1430 \times 500 \times 85\ voxel$). We showed, that with the straight-forward segmentation method, the Vesselness filter delivered the best results with respect to accuracy. However, this method had the longest computation time (approximately $26\ s$). The Adaptive Thresholding Methods performed quite well, with the Gaussian filter having the edge over the Mean and Median filter. Otsu's threshold and the Structure Tensor turned out to be unsuited for the implemented extraction algorithm. In future, we want to evaluate the methods also on scrolls. We also want to find the best scanning system parameters, such as X-ray energy, collimation and detector pixel shift. Furthermore, we want to try reconstructing different inks and papers, such as parchment or modern paper, with other reconstruction approaches.

References

[1] Kolar J, et al. "Historical iron gall ink containing documents - properties affecting their condition." *Analytica chimica acta 555.1*: 167-174, 2006.

[2] Mocella V, Brun E, Ferrero C, and Delattre D. "Revealing letters in rolled herculaneum papyri by X-ray phase-contrast imaging." *Nature communications*, vol. 6, 2015.

[3] Redo-Sanchez A, Heshmat B, Aghasi A, Naqvi S, Zhang M, Romberg J and Raskar R. "Terahertz time-gated spectral imaging for content extraction through layered structures." *Nature Communications*, vol. 7, p. 12665, 2016.

[4] Stromer D, Christlein V, Anton G, Kugler P and Maier A. "3-D Reconstruction of Historical Documents using an X-Ray C-Arm CT System." In *Proceedings of the 31th conference on Image and Vision Computing New Zealand 2016*, M. University, Ed., 2016.

[5] Stromer D, Schoen T, Holub W and Maier A. "3-D Reconstruction of Iron Gall Ink Writings." In *Proceedings of the 7th Conference on Industrial Computed Tomography*, Leuven, 2017.

[6] Seales W B, Parker C S, Segal M, Tov E, Shor P and Porath Y. "From damage to discovery via virtual unwrapping: Reading the scroll from En-Gedi." *Science Advances*, 2(9), e1601247, 2016.

[7] Budai A, Bock R, Maier A, Hornegger J and Michelson G. "Robust vessel segmentation in fundus images." *International journal of biomedical imaging, 2013.*

[8] Feldkamp L, Davis L and Kress J. "Practical cone-beam algorithm." *JOSA A*, vol. 1, no. 6, pp. 612–619, 1984.

[9] Maier A, Hofmann H, Berger M, Fischer P, Schwemmer C, Wu H, Müller K, Hornegger J, Choi J H, Riess C, Keil A and Fahrig R. "CONRAD - A software framework for cone-beam imaging in radiology." *Medical Physics*, vol. 40, no. 11, 2013.

[10] Otsu N. "A threshold selection method from gray-level histograms." *Automatica*, vol. 11, no. 285-296, pp. 23–27, 1975.

[11] Gonzalez R C and Woods R E, *Digital Image Processing* (2nd Ed). Prentice Hall, 2002.

[12] Powers, D M. "Evaluation: from precision, recall and F-measure to ROC, informedness, markedness and correlation." *Journal of Machine Learning Technologies 2 (1),* pp. 37–63, 2011.

IOP Conf. Series: Materials Science and Engineering **229** (2017) 012006 doi:10.1088/1757-899X/229/1/012006

Hydrogen Susceptibility of Pre-strained Type 316L Austenitic Stainless Steels in Aqueous Solutions

D Mati[1,*], A Takasaki[1] and S Uematsu[1]

[1]135-8548, Tokyo, koto 3-7-5 Toyosu, Koto-ku,Japan

*Md16056@shibaura-it.ac.jp

Abstract. This paper presents the observations made as a result of hydrogen effects in austenitic stainless steels which led to reduction of its original mechanical properties. This paper therefore seeks to understand the mechanisms and effects induced by hydrogen leading to embrittlement. The samples of Type 316L austenitic stainless steel with 20% pre-strain were charged with hydrogen through galvanostastic cathodic loading for 12, 24, 36, and 48 hours respectively, with sample as Cathode and the platinum wire as Anode. Afterwards, they were fractured under tensile test at a slow strain rates. The loss of ductility was evident as observed by the brittle nature of fracture. Small pieces were cut-off near the fracture region and examined through XRD and SEM analysis. The peak widening and slight shifting of the peak positions was observed. The surface cracking was also observed, an indication of surface-induced stresses. The severity of the effects increased with the hydrogen loading time, which is predictive of a real service conditions for austenite steels in aqueous environments.

1. Introduction

The hydrogen effects in metals and their alloys are rather complex and takes quite a deal to understand the interaction between hydrogen and constituent metals. Most of other forms of hydrogen degradation involving phase transformations are easily understandable and can therefore be remedied by simply identifying the actual cause of the damage causing process [1-4]. Though a concerted efforts have been dedicated to related research, comprehensive understanding of the damage process involved is much more intricate and overcoming the damage processes has proven quite difficult. This is in relation to its interdisciplinary attribute covering electrochemistry, materials science, and mechanics. Again, hydrogen being the lightest element, it has high mobility and insensitive to external excitations, thus making its detection difficult. Hydrogen and a surrounding stress fields have complex interactions in stainless steel materials [2]. More often, austenitic stainless steels find their various applications in service conditions that are frequently characterized by aqueous environments. The hydrogen source, the temperature and the duration of hydrogen exposure determines the potential severity of hydrogen embrittlement [3]. Consider the power plants for example, which comprises the bulk of austenitic stainless structural materials, and which run for long periods. Absorbed gases, some of which contains hydrogen elements like H_2S in geothermal power plants or hydrogen in water molecule and or the mass evolution of hydrogen gas in nuclear power plants are some of the sources of hydrogen that interact with the structural austenitic stainless steels in service conditions. Hydrogen is absorbed in metals as an atom or screened proton rather than as a hydrogen molecule. So, the absorption of hydrogen requires the presence of atomic (nascent) hydrogen on the metal surface [3-4]. Therefore, aqueous environments are rich source of hydrogen that absorbs into the metallic bulk over time.

Content from this work may be used under the terms of the Creative Commons Attribution 3.0 licence. Any further distribution of this work must maintain attribution to the author(s) and the title of the work, journal citation and DOI.
Published under licence by IOP Publishing Ltd

Austenitic stainless steels are less susceptible to hydrogen embrittlement because of low diffusivity and high solubility of hydrogen in the F.C.C. structures, although in severe environmental conditions, the brittle fracture associated with hydrogen embrittlement (HE) is observed [5-6].

The cathodic reaction for usual metallic corrosion is carried by the hydrogen reaction and/or the oxygen reaction. For the hydrogen reaction, there are two cathodic processes that produce hydrogen gas at metal electrodes. One is the reduction of hydrogen ions;

$2H^{+}_{aq} + 2e^{-}_{M} \rightarrow H_{2,\ gas}$

and the other is the reduction of water molecules.

$2H_20 + 2e^{-}_{M} \rightarrow H_{2gas} + 2OH^{-}_{aq}$

Pure diffusion does not sufficiently provide hydrogen penetration far from the surface of the sample because of the low diffusivity of hydrogen in AISI 316L. Therefore, the objective of this study is to identify and predict the progressive degradation involved in HE by cathodic charging for strained Type 316L stainless steel at 40° C.

2. Experiment

The apparatus used in this experiment included the galvanostatic cell for hydrogenation, Ringaku Ultma-IV X-ray diffractometer for XRD analysis, Kyowa ME-LUXZ optical microscope for surface observation, JSM- 7610 JOEL Field Emission scanning microscope and AG50KNX (Shimadzu) tensile testing machine. The investigations were performed on commercial type 316L austenitic stainless steel. The as- received sample was annealed at 1000°C for 1 hour. The test specimens were supplied in the form of 1mm thick plate. The hydrogen cathodic loading was carried out at 40° C by a chemical cell hydrolysis using 0.5M H_2SO_4 for 12, 24, 36, and 48 hours respectively, at current density of 4.5 mA/cm^2, with sample as Cathode and the platinum wire as Anode. The slow tensile deformation rates (0.02 mm/s) were employed for optimal results. The chemical composition of the 316L stainless steel used in this study is given in Table 1.

Table 1. Chemical composition of the test sample

C	Cr	Ni	Mo	Mn	Si	P	S	Co	N
0.017	18.0	14.0	3.0	2.0	0.75	0.007	0.03	0.15	0.1

3. Results and discussion

The resultant mechanical properties of the samples is given in figure 1 in the form of stress-strain curves and tabulated results as shown in table 2. The hydrogen uncharged and charged specimens including the un-strained/non pre-charged sample. This specimen which in this case is referenced, had zero strain and zero charge. It can therefore be clearly seen that hydrogen charging deteriorated the ductility property with increasing time of charge. The hydrogen charging decreased the elongation from 40.7% to 35% after 48 hours of charge. The hydrogen effect on tensile strength showed distinct decline with increased charging time though the results fluctuates between 24 hours and 36 hours of charge time.

Table 2. Mechanical properties before and after charging.

Specimen	UTS MPa	Yield Strength Mpa	Elastic Modulus GPa	Ductility %
0%S 0 hour charged	524.7	203	193	40.7
20% S 12 hour charge	506.1	197	218	39.3
20% S 24 hour charge	504.9	420	201	35.7
20 % S 36 hour charge	510.0	423	187	37.1

20% S 48 hour charge	488.7	405	170	35.0

On the other hand, hydrogen charging effects on the yield strength drastically increased with time, and the mixed results on Modulus of elasticity was observed. The results of the charged specimens fluctuated above and below that of the As-received annealed sample. The general tendency on the mechanical property deterioration increased with increasing hydrogen charging time. In many studies of the hydrogen embrittlement, the elongation and reduction of area loss (RA) are often used as the evaluation criteria, due to the well-established knowledge of hydrogen effect on the material ductility[7], though in this paper, the RA is not utilised owing to the geometry of the test samples used. The changes in tensile strength can be attributed to densities of dislocations and mechanical twins induced by pre-strain, different absorption rates of hydrogen into the material's voids partly occasioned by the pre-strain [6-7].

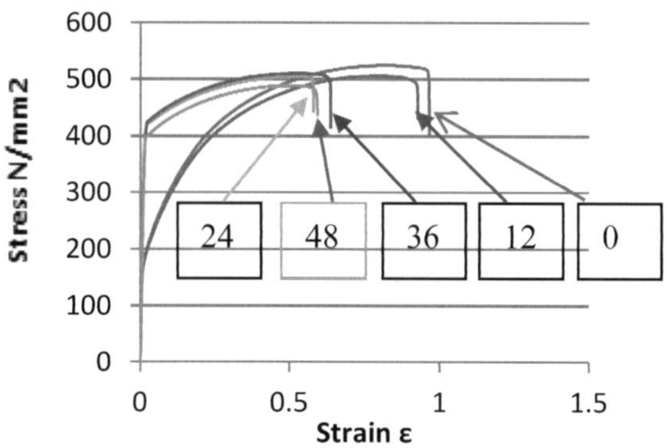

Figure 1. Stress-strain tensile curves of the tested Samples.

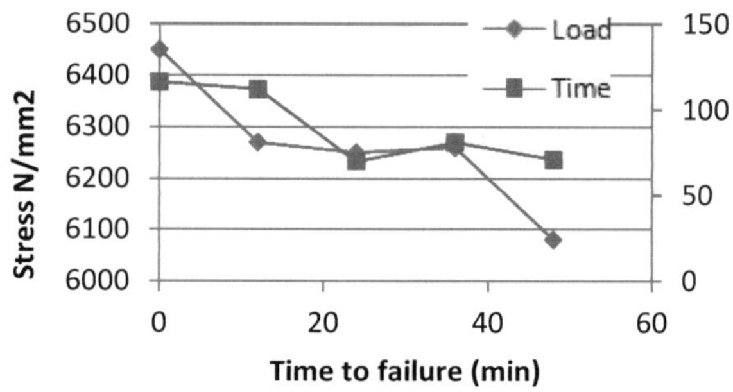

Figure 2. Stress to failure time.

The Modulus of elasticity and ductility tends to gradual decrease with time and agrees well with many research findings [7-9]. This is expected because of hydrogen accumulation at pores and microspores associated with the grain boundaries and micro structural defects caused by the pre-strain, leading to the hardening of the material. These results indicate that the overall resistance to HE deteriorated with the charging time. Hydrogen diffusion is well known to be activated by plastic deformation and that the activation is time dependent [8]. The time-to failure results supports that theory as is evidently shown in figure 2. This result is consistent with transport of hydrogen by dislocations occasioned by pre-strain [6, 9]. The presence of hydrogen in steel reduces tensile ductility and causes premature failure under static load that depends on the stress and time [10, 14].

To investigate the effects relating to ductility loss with hydrogen charging time and the pre-strained condition of the specimens, small pieces were cut off near the fractured region and the XRD and SEM tests conducted on them. All the xrd fractographs showed an austenite single phase without the reflections of ε-martensite peak. The diffraction patterns were analysed and plotted as shown in figure 3. Although the hydrogenation of 316L stainless steel seems to have caused the expansion of the austenite lattice, there were no significant reflections other than FCC that appeared in the diffraction patterns.

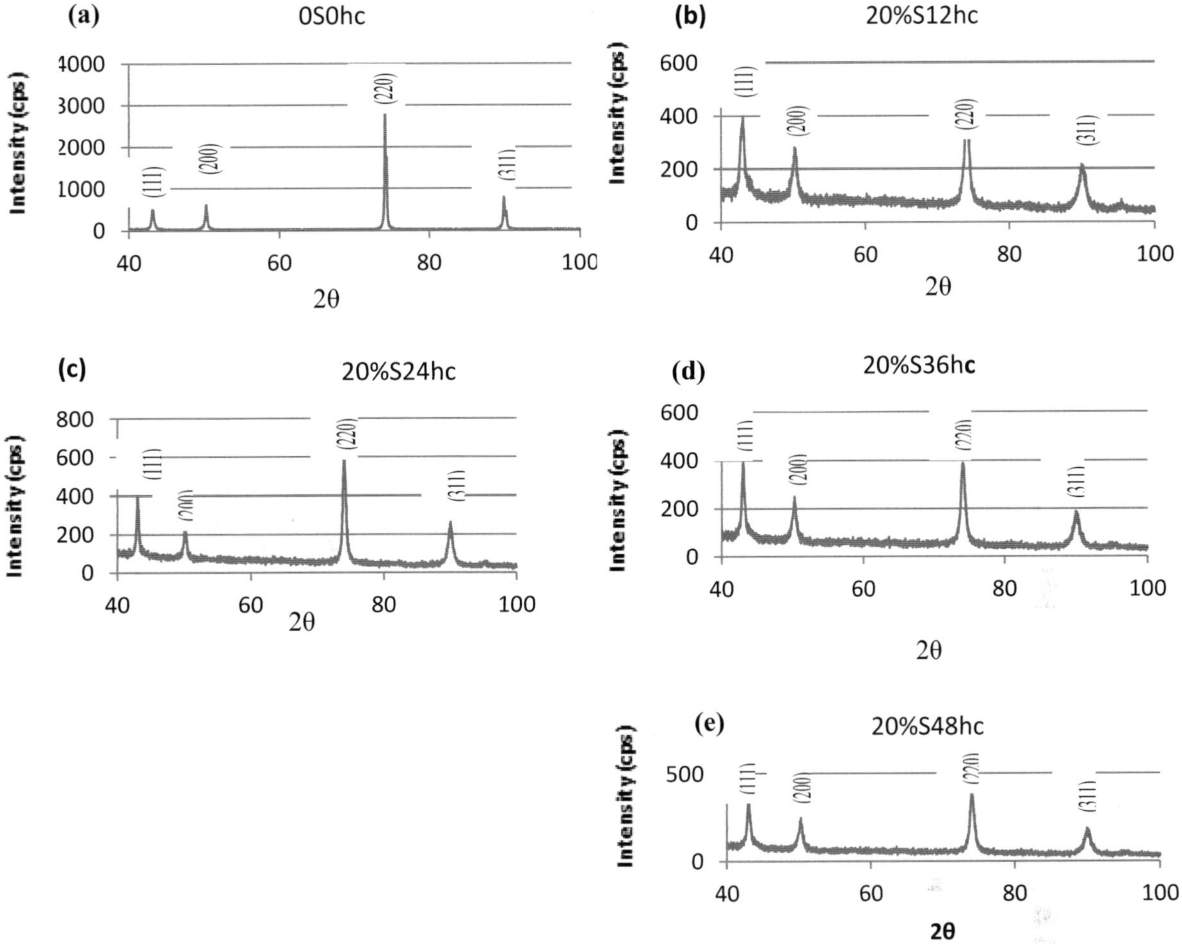

Figure 3. XRD diffraction patterns, (a) Unstrained and uncharged, (b) 20% strain with 12 h charge, (c) 20% strain with 24 h charge, (d) 20% strain with 36h charge and (e) 20% strain with 48h charge.

Based on investigations carried out on austenitic stainless steels type 304 and 310 [11], it was observed that there was continuous expansion of the austenite lattice due to hydrogen charging of 304 and 310 stainless steels. Slight shift of diffraction peaks and lines broadening were observed after hydrogen charging. Hydrogen penetration considerably expands the lattice and causes the significant shift of diffraction peaks toward the lower 2θ angles. After 48 hours of hydrogen charging, the 1 1 1 peak centroid was displaced by 0.153 of 2θ.

In the uniform solid solution, the intensity of the diffraction peaks decreases due to local lattice distortion. The reason for peak broadening is the formation of non-uniform solid solution of hydrogen in austenite [12].

SEM micrographs of the fracture surface after charging for 12h, 24h, 36h and 48h are given in Figure 4. Transition between brittle and intergrannular ductile fracture was observed in all the hydrogen charged samples, with the outermost edges bearing the brittle fracture, but the fracture mode changes to ductile in the inner depths from the surface. Near the brittle zone, small micro voids of low spatial density are observed (Figures 4c-d). This particular pattern corresponds to a decrease in ductility caused by hydrogen which is present in lower concentrations than near the surface of the

IOP Conf. Series: Materials Science and Engineering **229** (2017) 012006 doi:10.1088/1757-899X/229/1/012006

specimen where brittle fracture is prominent [12]. In the case of non-charged sample, the ductile failure is evident where the morphology of micro voids (decrease of density and size) is observed [13].

The presence of hydrogen atoms in a solid metal dissolve in the metal grid and accumulate in disturbed lattice regions resulting in the reduction of its ductility by decreasing the energy of cohesion and thus increasing its probability of brittle failure [14]. In Figure 4c, 4d and 4e, the ductile fracture surface is characterised by the presence of increasingly denser, shallower and smaller equi-axed micro voids, but in figure 4a and 4b, the dimples are larger and deeper. In all the charged samples, quasi-cleave tear ridges were also observed in different regions of the fracture profile, a confirmation of a significant degree of brittle fracture. These results agree well with the tensile test results in table 2. The decrease of the amount of hydrogen varied inversely with the depth from the surface of the samples to the center because of the lower diffusivity of hydrogen in the austenitic phase [1]. The presence of quasi-cleavage morphology from the brittle areas is a significant effect of the hydrogen transport by dislocations [15]. Shallow brittle multiple cracking were observed on all the sample surfaces. The inhomogeneous hydrogen distribution results in high multiaxial compressive stresses at the surface [16]. Although in this research hydrogen uptake has not been quantitatively analyzed, the results thus obtained cumulatively affirm that the 316L austenitic stainless steels stability to hydrogen embrittlement is significantly affected.

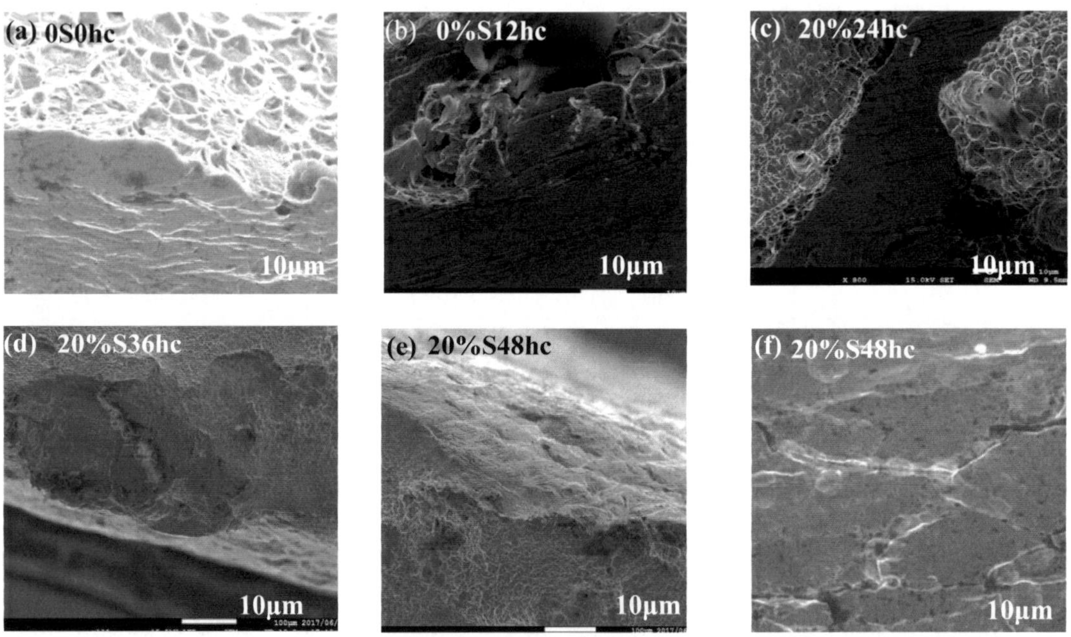

Figure 4. SEM images of the fracture morphologies: (a) non-charged, (b) 12h charge, (c) 24h charge, (d) 36h charge, (e) 48h charge and (f) the surface cracks developed at fracture.

4. Conclusion

The 316L austenitic stainless steels is mildly susceptible to hydrogen embrittlement and the effects increases with time of exposure. Warm temperatures coupled with pre-strain increases the hydrogen uptake and enhanced embrittlement as evidenced by reduction in ductility and brittle fracture morphologies. The 316L stability to HE may not be fully guaranteed especially if the material undergoes long periods of exposure to similar hydrogen environmental service conditions.The density, nature and size of micro voids present on dimpled fracture surfaces is a result of hydrogen effects on the material.

References

[1] Cayard, R K(1998) Roles of H2S the behavior of eng. alloys: InterCorr - CL1 International, Inc. Hassani A et al, Effect of Strain Rate on Stress Corrosion Cracking of 316L Austeniti Stainless Steel J. of Mat. Eng. and Performance: 2013, June 2013, Vol 22, Issue 6, pp 1783–1789, Lo K H et al , Recent developments in stainless steels, Mat. Scie. and Eng. R Reports, 2009 65(4-6):39-104, Fujita F, Sohmura T (1976) Hydrogen in stainless steel and Fe-Ni alloys, J. de Physique, Colloques, C6, Déc. 1976, Vol. 37, Numéro pp.C6-379-C6-383, Herms E et al Hydrogen embrittlement of 316L type stainless steel, Mat. Sci and Eng. Vol 272 issue 2, (1999)

[2] Song, Bonnie R Antoun and Bo (2009) Interaction of Hydrogen and Deformation in 316L Stainless Steel, Sandia National Laboratories, Livermore, CA, 94551-0969, USA.

[3] Louthan M R Jr.(2008) Hydrogen Embrittlement of Metals A Primer for the Failure Analyst J. of Failure Analysis and Prevention, Vol 8, Issue 3, pp 289–307

[4] Djukic M B et al, Hydrogen damage of steels. Article in Eng. Failure Analysis, 2015, Vol.58, pp 485-498

[5] Che^ne J et al, Solutions to Hydrogen Problems in Steels, American Society for Metals, 1981, pp. 263–271

[6] Che^ne J et al, Hydrogen transport by dislocations and hydrogen embrittlement in stainless steels, Bernstein I M, Thompson A W(Eds.): Effect of hydrogen on behavior of materials, publication of the Metallurgical Society of AIME, New York, 1976, pp. 455-466

[7] Xu H et al, Evaluation of hydrogen embrittlement susceptibility of temper embrittled 2.25Cr 1Mo steel by SSRT method, Eng Fail Anal 19(2012) 43–50 Ian M R et al Hydrogen Embrittlement Understood, The Minerals, Metals Materials Society and ASM International. 2015, Vol. 46B, June 2015—1085

[8] Herms E et al, Hydrogen embrittlement of 316L type stainless steel, *Mat sci. and Eng.* A272 (1999) 279–28

[9] ANTON J. et al: Dislocation Transport and Hydrogen Embrittlement, November 1979, Vol. 10, Issue 11, pp 1675–1682

[10] DAYAL R K and Parvathavarthini N, Hydrogen embrittlement in power plant steels, Springer link, Aqueous Corrosion and Surface Studies Section, June 2003, Volume 28, Issue 3, pp 431–451

[11] Mathias H, Y Katz and S Nadiv, *Met. Sci.* 12 (1978) 129 & Gavriljuk et al, *J. of Physics*. Condensed Matter, Vol 10, no. 31, Gavriljuk V G, Hanninen H, Tereshchenko A S, Ullakko K, Scripta ¨ *M etall. et Mater*. 28 (1993) pp 247-252

[12] Rozenak P et al Hydrogen effects on phase transform. *J. of Mat. Sci.* Feb. 1984, Vol 19, Issue 2, pp 567–573,Beer S, Israel, Narita N et al, *Met Trans*.Hydrogen-related phase transformations in austenitic stainless steels, 1982, Vol 13, issue 8, pp 1355-1365

[13] Caskey G R Jr., *Microstruct. Sci.* 9 (1981) 413–419, V.H. Desai, K. Scammon, Nace Annual International Conference Corrosion, 1995, paper 161, C.E. Price, L.B. Traylor, *Scr. Metall.* (1983) 901–904

[14] Fischer1 F D, Svoboda2J and Kozeschnik3 E (fracture Interstitial diffusion in systems with multiple sorts of traps1 Institute of Mechanics, Montanuniversit¨at Leoben, 8700 Leoben,Austria.

[15] Hazarabedian A, Ovejero-Garcia J, *Rev. Metall.* CIT, septembre 1993. Steel in hydrogen sulphide (H_2S) environment in SSRT conditions

[16] Narita N, Altstetter C J, Birnbaum H K *Met. Trans.* A 13A (1982) 1355.Vol. 13, Issue 8, pp 1355–136

IOP Conf. Series: Materials Science and Engineering **229** (2017) 012007 doi:10.1088/1757-899X/229/1/012007

Environmental Strategies for Sustainable Manufacturing Process of Composites

Maxim Kireitseu

Delft University of Technology, 1 Kluwerweg, Delft 2629HS, the Netherlands

m.kireitseu@gmail.com

Abstract. This research is focused on the strategic road mapping of composite manufacturing process and aims to understand the sustainability and related costs of composite part manufacturing. A manufacturing route of a serial automotive component is mapped and modelled using the following steps: (1) a holistic, cradle to grave product model for both manuflacturing and assembly operations, (2) development of life-cycle model and analytical tools, and (3) direct data collection and measure of environmental impacts of manufacturing. Besides the theoretical outcomes recommendations are given considering further recycling and recovery of materials so as to provide further direction for sustainability research in carbon and glass fibre composites.

1. Introduction

The manufacturing and assembly companies use a number of advanced technologies to fabricate parts and finished products from a range of materials such as metals, polymers and carbon or glass reinforced composites [1]. Environmental impact of such materials on landfill is significant and requires investigation and analytical modelling of product life cycle [2,3]. Therefore global engineering companies require further research and support with sustainable development, which includes the four domains: manufacturing, ecology, economics, politics and culture.

Metals, CF-reinforced plastics and ceramics used in automotive for example are energy intensive materials and ideal possible candidates for recycling [4]. However, these materials can be difficult to recycle and re-use in manufacturing for one or more of the following reasons: (1) cross-linking effects in thermosetting polymers prevent re-melting or remoulding; (2) mix of materials in composites are routinely combining many types such as glass and aramid fibres, core materials, paints and so producing multiphase waste; (3) costs and time of materials separation and its recycling are significant, not economic efficient [5-7].

Fundamental research with LCA modelling and manufacturing capabilities has continued to achieve better sustainability with composite materials and related carbon, plastics and metals. A set of new technologies such as chemical recycling, microwave heating, mechanical recovery, and fluidised bed, and pyrolysis [6-10] are being developed; however, a key question whether LCA modelling can capture and address the benefits of new technologies applications in manufacturing, what costs and time involved to replace an old process or supply chain [11-13]. Research at PUEB [14] suggested that using pyrolysis to recover carbon or glass materials from polymer waste will consume only 5-10 % of the energy required to produce a real carbon or glass fibre. For economic gains a cumulative impact from recycling should be less than a combined impact from alternative waste treatment and new material production [15].

Content from this work may be used under the terms of the Creative Commons Attribution 3.0 licence. Any further distribution of this work must maintain attribution to the author(s) and the title of the work, journal citation and DOI.
Published under licence by IOP Publishing Ltd

IOP Conf. Series: Materials Science and Engineering **229** (2017) 012007 doi:10.1088/1757-899X/229/1/012007

For recycled materials their benefits are usually compared with similar properties of engineered materials [16]. Further research using life cycle assessment (LCA) method is required to understand the benefits and limitations of new technologies such as lithography, deposition techniques [17-19]. Changes of properties in recycled materials, coupled with the potential inconsistency of supply and demand, prevent many Polish manufacturing companies from use of recycling technologies. Although the markets for recycled materials grow, a recent technology has identified the needs to speed-up applications for recycled glass fibres (GF), carbon fibres (CF) and also aluminium. Composites and steel products are usually compared in automotive applications but aluminium parts may win over composites in terms of emission benefits. Recycling via pyrolysis notably recovered 98% of carbon materials. Incineration with energy recovery and recycling (fluidised bed and mechanical) are applied for plastics and polymers as well as carbon waste [20].

In this research we aim to develop and demonstrate a holistic LCA modelling approach using the recent advancements in recycling and waste technologies of carbon and glass fibres and polymer materials for potential application in Polish manufacturing industry. This will also include the development of "Eco-design" tools aimed to enhance the modelling capabilities of Polish companies – both small and large manufacturing working in European or global supply chains and aimed to improve their sustainability, efficiency and productivity of operations. A demonstration and validation of research results are planned with industrial partners and academic collaborators as appropriate. In the present work, the LCA methodology is applied to assess the environmental benefits of recycling against other forms of treatment for materials waste, and to evaluate how potential benefits can increase material reuse.

2. Methodology

LCA is an internationally standardized methodological framework [21-23], a part of the ISO 14000 environmental management series, for estimating and assessing the environmental impacts associated with manufactured products over complete life cycles. A product life cycle typically consists of four individual phases: (i) raw materials, (ii) production, (iii) use, and (iv) an end of life. Any actions employed at each life cycle phase consume resources and release emissions that are quantified to determine a total impact of the complete product system as well as relative contribution of each phase [24,25]. The process of LCA involves a 4 step framework as follows:

1. Goal and scope definition: In this stage the aims of study are defined (i.e. to compare the LCA impacts of CFRP door panel Vs steel Vs aluminium).
2. Inventory analysis: In the second stage data is collected. Data is collected on the amount of materials and energy that is consumed to make each product. This data is entered into LCA software (such as software from Carbon Footprint), which becomes an LCA model. The model is built in accordance with standards on LCA, ISO 14040 and ISO 14044.
3. Impact assessment: With LCA software results can be processed and visualised.
4. Interpretation: Results are interpreted to draw conclusions (i.e. product with lowest impact)

Comparison of recycling of a carbon or glass fibres and epoxy composite part via pyrolysis against modern incineration with energy recovery, and disposal via landfilling as potential waste treatments for CFRP waste will be performed. The functional unit for the study is a standard automotive part manufactured at the selected Polish enterprises, related waste and recycling levels.

Life cycle models are compiled within open access software openLCA and commercially available LCA software such as SimaPro and Umberto [26] and the Impact 2002+ impact assessment method [27] is applied to estimate the environmental impacts. The software allows easy input of personalised data and access to the Eco Invent database. After discussing the approach it was decided that the functional unit would be on a "per kg of finished part" basis.

3. Results

Initially, a Process Flow Chart for automotive composite part (a body structure) as seen on Figure 1 is created for both additive manufacturing and traditional composite process technology to map the manufacturing route and further analyse the potential improvements in product life cycle and costs. The process mapping has also allowed the research project to put a boundary on what should and shouldn't be included in the assessment after discussions with the industrial partners. In further road-mapping processes that have less than a 1% influence on the final output have been left out of LCA modelling. Best practices are developed for LCA, including PAS2050.

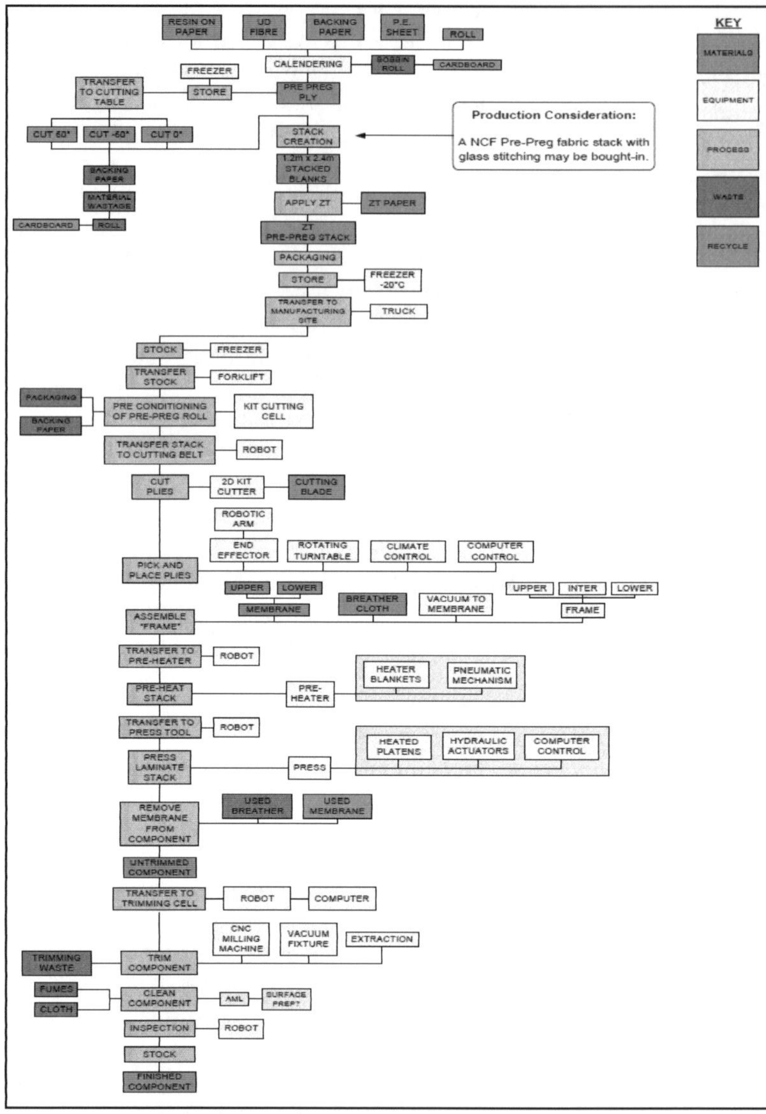

Figure 1. Manufacturing process road mapping for LCA.

Life cycle assessment (LCA) of pre-impregnated (prepreg) composite materials and ones manufacturing using additive manufacturing is presented. To understand the environmental impacts of manufacturing and recycling with carbon and glass fibres as opposite to aluminum a comparative study is done. The LCA is conducted on the current benchmark prepreg to obtain lifecycle results which will be shown in open reports. The benchmark materials are made from woven fabric carbon and thermosetting resins commercially supplied to manufacturing by Toray or similar suppliers. Studies also covered a uni-directional (UD) prepreg to understand the lifecycle properties which could be compared to the woven fabric benchmark. Both prepregs have the same carbon fibre and resin quantity measured in grams per square metre (GSM). Both of these prepregs are being studied to understand if they can be used on a future body design for body parts of automotive components.

Several steps of LCA have been clarified; the energy consumption of the prepreg production, the type of raw materials, where the raw materials are located, how to transport them, and the waste generated during the production. The entire life cycle assessment required data collection to be done using either measurement or modeling/calculation approach as shown in Fig. 2. This provided data for waste generation and energy consumption during prepreg production.

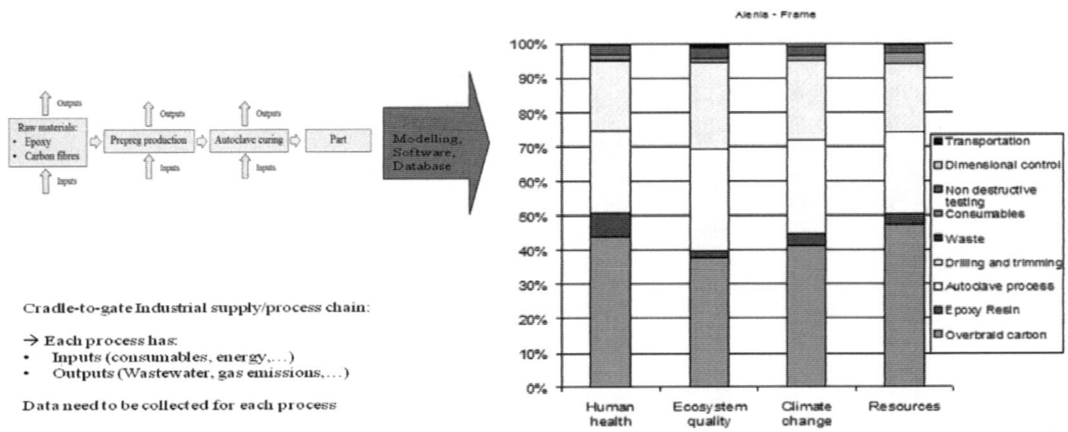

Figure 2. Life cycle assessment and data collection approaches used.

There are two scenarios assumed for carbon fibre composite recycling and re-use. The first approach corresponds to the production and use of carbon fibre-reinforced composites, which is either disposed, through land filling, or incineration, or recycled. Impacts related to each scenario are quantified and compared to determine the most appropriate.

For each waste treatment scenario, the starting point is 1kg of materials waste. The sub-processes included for modelling each making and waste technology route is shown in Figures 1, for land filling, incineration, and recycling respectively. Life cycle inventory (LCI) data will be obtained from Ecoinvent 3.1 LCI database [28] with over 10,000 processes. In each case, waste is first transported with a truck, and then reduced in size by an industrial shredder consuming 0.0025 MJ/kg of electrical energy [29,30]. Transport distances are to be assumed. Electrical energy is produced with the European wide average power mix (nuclear 29%, fossil 52%, Hydro 12%, other renewable 6%) [31-33]. Subsequently this is used in mapping the entire manufacturing and recycling processes and quantify contributions of each sub-process using graphical tools (Fig. 3-4).

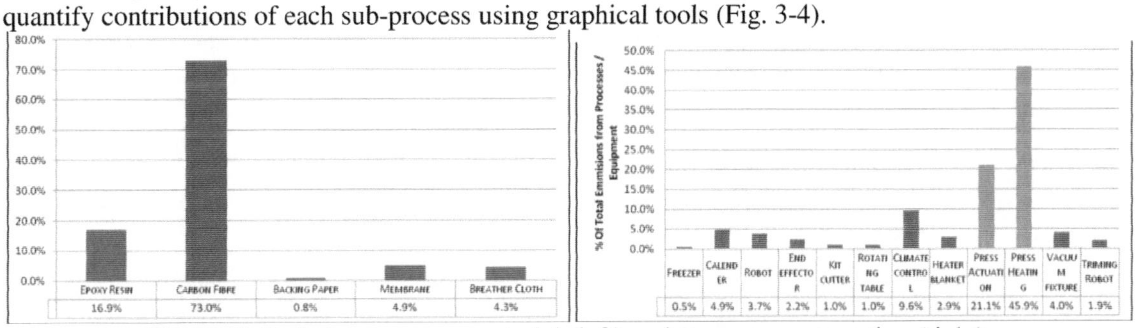

Figure 3. Emissions per raw material (left) and per process operation (right).

The embedded energy in carbon fibre is the biggest contributor to the total emissions (Fig. 3). Depending on the quantity of carbon manufactured, the $kgCO_2e/kg$ will decrease as the quantity increases. According to Harper Beacon, this figure can be as high as 24 for runs as low as 500 MTPY to 16 for runs as high as 2250 MTPY. It is clear that if a reduction in the $kgCO_2e/kg$ is sought after, then focussing on the CF manufacturing is a good place to start.

The biggest contributor to the total emissions by process / equipment (Fig. 3 right) comes from the press platen oil heaters. This is based on 2 heaters each using 48 kW for 5 minutes. This may not be truly representative and physical measurements would be recommended as the heaters may use significantly less power once the platens and tool are up to temperature. During the production implementation newer designs of heated platens are available and significantly decrease the energy consumption. The majority of total emissions generated by using the manufacturing technique come from the raw materials. The total emissions from the traditional composite manufacturing technique are 25.8 $kgCO_2/kg$. 79% of emissions come directly from the raw materials, 20% from the actual manufacturing of the components and only 1% from transportation (Fig. 4).

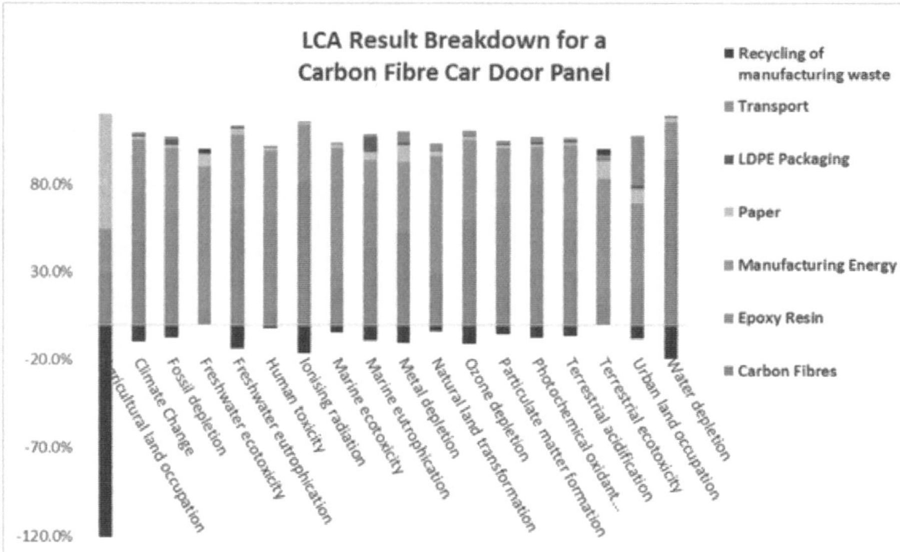

Figure 4. Contributions to each environmental impact category for manufacturing of composite part.

To compare the carbon fibre component to an equivalent stamped aluminium component would require going through the same process as was demonstrated. From the analysis, the embedded energy in the carbon fibre and resin are the biggest contributors to the overall $kgCO_2$. This is 18.6 $kgCO_2e/kg$. To compare this to an equivalent aluminium component, the embedded energy would be 3.3 $kgCO_2e/kg$. From the analysis, the embedded energy in the carbon fibre and resin is used as comparison criteria contributors to the overall $kgCO_2$.

For the recycling strategies of composites there are several scenarios assumed. In first scenario, disposal of waste occurs in a sanitary landfill to be constructed on site. Landfill gas is extracted and used for heating in this case without energy recovery. Other emissions could be from construction and related transport equipment at this site. LCI data for landfill of mixed plastics is used as compared to specific materials data from manufacturing. Landfill construction, wastewater treatment and emissions from operation will be considered in the LCA approach.

In a second scenario, a thermal waste treatment process such as incineration is assumed to convert waste material into flue gas, ash, heat, enabling embodied energy to be reduced.

In a third scenario, fibre recovery is achieved via pyrolysis technology to be added to LCA modelling. LCI for carbon or glass fibre recycling has not been found [34]; however, process specific assumption can be made based upon literature and the technical specifications of a commercial pyrolysis process [35-37]. Energy requirements for the recovery of 1 kg of CF are reported to be approximately 10% of that required to produce carbon fibres originally. Emissions from matrix combustion are assumed to be the same as from the incineration of plastics waste [38, 39]. The study has not expanded to the use phase of a vehicle made from recycled materials either carbon or glass fibre composites but other studies may be needed [40-42].

Acknowledgments
Project is supported by European funds from FP7-PEOPLE-2013-IIF – Marie Curie Action Ref. Nr. 913974 and hosted at TU Delft.

References
[1] Raugei M 2015 *J. Cleaner Production* **108** 1168.
[2] Council directive 2008/98/EC (Waste Framework Directive)
[3] Witczak J 2014 *Int J Life Cycle Assess*ment **19** 891.
[4] Witczak J 2016 *In: Blaszczyk A. (ed.), Product Proc. Mgt. Envir.&Tech.*, PUEB, Poznan, p. 41.
[5] Pickering S 2006 *Comp. Mat Part A*; **37(8)** 1206.
[6] Akesson D Foltynowicz Z 2013, *Polimery* **7-8** 543.
[7] Åkesson A Foltynowicz Z 2013 *Polymers & Polymer Composites* **21 (6)** 34.
[8] Palmer J Ghita O Savage L Evans K 2009 *Composites Part A* **40 (4)** 490.
[9] Pickering S 2000 *Composites Science and Technology* **60 (4)** 509.
[10] Ingarao G 2016 *J. Cleaner Production* **126** 493.
[11] Muradin M Foltynowicz Z 2014 *Sustainability* **6** 5065.
[12] Turner T Pickering S Warrior N 2011 *Composites Part B* **42 (3)** 517.
[13] Giannadakis K 2010 *Experim. Mech.* **51(5)** 767.
[14] Foltynowicz Z 2012 *J. Reinf. Plastics and Comp.* **17** 1136.
[15] Kurczewski P Lewandowska A 2012 *Int J Life Cycle Assess* **15** 777.
[16] Pimenta S Pinho S 2011 *Waste Manag.* **31(2)** 378.
[17] Witik R 2011 *Composites Part A* **42(11)** 1694.
[18] Das S 2011 *Int. J of life cycle assessment* **16 (3)** 268.
[19] Duflou J 2009 *Manufacturing technology* **58(1)** 9.
[20] Lewandowska A 2015 *Int. J Life Cycle Assess.* **20** 9.
[21] La Rosa A.D. 2016, *Composites Part B: Eng.* **104** 17.
[22] Clift R Doig A Finnveden G 2000 *Process Safety and Environmental Protection* **78(4)** 279.
[23] Pryshlakivsky J Cory S 2014 *Journal of Cleaner Production* **23(1)** 345.
[24] ISO 2006. Environ. Mgt. – LCA – principals and framework. ISO 14040, Geneva.
[25] Wachowski L Kirszensztejn P Foltynowicz Z 2001 *Polish J. of Environ. Studies* **10.6** 415.
[26] Simapro Software. web: www.pre-sustainability.com (Accessed Sept 2016)
[27] Liu Y Yi L Jining C 2015 *Journal of Cleaner Production* **15**(2) 102.
[28] Ecoinvent Centre. Ecoinvent data v2.1, Ecoinvent reports No. 1-25. CD-ROM, Swiss Centre for Life, Cycle Inventories: Duebendorf, Switzerland; 2009.
[29] Spielmann M Dones R Bauer C 2007 *Report ecoinvent v2.0* **14**. Swiss LCA Centre, Dübendorf.
[30] Duflou J 2012 *Bulletin* **37.04** 374.
[31] Witik R 2012 *J. of Cleaner Production* **29** 91.
[32] Li X Ruibin B 2016 *Journal of Cleaner Production* **127** 451.
[33] European Commission 2010. Inst. for Environment and Sustainability: Int-l Ref. LCA Data System Handbook -General guide for LCA. Luxembourg. Publications Office of the European Union.
[34] Witik R 2013 *Composites Part A: Applied Science and Manufacturing* **49** 89.
[35] Simões C 2017 *Int. J. of Materials and Product Tech.* **52(1-2)** 162.
[36] Gill A 2017, *J. of Manufacturing Science and Eng.* **139.4** 41.
[37] Foltynowicz Z Jakubiak P 2002 *Polimery* **47(11-12)** 769.
[38] Lewandowska A Foltynowicz Z 2004 *Int-l Journal of Life Cycle Assessment* **9(2)** 86.
[39] Doka G 2009 Final report ecoinvent v2.1 **13**. Swiss Centre for Life Cycle Inventories, Switzerland.
[40] Mayyas A 2012 *Energy* **39.1** 412.
[41] Lewandowska A 2008 *Int. J. of Life Cycle Assessment* **13(4)** 319.
[42] Koffler C Rohde-Brandenburger K 2010 *Int. J. of Life Cycle Assessment* **15(1)** 128.

IOP Conf. Series: Materials Science and Engineering **229** (2017) 012008 doi:10.1088/1757-899X/229/1/012008

Influence of Ga/Hg Ratio on Phase Constituents and Electrochemical Performance of Mg-Hg-Ga Anode Materials

Libo Wang[*1,2]**, Peijie Li**[2] **and Liangju He**[3]

[1] China Special Equipment Inspection and Research Institute, Beijing 100029, China
[2] Department of Mechanical Engineering, Tsinghua University, Beijing 100084, China
[3] School of Aerospace, Tsinghua University, Beijing 100084, China

*wanglibo@188.com

Abstract. This study investigated the impacts of Mg-Hg-Ga alloys of various Ga/Hg ratios on phase constituents and electrochemical performance. The relationship between composition and phase constituents of the casting alloys were investigated by SEM and XRD Potentiodynamic polarization curves and the galvanostatic curves of the alloys in 3.5wt% NaCl solution were obtained. With a Ga/Hg ratio greater than 0.97, the second phase changes from Mg_3Hg to Mg_5Ga_2, and the normal eutectic becomes a divorced eutectic. Additionally, corrosion is inhibited and passivation appears with an increase in the Ga/Hg ratio increase. With a starting Ga/Hg ratio of less than 0.68, the discharge process becomes steadier and discharge time simultaneiously increases with the Ga/Hg ratio. Mg-Hg-Ga alloys with a 0.68 Ga/Hg ratio are suitable as the anode material for seawater batteries.

1. Introduction

Magnesium alloys are used widely as anode sheets for seawater batteries due to their high specific energy, long-term storage, stable discharge performance and fast activation [1-3]. For example, magnesium alloy seawater batteries are extensively used in torpedoes and some kinds of life jackets [4].

Mg-Hg alloys of 1~3wt% Hg are typically regarded as the best known seawater battery anode materials. The specific energy of a seawater battery can reach 150Wh/kg. The addition of an suitable amount of Hg may inhibit microgalvanic corrosion, and thus reduce the thickness of corrosion film on alloy surfaces; Mg_3Hg as the second phase may also activate a magnesium matrix [5,6]. However, these properties are often accompanied by disadvantages such as significant self-corrosion and short discharge times [7,8].

Previous research has attempted to addition of Ga to the Mg anode material, resulting in improved electrochemical performance of the anode material due to the activation effect of Ga on the Al anode [9,10]. However, the influence of Ga on the alloy solidification structure and the discharge process remains unclear. This study prepared Mg-Hg-Ga alloys of various Ga/Hg ratios to analyze the effect of the ratio on the microstructure and electrochemical performance of Mg-Hg-Ga anode materials, such as those utilized in seawater batteries.

2. Experimental Method

Hg and Ga are volatile elements at high temperatures, requiring the pre-preparation of Mg–Hg and Mg-Ga master alloys; other alloys were obtained by dilution. Pure magnesium (99.99%) was melted in

Content from this work may be used under the terms of the Creative Commons Attribution 3.0 licence. Any further distribution of this work must maintain attribution to the author(s) and the title of the work, journal citation and DOI.
Published under licence by IOP Publishing Ltd

a resistance furnace under the protection of gaseous CO_2 and SF_6. The pre-prepared Mg-Hg and Mg-Ga master alloys were then added into the melt at 600℃. The melt was stirred vigorously, and then cast into a water-cooled metallic mold. Table 1 lists the chemical compositions of tested alloys, which determined by ICP-AES.

Table 1. Chemical compositions and Ga/Hg ratios of alloys (wt%).

Samples	Hg	Ga	Mg	Ga/Hg
MHG1	1.77	0.83	Balance	0.47
MHG2	1.70	1.16	Balance	0.68
MHG3	1.72	1.67	Balance	0.97
MHG4	1.74	1.95	Balance	1.12

The microscopic structures of the magnesium alloys were observed by SEM (JEOL JSM-6301F) equipped with EDS. Electrochemical experiments were conducted with a potentiostat/galvanostat model 273A in 3.5wt% NaCl solution. The counter electrode consisted of a platinum plate, and the reference electrode consisted of a SCE. All potentials were referred to the SCE. The potentiodynamic test employed a sc anning rate of 5mV/s. The current density of the galvanostatic measurements was $100mA/cm^2$. All samples were ground to 1200 grit, and then polished using 1μm diamond polishing paste.

3. Results and discussion

The alloy microstructures are displayed in Figure 1. All alloys consist primarily of an α-Mg matrix with eutectic phase distributed in the grain boundaries. All alloys also display well-developed α-Mg dendrites and sixfold symmetry due to their rapid solidification when cast in the water-cooled metallic mold, as shown in Figure 1(c).

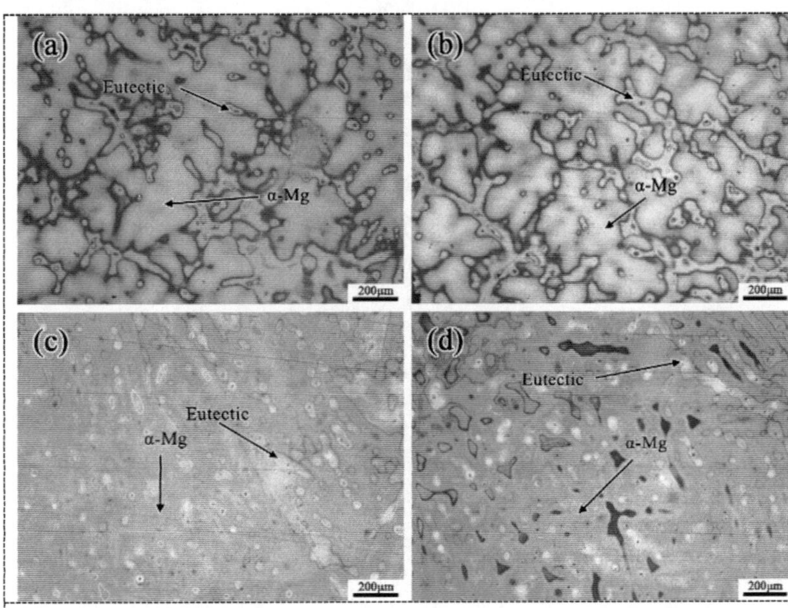

Figure 1. As-cast structures of (a) MHG1; (b) MHG2; (c) MHG3; and (d) MHG4 alloys

The X-ray Diffraction (XRD) spectral lines of four prepared alloys are displayed in Figure 2. Alloys MG1 and MG2 displayed the Mg_3Hg phase, while the Mg_5Ga_2 phase was found in alloys MG3 and MG4. These results indicate that the secondary phase of Mg-Hg-Ga alloys changes from Mg_3Hg to Mg_5Ga_2 with an increase in the Ga/Hg ration. As shown in Figures 1(a) and (b), when the Ga/Hg ratio is less than 0.68, the microstructures of Mg-Hg-Ga alloys are composed primarily of eutectic α-Mg and α-Mg+Mg_3Hg, which is similar to the structure of the Mg-Hg alloy without Ga.[11,12]. However, when the Ga/Hg ratio is increased to 0.97, the Mg_5Ga_2 phase appeared in alloys MHG3 and MHG4, and the Mg_3Hg phase was absent. Thus, experimental results indicate that Mg_5Ga_2 inhibits the phase formation of Mg_3Hg.

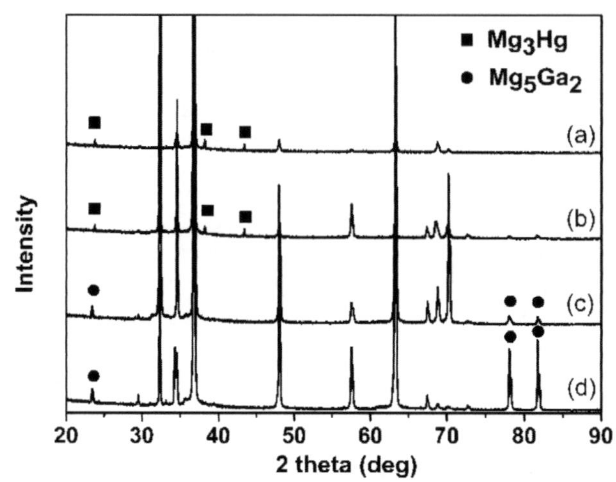

Figure 2. XRD spectral lines of Mg_3Hg and Mg_5Ga_2 alloys

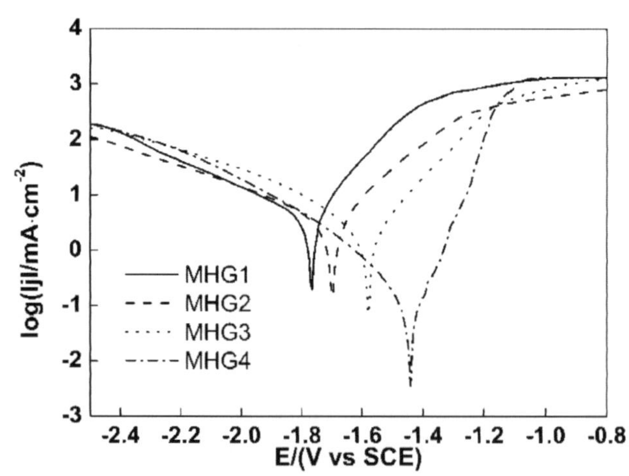

Figure 3. Potentiodynamic polarization curves of alloys in 3.5 wt% NaCl.

Figure 3 depicts the potentiodynamic polarization curves of alloys in 3.5wt% NaCl. Table 2 lists the experimental curves as fitted by the Tafel method with CorrView software. When the ration of Ga/Hg was increased from 0.47 to 0.97, the corrosion current density (icorr) decreased from 5.23 to 1.95; when the Ga/Hg ratio reached 1.12, the corrosion current density rapidly decreased to 0.025 mA/cm^2. According to Faraday's Law, the electrochemical corrosion rate increases linearly with increasing corrosion current density [13]. The MHG4 alloy demonstrated the highest corrosion resistance of all tested alloys, indicating the inhibition of corrosion with an increasing Ga/Hg ratio. This result is consistent with the fact that Ga is a high hydrogen evolution overpotential element [14].

Alloy MHG4 demonstrated the minimum corrosion current density, and the most positive corrosion potential (Ecorr). Alloy MHG2 demonstrated a more negative corrosion potential and a higher corrosion current density than MHG3, caused by the high activity of Mg_3Hg. Mg_3Hg could also reduce the formation of a corrosion product film on the alloy surfaces [6]. The passivation of the polarization curves of MHG4 can be observed in Figure 3, passivation disappeared as the Ga/Hg ration decreased.

Table 2. Potentiodynamic polarization curves of experimental results.

	Ecorr (V/SCE)	icorr (mA/cm^2)	ba (V/decade)	bc (V/decade)
MHG1	-1.767	5.23	0.180	0.589
MHG2	-1.698	3.43	0.195	0.474
MHG3	-1.578	1.95	0.178	0.299
MHG4	-1.437	0.025	0.071	0.106

Figure 4 displays the galvanostatic curves of the tested alloys. Experimental current density was 100mA/cm^2, with an electrolyte of 3.5wt% NaCl. The MHG1, MHG3, and MHG3 alloys reached maximum voltage in 10 seconds, whereas MGH4 obtained maximum voltage in 60 seconds. This result indicates that the activation times of the first three alloys are shorter; thus, the MHG1, MHG2, and MHG3 alloys have been identified as potential anode materials for seawater batteries.

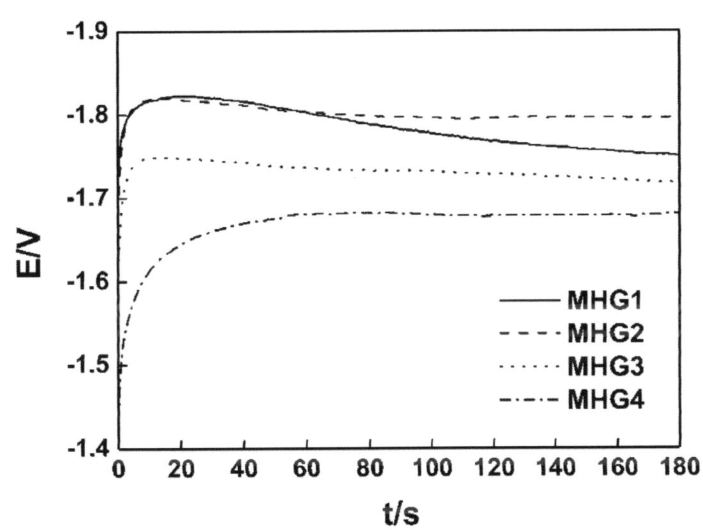

Figure 4. Galvanostatic curves of alloys in 3.5 wt% NaCl solution.

Seawater batteries require negative potential. As the Ga/Hg ratio increases, the electrode potential shifts to less negative values, thus decreasing the driving potential. The MHG4 alloy with a Ga/Hg ratio of 1.12 demonstrated the most positive potential of -1.68V (vs. SCE). Thus, we determined that alloys with a Ga/Hg ratio that exceeds 0.97 cannot meet the functional requirements of a seawater battery anode. The demonstrated electrode potentials of MHG1 and MHG2 were both approximately -1.83V. When the Ga/Hg ratio increases to 0.68, the discharge process of MHG2 became steadier and discharge time increased; this high voltage and short activation time indicate that the MGH2 alloy may be a suitable material for seawater battery anodes.

IOP Conf. Series: Materials Science and Engineering **229** (2017) 012008 doi:10.1088/1757-899X/229/1/012008

When the Ga/Hg ratio increases above 0.97, the effect of Ga becomes more apparent as the second phase changes to Mg_5Ga_2. With an increasing Ga/Hg ratio, the electrode potential shifts, passivation appears and corrosion is inhibited. Contrarily, with a Ga/Hg ratio of less than 0.68, the effect of Hg becomes more apparent as the second phase shifts toward Mg_3Hg. As the Ga/Hg ratio increases, the discharge process becomes steadier and the discharge time increases.

4. Conclusion

An increasing Ga/Hg ratio results in the inhibition of second phase Mg_3Hg. As the Ga/Hg ratio reaches 0.97, the eutectic phase changes from eutectic α-Mg+Mg_3Hg to eutectic α-Mg+Mg_5Ga_2. The normal eutectic morphology simultaneiously transforms to a divorced eutectic structure.

As the Ga content of the alloy increases, the corrosion density decreases, and anode corrosion is prevented. When the Ga/Hg ratio increases above 0.97, the corrosion current density and the discharge voltage decrease, and the activation time increases. With a Ga/Hg ration of less than 0.68, the corrosion current density remains stable, resulting in a steadier discharge process. Results indicate that the Ga/Hg ratio should not exceed 0.68 in order to obtain optimum electrochemical performance.

When the Ga/Hg ratio increases above 0.97, the electrode potential shifts, inhibiting anode corrosion and resulting in passivation.. When the Ga/Hg ratio is less than 0.68, the discharge process becomes steadier and the discharge time increases with an increase in the Ga/Hg ration.

References

[1] Shinohara M, Araki E, Mochizuki M and Kanazawa T 2009 *J. Power Sources* **187** 253
[2] Shi Y, Peng C, Feng Y, Wang R and Wang N 2017 *Mater. Des.* **124** 24
[3] Renuka R 1999 *Materials Chem. Phy.* **59** 42
[4] Hahn R, Mainert J, Glaw F and Lang K 2015 *J. Power Sources* **288** 26
[5] Zhao J, Yu K, Hu Y, Li S, Tan X, Chen F and Yu Z 2011 *Electrochim. Acta* **56** 8224
[6] Wang L, Li P and He L 2011 *Russ. J. Electrochem.* **47** 900
[7] Kim J G and Koo S J 2000 *J. Mater. Sci. Lett.* **19** 477
[8] Kim K, Hwang S, Park J, Han J and Kim Y 2016 *J. Power Sources* **313** 46
[9] He T, Wang W, Chen W, Chen D and Yang K 2017 I*nt. J. Hydrogen Energy* **42** 5627
[10] Feng Y and Wang R 2009 *J. Alloy Compd.* **473** 215
[11] Zhu M, Zhang L, Zhao H and Stefanescu D 2015 *Acta Mater.* **84** 413
[12] Lipchin T, Mushkov S and Vyatkin I 1974 *Met. Sci. Heat Treat.* **16** 1037
[13] Scully J C 1975 *The Fundamentals of Corrosion* (Oxford: Pergamon Press)
[14] Song G 2006 *The Corrosion and Protections of Magnesium Alloys* (Beijing: Chemical Industry Press.)

IOP Conf. Series: Materials Science and Engineering **229** (2017) 012009 doi:10.1088/1757-899X/229/1/012009

The Partial Density of States of CO_2 Molecules Adsorption on the Fe (111) Surface

Junfang Wu

Department of Physics, School of Science, Xi'an Polytechnic University, Xi'an 710048, China

1074897770@qq.com

Abstract. The state of CO_2 molecules adsorption on Fe (111) surface is studied by simulation with the software, the partial density of states the adsorption is obtained. Through the graphical distribution, the pseudogap and the partial density of states at the Fermi level of the CO_2 molecules adsorption on the Fe (111) surface is analyzed and compared. The key mechanism of CO_2 molecules adsorption on the Fe (111) surface is revealed. The results showed that the CO_2 molecules adsorption on the bridge position of Fe (111) surface is stable. The main reason of O atom and Fe atom combining with the bonding is that the resonance of the density of states happed between the O 2p orbital and Fe 3d orbital.

1. Introduction

In solid-state and condensed matter physics, the density of states (DOS) and the partial density of states (PDOS) of a system describe the number of states per interval of energy at each energy level that are available to be occupied [1]. A high DOS of molecules at energy level means that there are many states available for occupation. The zero DOS means that there are no states can be occupied at the energy level. In general the DOS is an average occupied by the system over the time and space domains. The DOS is the main index of adsorptions' properties, the DOS is defined as follows [2]:

$$\mathrm{DOS}\ (E)\ \mathrm{d}E = \text{number of levels between } E \text{ and } E + \mathrm{d}E \qquad (1)$$

E is the energy. The DOS for CO_2 adsorption on Fe (111) surface are obtained by the different position adsorption on Fe (111) surface [3,4].

2. Model building

The figure 1 is the unit cell structure of CO_2 molecule, the length of the cell is a=b=c=8.00 Å, the bond length of C-O is 1.16Å in the CO_2 molecule, the bond angle is 180°.

Content from this work may be used under the terms of the Creative Commons Attribution 3.0 licence. Any further distribution of this work must maintain attribution to the author(s) and the title of the work, journal citation and DOI.
Published under licence by IOP Publishing Ltd

IOP Conf. Series: Materials Science and Engineering **229** (2017) 012009 doi:10.1088/1757-899X/229/1/012009

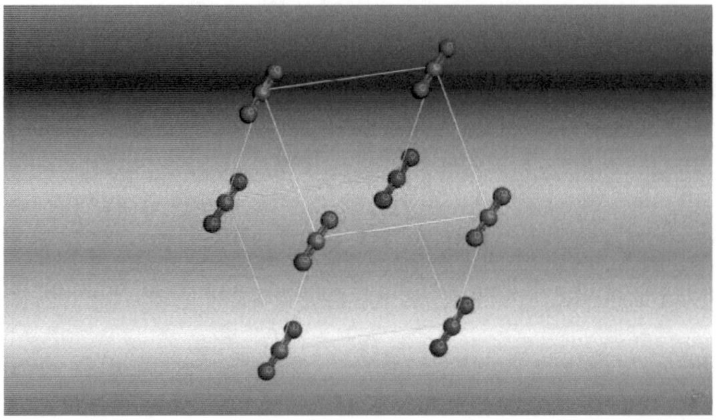

Figure 1. The unit cell structure of CO_2 molecule.

The unit cell structure of Fe is the figure 2, the length of the cell is a=b=c=2.8664 Å.

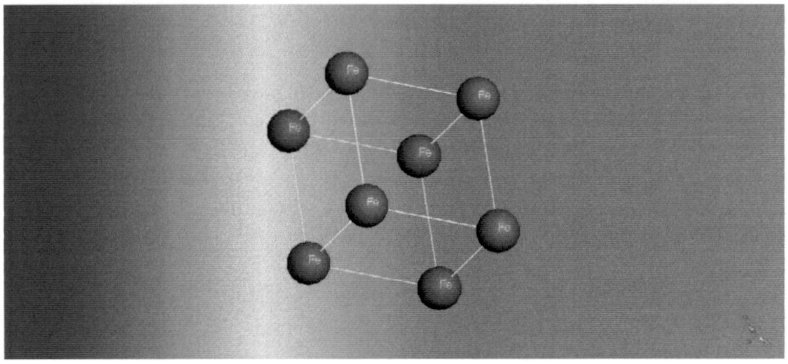

Figure 2. The unit cell structure of Fe.

The figure 3 is Fe (111) surface, the fractional top is 1.5 Å and the thickness is 1.5 Å, the vacuum thickness is 10 Å.

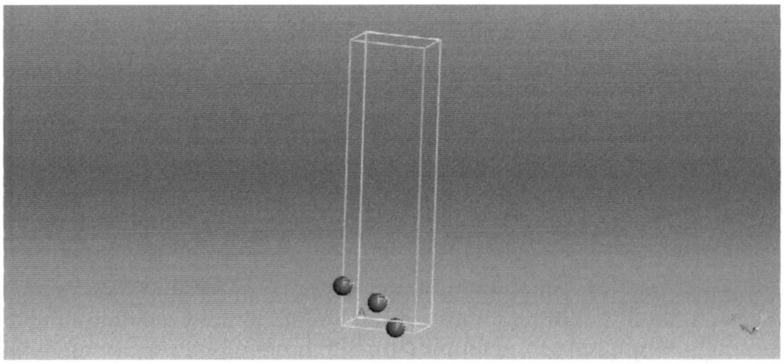

Figure 3. The Fe (111) surface.

The bond length of Fe-O is 1.887Å , the coordinates of O atom are x=1.8152, y=0.3077, z=2.9512. The 3D model of the CO_2 molecule adsorption on the Fe (111) surface is figure 4,the figure 4 (a) and the figure 4 (b) are different angles of the Fe (111) surface.

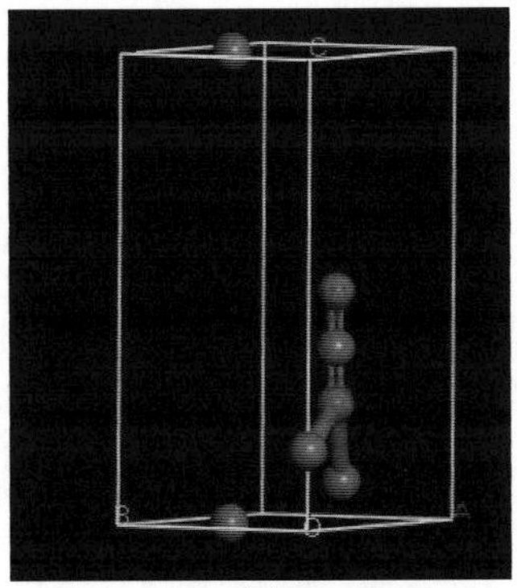

(a)

(b)

Figure 4. The Fe (111) surface.

The PDOS of CO_2 molecule is figure 5:

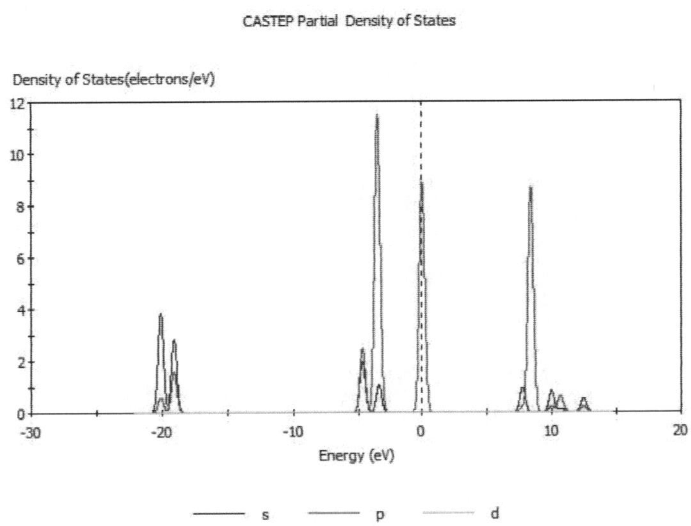

Figure 5. The PDOS of CO_2 molecule.

The figure 6 is the PDOS of Fe (111) surface:

IOP Conf. Series: Materials Science and Engineering **229** (2017) 012009 doi:10.1088/1757-899X/229/1/012009

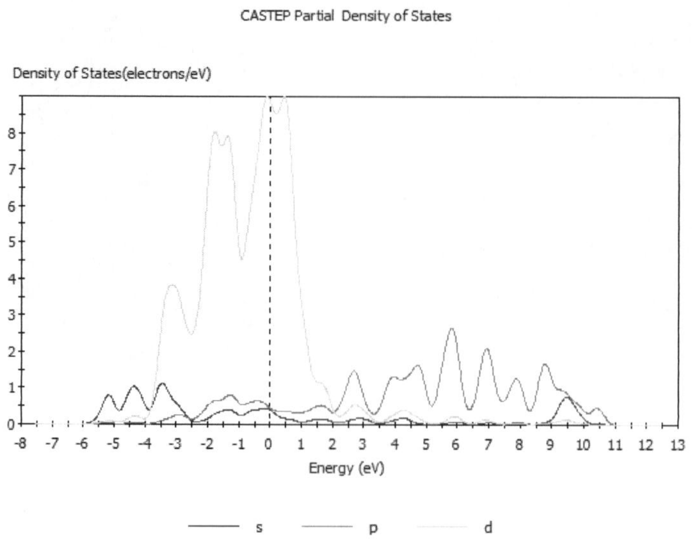

Figure 6. The PDOS of Fe (111) surface.

The figure 7 is the PDOS of CO_2 molecule adsorption on the Fe (111) surface:

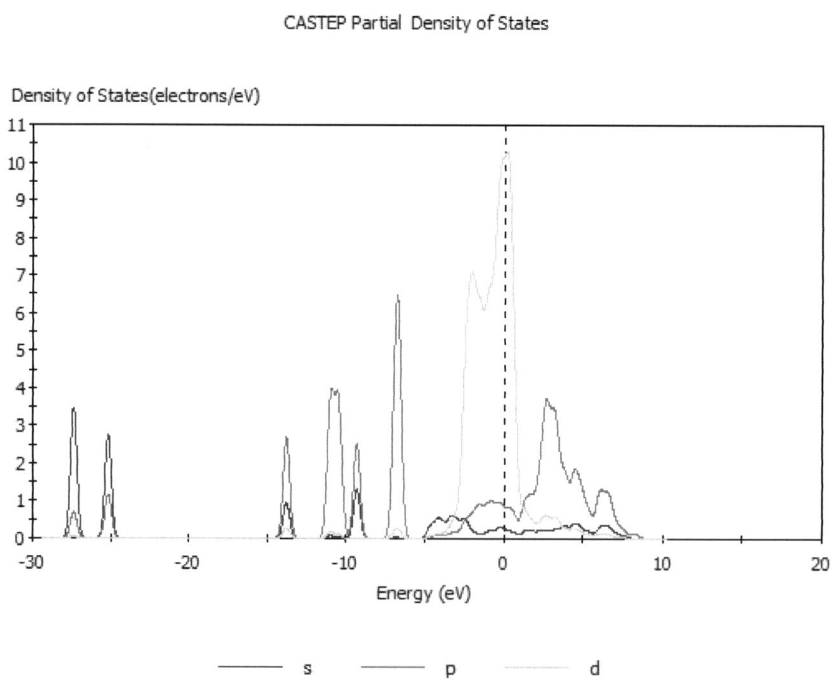

Figure 7. The PDOS of CO_2 molecule adsorption on the Fe (111) surface.

The figure 8 is the PDOS of CO_2 molecule after adsorption on the Fe (111) surface:

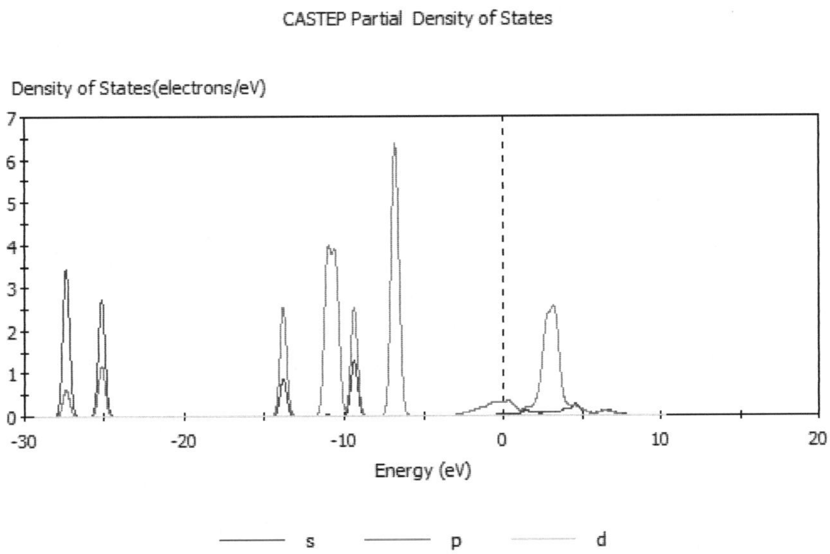

Figure 8. The PDOS of CO_2 molecule after adsorption.

The figure 9 is the PDOS of Fe after adsorption:

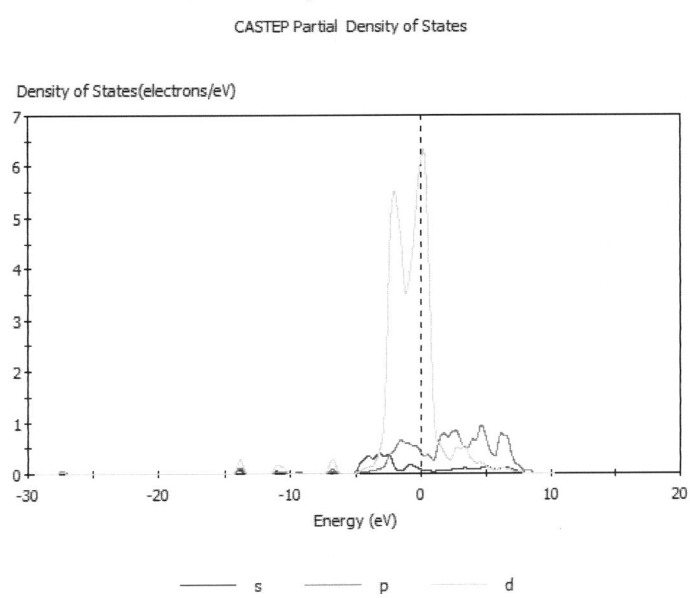

Figure 9. The PDOS of Fe after adsorption.

3. Results and discussions

The dotted line is the Fermi level in figure. From the figure 7, the DOS is very uneven in the whole energy interval distribution, it shows that the delocalized properties of electronic are very weak [5]. The values of PDOS in Fermi level are not zero, there are partial wave DOS across the Fermi level. These characteristics indicate that the system is analogous metal conductor. Atom Fe is itself a metal, the nature of the metal is due to the surface state of the Fe. In the Fig.7, there have respectively two peaks on both sides of the Fermi level [6,7]. The DOS between the two peaks is not zero. The energy discrepancy between the two peaks is 5.57eV, that is the pseudogap.

From the figure 5 and figure 8, the s orbital and p orbital of CO_2 move toward to low energy level. From the figure 6 and figure 9, the PDOS of the 3p orbital ,4s orbital and 3d orbital for atom Fe decreases. From the PDOS, the 4s orbital and 3d orbital of atom Fe contribute to the PDOS of O 2s for CO_2 near the -25eV.The 3d orbital of Fe and O 2p orbital of CO_2 resonate at the -14eV,-10eV,-7eV.

4. Conclusions

By simulation the CO_2 molecule adsorption on the Fe(111) surface, the CO_2 molecule adsorption on the Fe(111) surface. The pseudogap is 5.57eV in the PDOS of the CO_2 molecule adsorption on the Fe(111) surface.From the PDOS of the CO_2 molecule adsorption on the Fe(111) surface, the bond is strong covalent, the CO_2 molecule adsorption on the Fe(111) surface is stable. The values of PDOS in Fermi level are not zero, there are partial wave DOS across the Fermi level. These characteristics indicate that the system is the analogous metal conductor. The PDOS of the 3p orbital ,4s orbital and 3d orbital for atom Fe decreases. From the PDOS, the 4s orbital and 3d orbital of atom Fe contribute to the PDOS of O 2s for CO_2 near the -25eV.The 3d orbital of Fe and O 2p orbital of CO_2 resonate at the -14eV,-10eV,-7eV.

Acknowledgments

This work was supported by the financial support of the Natural Science Foundation of Shaanxi Province of China (2015JM1035) and the Special Foundation Item of the Key Academic Subjects Development of Shaanxi Province of China [Shaanxi Education Supporting (2008)169].

References

[1] Wang, Fugao, Landau, D. P. (Mar 2001) 2001 *Phys. Rev. Lett* **86** 2050
[2] J. F. Wu 2015 *Advanced Materials Research* **1061-1062** 315
[3] Y.C. Ding, M. XU, Y.B. Shen 2006 *Materials Review* **20** 116
[4] Jun-Fang Wu, Ying-Tang Zhang 2013 *Optik* **124** 2667
[5] K. Kimura, K. Oh shima, M. Mannami1995 *Phys Rev B* **5** 5737
[6] Jun-Fang Wu, Ying-Tang Zhang 2013 *Optik* **124** 2667
[7] Wu JF, Zhang CM 2010 *Chinese Physics B* **19** 034201-1

Study on Silver-plated Molybdenum Interconnected Materials for LEO Solar Cell Array

Jia-jun Zhu[1,2,*], Yu-hao Hu[1,2], Meng Xu[1,2], Wu-lin Yang[1,2], Li-cai Fu[1,2], De-yi Li[1,2] and Ling-ping Zhou[1,2]

[1] College of Materials Science and Engineering, Hunan University, Changsha 410082, China
[2] Hunan Province Key Laboratory for Spray Deposition Technology and Application, Hunan University, Changsha 410082, China

*spolestar@163.com

Abstract. Atomic oxygen (AO) is one of the most important environmental factors that affected the performance of low earth orbit spacecraft in orbit. In which, silver was the most common materials as the interconnected materials. However, with the poor AO resistance of silver, the interconnectors could be failure easier , and the lifetime of the spacecraft was also reduced. In this paper, the silver-plated molybdenum interconnected materials made by Ag thin films deposited on the Mo foils by vacuum deposition methods was studied. And the effects of the preparation process on the micro-structure of the Ag thin films, the interfacial adhesive strength and the electrical conductivity of the composites were investigated. It was found that the Ag thin films deposited on the Mo substrates coated the Ag thin films by ion beam assisted deposition(IBAD) methods exhibited a perfectly (200) preferred orientation. The interfacial adhesive strength had been increased to 18.58MPa. And the composites also have excellent electrical performance .

1. Introduction

All spacecraft in low Earth orbit (LEO, defined for altitudes between about 200-1000km), such as space shuttle, International Space Station, reconnaissance satellite, navigation satellite, are affected by different components of the LEO environment. The LEO space is a complex and dynamic environment that was consisted of high vacuum, radiation, temperature extremes, atomic oxygen(AO), charged particles, micrometeoroids, and man-made debris, which may severely affect the lifetime of the spacecraft. Among them, AO is the predominant and the most active component of LEO atmosphere. When the spacecraft is flying in LEO at an orbital velocity of 7.8 $km \cdot s^{-1}$, AO flux has impingement kinetic energy of about 3-5eV. At this energy, AO would initiate some chemical and physical reaction with the surface materials on spacecraft.[1-3].

So that, for LEO applications, the solar cell interconnectors have to sustain extremely high fatigue cycles and be subjected to the AO environment. It is well known that silver was one of the earliest interconnected materials for space solar cell panels. And the thickness of silver foil is approximately 30μm. However, in the case of silver, the AO reaction rate is about $10.5\times10^{-24} cm^3 \cdot atom^{-1}$[4], so a very high fraction of the incident AO reacts with the silver to form silver oxide. As figure 1 shows that the silver interconnectors from Eureca Program, the silver layer oxidized by atomic oxygen was flaked off. Due to the thermal stresses, the oxide layer exposing underlying fresh material, it continuously breaks

Content from this work may be used under the terms of the Creative Commons Attribution 3.0 licence. Any further distribution of this work must maintain attribution to the author(s) and the title of the work, journal citation and DOI.
Published under licence by IOP Publishing Ltd

up. One way to protect the silver of being oxidized by AO is by plating with gold. But as illustrated by figure 2, still a lots of pin-holes under the gold plating were created and silver-oxide was found at the surface[5]. So the average lifetime of the spacecrafts was only about 1-3 years.

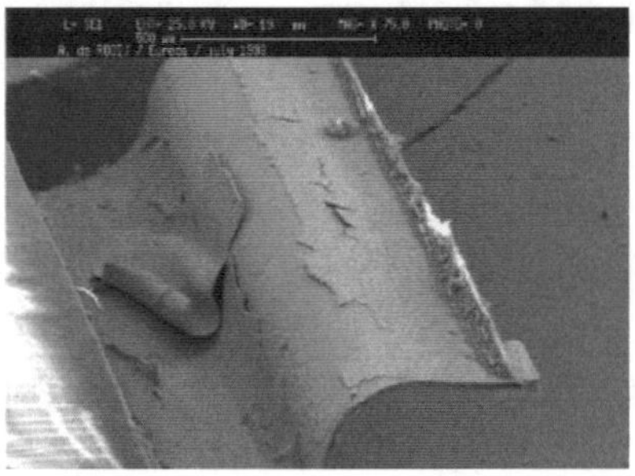

Figure 1. Silver interconnected materials from Eureca.

Figure 2. The pin-holes under the gold-plated silver foils.

For such extreme conditions, the interconnectors consist preferably of Kovar/lnvar [6-7]or molybdenum with applied silver coatings to allow welding, because of the good match of coefficients of expansion to silicon and, consequently, the relative lower incidence of cracks in the joints. But Kovar or Invar is ferromagnetic, which could induce some negative effects for space application. In comparison, molybdenum is still AO-resistant for its AO reaction rate about $1.4\times10^{-27}cm^3\cdot atom^{-1}$. The European Space Research Organization had performed resistance welding (parallel-gap) using silver-plated molybdenum(Ag/Mo) interconnectors, and no damage was found in temperature cycling test[8].

However, in the case of binary immiscible Ag-Mo system, which has an extremely large positive ΔHf being +57 kJ/mol. The interfacial adhesive strength between the Ag films and Mo substrate made by electroplating directly was very low. In this case, the metal transition layer was often needed to improve the adhesive strength, hence the process was complicated and the reliability of the materials was not high. Yuan HUANG[9] had researched that Ag/Mo composites was prepared by irradiation damage alloying method on Mo base. And the amorphous alloy layer was formed on the surface of the Mo base. In this paper, the Ag thin films deposited on the Mo substrates coated the Ag thin films by ion beam assisted deposition(IBAD) methods exhibited a perfectly (200) preferred orientation, while enhancing the adhesive strength of Ag/Mo interface greatly.

2. Experimental details

2.1. Deposition conditions

The deposition of Ag films was performed on homemade deposition system. Two Kaufman ion sources were set in the chamber as shown in figure 3(a), one of them called sputtering ion resource was used for sputtering an Ag target (purity >99.99%) with 1200eV Ar^+ ions, the other called assisted ion resource was used for bombarding the growing films with 30keV Ar^+ ions. And the assisted ion beam was incident along normal direction of film surface. The Magnetron sputtering devices were also installed in the same chamber as shown in figure 3(b).

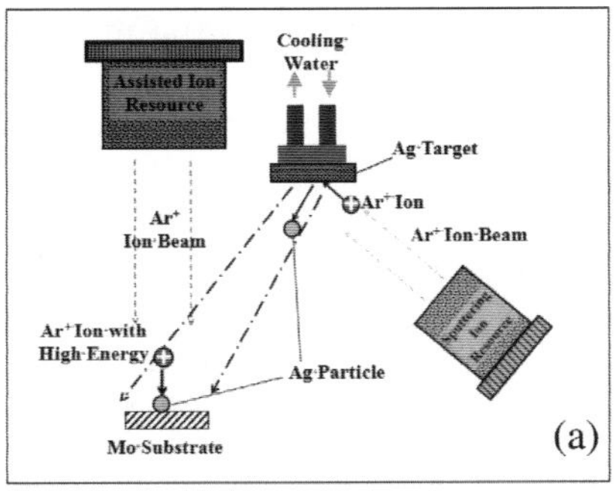

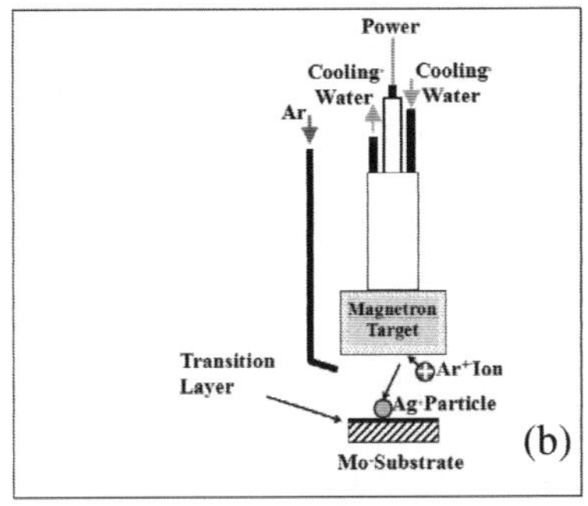

Figure 3. Schematic diagram of the Ag thin films deposited on Mo fiol by （a） ion beam assisted deposition （b） magnetron sputtering deposition

The Ag thin films was deposited on 12μm-thick Mo foil substrates by three sputter methods, which are ion beam sputtering deposition(IBD), ion beam assisted bombardment deposition(IBAD) and magnetron sputtering deposition(MD), respectively. The deposition method of Ag thin films only using sputtering ion resource without assisted ion resource was named IBD. The deposition of Ag thin films using sputtering ion resource, while the growing films were bombarded by assisted ion beam, which was named IBAD. Pure Ag (purity>99.99%) was used as a target materials, and the purity of argon as working gas was about 99.999%. The deposition chamber was evacuated to a background vacuum pressure of less than 5×10^{-4} Pa. The process parameters of deposited Ag thin films are shown in table 1. Before deposition, the surface of Mo foils was cleaned by 0.7keV and 60mA low energy ion beam about 15 minites.

Table 1. The process parameters of deposited Ag thin film

Assisted ion resouce		Sputtering ion resource			Magnetron sputtering			The background vacuum Pressure (Pa)
plate voltage (keV)	beam current (mA)	plate voltage (keV)	beam current (mA)	Working Gas Pressure (Pa)	Power (W)	Working Gas Pressure (Pa)	Temperature on substrates (K)	
15-35	2	2.5	60	2.2×10^{-2}	200	1.1	Self-heating	5×10^{-4}

2.2. *Evaluation*

Hitachi S4800 scanning electron microscope was used to observe the morphology of the composites. The orientation of Ag films was characterized by D5000 X-rays diffraction (XRD) instrument, with a CuKa radiation source.

In the case of the silver-plated molybdenum foils, the adhesive strength P between films and substrate was expressed as

$$P=F/A \tag{1}$$

Where, ***A*** and ***F*** are the area of the flaking Ag thin films and the maximum force, respectively, when the Ag thin films had been pulled off from Mo substrate.

The electrical properties of the samples were tested by SB2232 DC digital resistance test instrument. By measuring the resistance ***R***, area ***A*** and thickness ***L*** of the samples, the electrical conductance $\boldsymbol{\kappa}$ can be calculated according to the equation (2).

$$\kappa = 1/\rho = \mathrm{L}/(R \cdot A) \tag{2}$$

3. Results and discussions

3.1. *Orientation of the Ag thin films*

The XRD spectra of the Ag thin films on Mo(200) substrate by different deposition methods was shown as figure 4. It can be seen from figure 4(a) that the Ag thin films deposited by MD method, exhibited (111) and (200) mixed orientation, while the (111) diffraction peak intensity was about two times than (200); When the films was deposited by IBD method, the (200) peak was greatly weakened and exhibited high preferred (111) orientation, and the (111) diffraction peak intensity was about five times than (200), as seen in figure 4(b). However, as seen in figure 4(c), when the Ag thin films deposited by IBAD method, its (200) peak was greatly enhanced and the (200) diffraction peak intensity was about ten times than (111) . And on this basis, the Mo substrate coated the Ag thin films by IBAD method, was deposited a layer of Ag thin films again by MD method. It could be found that the Ag thin films was showed a perfectly (200) preferred orientation, as shown in figure 4(d).

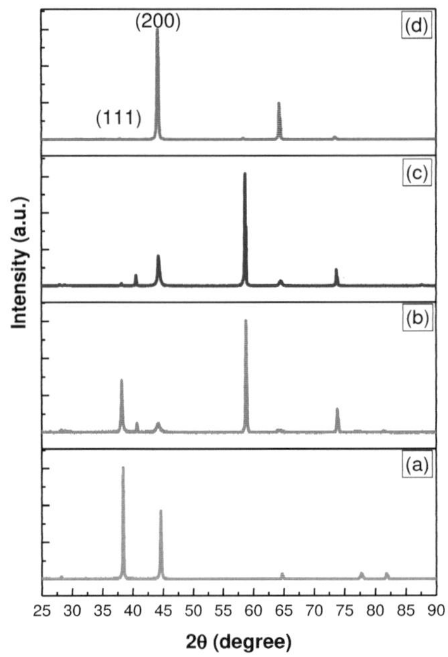

Figure 4. X-ray diffraction spectra of the Ag thin films on Mo(200) substrate (a) Deposited by MD(about 40 minutes), (b) Deposited by IBD(about 40 minutes), (c) Deposited by IBAD(about 40 minutes), (d) Deposited by MD(about 30 minutes) after IBAD(about 40 minutes).

The effect of IBAD on the preferred orientation of deposited films had been intensively studied[10-13]. In general, the prevailing model in which the texture transformation is attributed to a competition between anisotropic interface and strain energies, channeling effects, surface energy of the thin films, micro-structure and stresses of the substrate surface, and so on.

Usually, based on the thermodynamic stability theory, the thin films deposited by PVD (physical vapor deposition) grows with the plane of highest atomic density parallel to the substrate or the plane with minimum strain energy. For FCC (face-centered-cubic) materials such as Ag, its (111) crystal plane was the plane of highest atomic density as well as the lowest surface energy, so the thin films were highly preferred (111) orientation. Only if thought about surface energy, the preferred orientation were mainly (111) crystal planes, which had been reported for FCC metal films such as Au, Al, Ag, Cu and so on.

Compared to MD or IBD, the growing films were continuously bombarded by more high-energy Ar^+ ions with the assisted ion resource during IBAD. Namely, the energy of the ions was strong enough to clean the substrate's surface. Even if the assisted ion beam could induce atomic sputtering from surface, it could induce damage were significantly reduced when the ion beam was oriented along channeling directions of the films, which were named channeling effects. According to the study results about channeling effects[14-16], the sputtering yields in [111] direction is greater than that of in [100] direction in FCC materials. So, the growth speed of the films with (100) orientation was enhanced relatively, and the Ag films exhibited weak (111) and strong (200) mixed orientations. What is more, this layer of thin films would play an important role to induce the growth of the thin films deposited by MD along (200) orientation. Finally, the Ag thin films showed a perfectly (200) preferred orientation. Therefore, in the process of IBAD, the role of the channeling effects was more important than surface energy .

3.2. The adhesive strength between films and substrate

For silver-plated molybdenum interconnection-materials, because of the difference of the CTE (coefficient of thermal expansion) between Ag and Mo (Ag $19.2\times10^{-6}K^{-1}$, Mo $5.4\times10^{-6}K^{-1}$), the higher thermal stress was produced at the interface between the Ag films and the Mo substrates. Therefore the interconnectors would be bent over, broken and even distorted, if this thermal stress was higher enough. Based on the previous study [17], the interconnectors with sandwich-structure formed by the Mo foils deposited Ag thin films two sides should been designed for reducing some thermal stress. The cross-section micro-graph of this kind of Ag/Mo/Ag composites was shown as figure 5. It can be seen that the Ag films had very uniform thickness, and the cohesion of films/substrate were ideal.

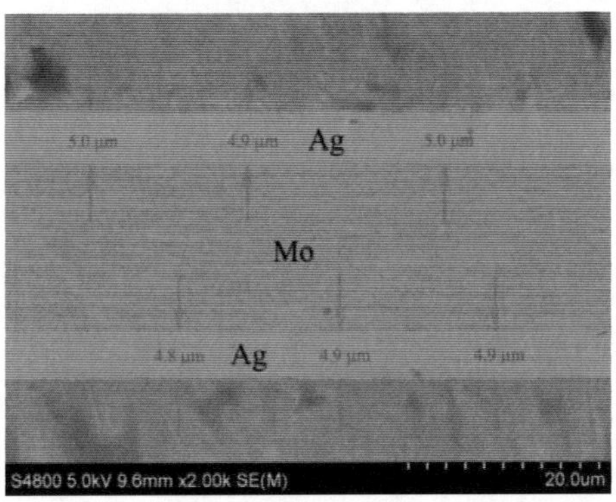

Figure 5. Cross sectional SEM image of the Ag/Mo/Ag composites made by MD after IBAD Ag thin films on the Mo foils.

The adhesive strength between the substrates and the films deposited by IBAD was significantly improved, which was attributed to the change of the interface bond model between the films and the substrate [18] mainly. It was found that the Ag thin films made by MD directly very easily desquamated from the substrates, because the combinative state between the films and the substrate could be physical mainly and the adhesive strength was lower. But during the IBAD process mentioned above, the growing film was continuously bombarded by a lots of high-energy Ar^+ ions, and a similar "ion implantation" effects was produced. So a part of Ag atoms could enter the Mo substrates to a depth of several nanometers from the surface, while a transition layer was made through the mixture of Ag and Mo atoms under the influence of ion beam. And the interficial combinative state between the films and the substrate was changed to metallurgical also. Moreover, the transition layer has a good thermal match with Ag or Mo because its CTE was between of them. The adhesive strength was significantly improved also. Moreover, the energy of ion beam had a profound impact on the adhesive strength during the IBAD process. The data including ion beam energy, the test results of the bond strength between the films and substrate, the area of the flaking Ag thin films and the maximum tensile force when the ion beam energy was changed, were given in table 2. Where, the interfacial adhesive strength was calculated by equation (1).

Table 2. Interfacial adhesive strength at different assisted ion beam energies

Samples' number	Assisted ion beam energy /keV	the area of the flaking Ag thin films /mm^2	Maximum tensile force /N	interfacial adhesive strength /MPa
1	15	10.22	58.8	5.75
2	20	9.72	63.7	6.55
3	25	9.87	121.03	12.26
4	30	9.23	171.5	18.58
5	35	11.21	117.6	10.49

According to the results of table 2, it can be found that the interfacial adhesive strength had been increased to 18.58MPa when the energy of ion beam was 30keV. With the increase of energy, the depth of bombardment and the thickness of the Ag-Mo mixed-layer were increased respectively. Therefore, the interfacial adhesion strength was increased with them. However, with the further increase of the ion beam's energies, the Ag particles entering the Mo substrate maybe too deep to form a effective transition layer on the substrate's surface, which made the interfacial adhesive strength decreased.

3.3. Electrical properties

The electrical performance of the silver-plated molybdenum composites was studied in this paper. The test results of the composites' electrical conductivity was shown as table 3. Where, the electrical donductivity of the composites was calculated by equation (2).

Table 3. Test results of electrical conductivity of composites

Thickness of the Mo foil /μm	Total thickness of Ag films both sides /μm	Resistance of the composites /mΩ	Electrical conductivity of the composites /$\times 10^6 S\cdot m^{-1}$
12	10	10.60	31.2
17	10	9.16	29.4

It can be seen that the increase of the electrical conductivity of composites along with the increases of Ag layer thickness ratio. So using the 12μm thick Mo foils, the current transmission loss would be reduced on some level. With 12μm thick Mo foil for example, according to the different current transmission modes in series or parallel, the calculated value of electrical conductivity of the composites was from $26.77\times 10^6 S\cdot m^{-1}$ to $38.51\times 10^6 S\cdot m^{-1}$. And the test results was $31.2\times 10^6 S\cdot m^{-1}$, which illustrated that the interfacial contact resistances were very small and further explained that the combination between the films and the substrate was very close.

4. Conclusions

For interconnected materials of solar cell array, compared to the pure silver foils or gold-plated silver foils, the silver-plated molybdenum composites have more excellent AO resistance performance. The preparation and performance of the silver-plated molybdenum composites was studied in this paper. It was found that the Ag thin films deposited by different preparation process showed different crystal orientation. Using of IBAD methods, the interfacial adhesive strength could been increased dramatically, and hence the electrical conductivity of the composites was more than $30\times 10^6 S\cdot m^{-1}$. Which could meet the requirements of the longtime for the LEO spacecraft.

Acknowledgments

This work is supported by the National Natural Science Foundation of China under grant No. 51401080.

References

[1] S W Samwel 2014 Space Research Journal **7** p 1
[2] Dooling D and Finckenor MM 1999 NASA **TP-1999-209260** p 1
[3] LEE A and G Rhoads 1985 20th Thermophysics Conference p 1065
[4] M Raja Reddy 1995 Journal of Materials Science **30** p 281
[5] de Rooij A 2010 ESA/ESTEC EUROCORR (Moscow) p 1
[6] George C 1976 NASA-CR-**144312** p 1
[7] Silverman E M 1995 NASA-CR-**4661** p 1
[8] W Luft 1971 Transactions on aerospace and electronic systems p 781
[9] Yuan HUANG, De-yue KONG and Fang HE 2012 Acta metallurgica sinica **48** p 1253
[10] Jian-min ZHANG and Ke-wei XU 2003 ACTA PHYSICA SINICA **52** p 145
[11] Tao FENG and Bing-yao JIANG et al. 2008 Applied Surface Science **254** p 1565

[12] Huan Ma and Yu Zou Alla et al. 2015 Acta Materialia **98** p 17
[13] V A Gasparov and M Riehl-Chudoba 2007 Surface Science **601** p 5403
[14] Lindhard J 1965 Kgl. Dan. Vidensk. Selsk., Mat.-Fys. Medd **34** p 64
[15] J W Mayer and E Rimini 1977 *Ion Beam Handbook for Material Analysis*(New York: Academic Press) p 67
[16] Matteo Seita and Christina M Pecnik et al. 2010 Acta Materialia **58** p 6513
[17] Pei QIU and Jia-jun ZHU 2015 Materials Review **29** p 143
[18] Ling-ping ZHOU and Ming-pu WANG et al. 2008 Transactions of Nonferrous Metals Society of China **18** p 372

Micromilling of Ti-6Al-4V Titanium Alloy Using Ball-end Tool

T Pratap[1*] and K Patra[2]

[1, 2] Micro-fabrication Lab, Department of Mechanical Engineering, Indian Institute of Technology Patna, Bihta, Patna-801103, Bihar, India

* tejpratap.pme13@iitp.ac.in

Abstract. The present work experimentally investigates the characteristics of micromilling of Ti-6Al-4V workpiece using ball-end tool. Micromilling were performed using 400 μm ball-end milling tool with the consideration of effective tool diameter. The effective tool diameter in ball-end milling is different from flat-end milling due to variation in cutting zone with axial depth of cut which accounts actual cutting speed and feed the machining. Process input variables includes cutting speed, feed per tooth and axial depth of cut, each with three levels, whereas cutting forces and surface roughness as process responses. Response surface methodology (RSM) based quadratic regression analysis was utilized to develop the correlation between process inputs and responses. Experimental results of present study on micromilling of Ti-6Al-4V using ball-end tool showed that axial depth of cut and feed per tooth could significantly influence the cutting forces and surface roughness. However, present study shows cutting speed has least effect on process responses. Finally, suitable set of process variables were proposed based multi objective optimization approach to obtain minimum cutting forces and surface roughness which were also validated with conformation test results.

1. Introduction

The applications of highly precise complex three-dimensional micro-features have grown over the years to enhance the performances of micro-fluidic devices, electronic chips, biomedical devices, aerospace components, optical lens, etc [1, 2]. Effective techniques to fabricate these micro-features are highly dependent on the applications, the materials to be used and the geometrical complexity [2]. However, with the ease of material processing, applicability on wide range of materials processing, higher material removal rate (MRR), operational flexibility, good accuracy, etc., tool based micromachining is considered as a effective technique among available micro-fabrication techniques for mass production of such micro-features [1-3]. Micromilling is one of the better choice among tool based processes for producing such complex features with three dimensional geometries and its applicability even enhanced by the selection of ball-end milling tool. Because of its hemispherical geometry it provides additional ability to produce complex edge free surfaces, good finish, fewer burrs, mold cavities and various surface-textures which is highly demanding nowadays [1-5].

Ti-6Al-4V titanium alloy is an extensively applied material in biomedical implants, automotive suspensions and several aerospace equipment for its superior physical and chemical properties like higher specific strength, non-toxic nature, biocompatibility and good fatigue and corrosion resistance

Content from this work may be used under the terms of the Creative Commons Attribution 3.0 licence. Any further distribution of this work must maintain attribution to the author(s) and the title of the work, journal citation and DOI.
Published under licence by IOP Publishing Ltd

even at extremely high temperatures [3, 6-7]. However, Ti-6Al-4V machining is challenging due to its unfavourable machining natures like: lack of elastic modulus, low heat conductivity, etc., which promotes higher temperature generation leading to influence in reactivity with tool material, promotes tool wear and failure, etc. [6-9]. In addition to these factors, scaling of micro-tools (Size effect) added further challenges in machining [7, 10]. Moreover, downscaling of the tool size increases aspect ratio, which leads to breakage of the tool even for very small discrepancy in tool overhang [7, 10, 11].

Along with the aforementioned micromilling issues, selection of ball-end milling tool produces ball tip rubbing due to variation in cutting speed along hemispherical portion (zero at the ball tip to highest at extreme of tool workpiece contact). The rubbing phenomenon will be more predominant for smaller axial depth of cut, which is generally required for hard and brittle materials [11, 12]. It significantly affects the chip formation mechanics that causes variation in cutting forces, surface quality and reduced tool life. To overcome its adverse effect, inclined milling is generally preferred which reduces the actual cutting time and allows the proper cooling, thus, decreases the rise in temperature and tool wear [12, 13]. However, tool inclination changes the machined profile which significantly affects the application and used based on the requirement. Berestovskyi *et al.* [11] showed how profile of the machined micro channel varied with the flute length of micro ball-end milling tool. Arif *et al.* [12] also reported that surface profile varies with any discrepancy in workpiece flatness or uneven mounting with ball-end tool. Qui *et al.* [13] observed the behaviour of ball-end tool inclination on surface finish and cutting forces in milling of quartz glass. Another challenging issue with micromilling is burr formation control, which reduces the quality and functionality [14]. However, from the literature it was found that smaller size top burr was produced by selection of ball-end milling tool compared to flat-end milling tool due to the burr suppressed by ball-end tool [15].

However, from literature, it was found that selection of suitable surface profile depends upon the application and to fabricate such profiles required proper selection of tool geometry and its position with respect to workpiece. It was also observed that, by the consideration of effective tool diameter of ball-end milling tool, micromilling performance characteristics such as; cutting forces, surface roughness, machined surface profile, burr formation are not optimized. This change in effective tool diameter is due to the variation in axial depth. Therefore, contrary to micro-flat-end milling where cutting speed does not vary with cutting depth at a particular spindle rotation, effective cutting speed of micro-ball-end tool varies with the axial direction cutting depth. Therefore, the present work targets the experimental investigation of micromilling of Ti-6Al-4V using ball-end tool considering effective tool diameter. Process responses are observed in terms of cutting forces and surface roughness whereas process inputs are cutting speed (V), feed per tooth (f_t) and axial depth of cut ($ADOC$). Process responses are predicted using quadratic regression models and RSM based plots are utilized to describe the parametric effects on individual response. Finally, multi-objective optimization has been performed by desirability function approach to select the optimal combination.

2. Experimental methodology

2.1. Experimental details

Test samples were fixed on the top surface of dynamometer and facing is performed using flat-end milling at micromachining centre (DT-110, Mikrotools Pte. Ltd.). It reduce the chance of any flatness error which is further ensured by the use of contact probe. After the face milling, fresh solid carbide two fluted micro ball-end milling tool of diameter 400 μm is used to perform the experiments under dry condition with three repetitions considering effective tool diameter as shown in the figure 1. Positional accuracy of ±1 μm per 100 mm of travel is achieved by its programmable multi axis controller. The cutting forces (F_x, F_y and F_z) are acquired in time domain using piezo electric dynamometer (MiniDyn Type 9256C2) at sampling rate of 50 kHz and average of three experimental root mean square (RMS) values are considered for further analysis. Unwanted chips and wear debris of the tool presented are ultrasonically cleaned under acetone solution and dried with compressed air. Surface roughness (R_a) was measured by digital type surface analyzer (Mitutoyo, SJ-400) at three different locations along the slot base and average value has been reported. Machined surface profile was also investigated by zeta-20, 3D optical profiler at three different locations to see the consistency of the machined surface profile.

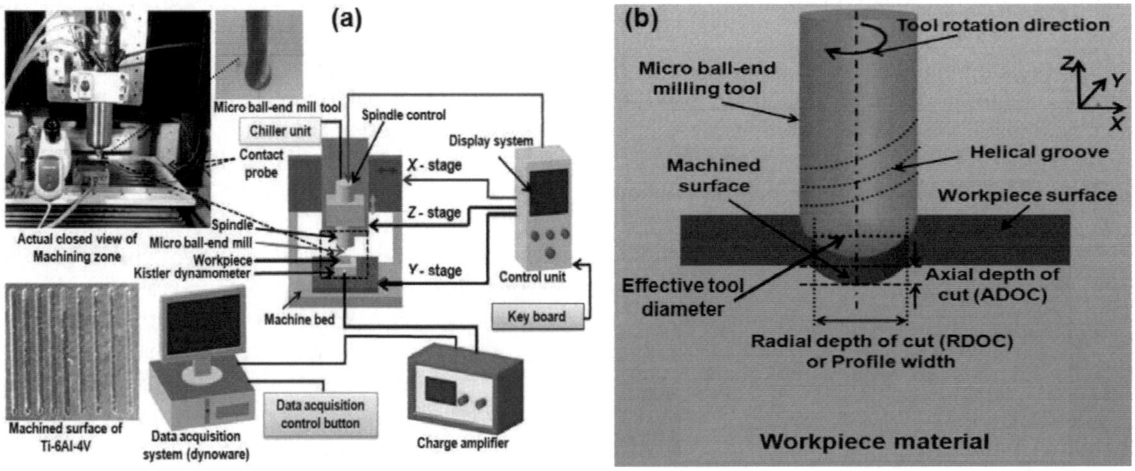

Figure 1. (a) Experimental setup used for micromilling of Ti-6Al-4V (b) Schematic representation of effective tool diameter in ball-end milling tool for given axial depth of cut

2.2. Effective diameter calculation and modified process inputs

Effective tool diameter (D_e) are calculated based on tool geometry and *ADOC* using the equation (1). This effective diameter is used to calculate spindle speed (N) for designed value of cutting speed (V) by equation (2) and finally, modified spindle speed is multiplied by selected feed per tooth (f_t) to achieve actual feed rate as per equation (3).

$$D_e = 2\sqrt{R^2 - (R - ADOC)^2} \quad (1)$$

$$V = \pi D_e N \quad (2)$$

$$f = z f_t N \quad (3)$$

where, D_e = effective tool diameter (μm)
R = radius of ball end tool (μm)
$ADOC$ = axial depth of cut in (μm)
V = cutting speed (m/min)
N = modified spindle speed in rpm
f = feed rate (mm/min)
z = number of tooth and
f_t = feed per tooth (μm/tooth)

2.3. Design of Experiments (DOE) and process responses

To select suitable process variables, experiments are designed by DOE with three factor (V, f_t and *ADOC*) and three levels. Process responses for the given set of experiments are measured as cutting forces F_x, F_y and F_z along tangential, feed and thrust direction, respectively, and surface roughness (R_a) along the slot base as shown in Table 1. RSM based quadratic regression expressions are also developed to relate the process input to each of the response and compared with experimental results to check the maximum possible error present in the developed model. Design-Expert 9 software package (Stat-Ease Inc., USA) has been utilized to plot response surface which provides the idea of influencing factor for the given sets of experiments.

Table 1. Experimental outcomes of micromilling of Ti-6Al-4V using ball-end milling tool

Process inputs with their assigned levels				Process responses			
Cutting condition	V (m/min)	f_t (μm/tooth)	$ADOC$ (μm)	F_x (N)	F_y (N)	F_z (N)	R_a (μm)
1	8	0.5	10	0.2081	0.3356	1.7112	0.321
2	8	1.5	20	0.4511	0.6010	2.1386	0.192
3	8	2.5	30	0.8308	0.9142	2.5735	0.167
4	16	1.5	30	0.6016	0.7937	2.3484	0.159
5	16	2.5	10	0.6873	0.6904	2.1735	0.191
6	16	0.5	20	0.3606	0.5989	1.8772	0.269
7	24	2.5	20	0.7761	0.7797	2.2329	0.174
8	24	0.5	30	0.5526	0.7863	2.1829	0.177
9	24	1.5	10	0.3822	0.5193	1.9772	0.262

3. Experimental results and discussions

Quadratic regression expressions of the predicted cutting forces and surface roughness are given in equation (4), (5), (6) and (7), respectively. Predicted results of each of the response are compared with the experimental results and shown in figure 2 (a-d). Observation also indicates that predicted values are well suited with the error bar plot for all the responses. From the all proposed mathematical expression, it was observed that cutting velocity (V) term have very smallest co-efficient and hence, less contribute to the prediction of given responses. Therefore, influences of the significant process inputs f_t and $ADOC$ on the process responses F_x, F_y, F_z and R_a are described with the help of three dimensional response surface plots shown in figure 3 (a-d). figure 3 (a) and (b) show the variation of F_x and F_y with $ADOC$ and f_t, respectively. Response surfaces in both plots indicate that with increase in $ADOC$ and f_t both tangential force (F_x) and feed force (F_y) increases. The increase in force is due to the increase in effective chip area with increasing value of $ADOC$ and f_t. Contrary to micro-flat-end milling, thrust force (F_z) in micro-ball-end milling is one order more than tangential and feed forces due to rubbing of the ball-end of the tool with respect to workpiece. Figure 3 (c) shows that thrust force (F_z) increases with increasing $ADOC$ and f_t caused by increase in effective cutting zone. The variation in effective tool diameter by the change in $ADOC$ is the major contributor for the variation in cutting zone. Response surface plot shown in figure 3 (d) shows that with increase in f_t and $ADOC$ surface roughness decreases. It shows that with the increase in f_t and $ADOC$, ploughing and rubbing effect reduces and results the smoother surface [6, 10-12]. Similarly for low axial depth of cut, rubbing effect is more dominant than cutting and it leads to more roughness of the workpiece surface.

$$F_x = +0.00784 + 0.00234\,V - 0.00493\,f_t + 0.01722\,ADOC - 0.00362\,f_t\,ADOC + 0.09095\,f_t^2 \tag{4}$$

$$F_y = +0.12768 + 0.00198\,V + 0.03842\,f_t + 0.02146\,ADOC - 0.00399\,f_t\,ADOC + 0.04952\,f_t^2 \tag{5}$$

$$F_z = +1.61994 - 0.0028\,V + 0.27091\,f_t - 0.00536\,ADOC - 0.00347 f_t\,ADOC + 0.00078\,ADOC^2 \tag{6}$$

$$R_a = +0.47233 + 0.00076\,V - 0.14487\,f_t - 0.00971\,ADOC + 0.00346\,f_t\,ADOC + 0.01217\,f_t^2 \tag{7}$$

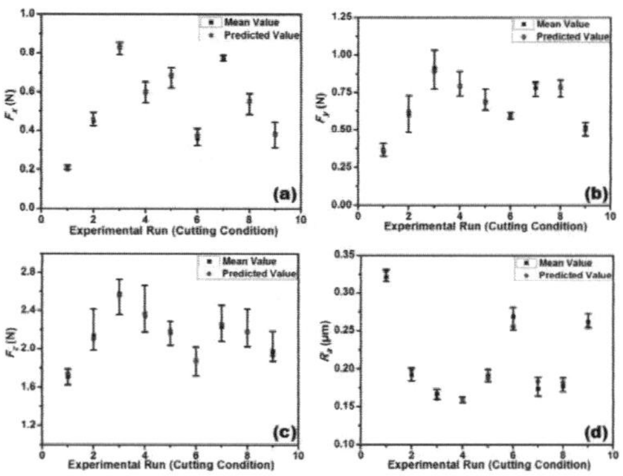

Figure 2. Predicted and experimental values with error bar (a) F_x (b) F_y (c) F_z and (d) R_a

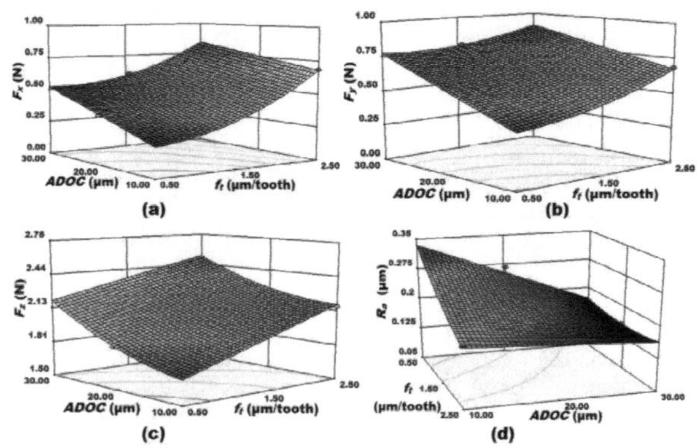

Figure 3. Effects of *ADOC* and f_t on cutting performances (a) F_x (b) F_y (c) F_z and (d) R_a

Machined surface profile variations are also investigated by measuring surface profile at three locations along the length for all the channels to check the accuracy of the produced surface. Figure 4 shows the machined surface profile variation along with their respective three dimensional views. It can be observed from the machined surface profiles that cross-sectional features are almost similar in nature with minor variation and there is almost no burr on the top surface of the workpiece. This shows that micro-ball-end mill is capable to produce surfaces of repeatable profiles along with minimum top burr. Ball-tip rubbing zone along the centre line are also observed which significantly effects on the surface roughness as observed in the experimental results.

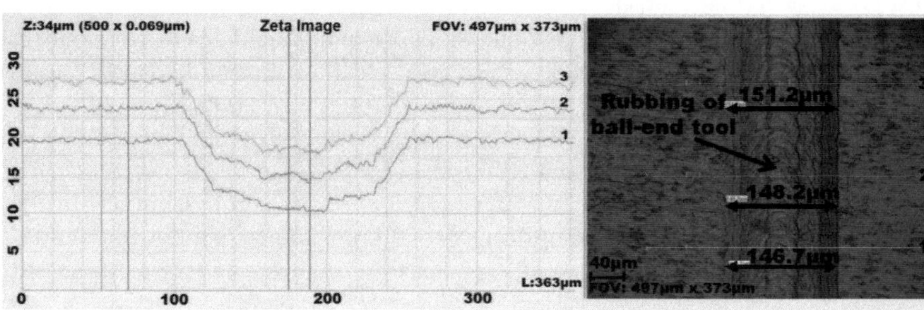

Figure 4. Machined surface profile measurement using Zeta-20 optical profiler for cutting condition: V = 8m/min, f_t = 0.5 µm/tooth and *ADOC* = 10 µm

From the response surface analysis we have also found that cutting forces increases with increase in f_t and *ADOC* while surface roughness reduces. As the nature of cutting forces and surface roughness are reverse in nature, there should be a compromise to fulfil the condition of minimum cutting forces along with smother surface quality. This can be performed by multi-objective optimization using desirability function approach of Design-Expert 9 software package. The constraints of the process inputs are selected within the performed range and process response are of minimization type. From the set of optimized solution, minimum cutting forces (F_x, F_y and F_z) are obtained at V = 8 m/min, f_t = 1.34 µm/tooth and *ADOC* = 10.79 µm. However, for minimum surface finish (R_a) optimum condition is obtained at V = 8 m/min, f_t = 1.28 µm/tooth and *ADOC* = 12.62 µm. To validate the optimized results, fresh experiments were conducted at V = 8m/min, f_t = 1.34 µm/tooth and *ADOC* = 10.79 µm and compared with the predicted optimized results. The maximum error for cutting forces (F_x, F_y and F_z) are found to be 4.02%, 2.49% and 2.09%, respectively, whereas for surface roughness (R_a) maximum error is 4.71%.

4. Conclusions

IOP Conf. Series: Materials Science and Engineering **229** (2017) 012011 doi:10.1088/1757-899X/229/1/012011

The present work, effects of process inputs on the cutting forces (F_x, F_y and F_z) and surface roughness (R_a) have been investigated through micromilling of Ti-6Al-4V using ball-end milling tool considering effective tool diameter. Research findings are concluded with the following points.

- For axial depth of cut (*ADOC*) below tool radius, cutting speed (*V*) and profile width changes accordingly due to variation in effective tool diameter (D_e). To maintain constant V and f_t, N and f must be modify.
- Based on the developed response surface plots, it is observed that f_t and *ADOC* significantly influence the cutting forces and surface roughness, whereas they are found to of opposite in nature. The selected range of cutting speed (*V*) is least significant for given responses.
- Machined profile is found to be similar in nature.
- Micromilling with ball-end tool is capable to reduce the top burr which is a major challenge with flat-end tool. However, rubbing of ball-end tip has also been observed along slot base.
- The validation experiments results are found to be comparable with optimized predicted values.

References

[1] Kumar SPL, Jerald J, Kumanan S and Prabakaran R 2014 A review on current research aspect in tool-based micromachining processes *Materials and Manufacturing Processes* **29** 1291-1337.

[2] Camara MA, Rubio JCC, Abrao AM and Davim JP 2012 State of the art on micromilling of materials, a review *Journal of Materials Science & Technology* **28(8)** 673-685.

[3] Mamedov A and Lazoglu I 2015 Micro ball-end milling of freeform titanium parts *Advances in Manufacturing* **3(4)** 263-268.

[4] Özel T and Liu X 2009 Investigations on mechanics-based process planning of micro-end milling in machining mold cavities *Materials and Manufacturing Processes* **24(12)** 1274-1281.

[5] Baburaj M, Ghosh A and Shunmugam MS 2016 CAD based simulation of ball end mill manufacturing *Computer-Aided Design and Applications* doi: 10.1080/16864360.2016.1199754

[6] Hassanpour H, Sadeghi MH, Rezaei H and Rasti A 2016 Experimental study of cutting force, microhardness, surface roughness, and burr size on micromilling of Ti6Al4V in minimum quantity lubrication *Materials and Manufacturing Processes* doi: 10.1080/10426914.2015.1117629

[7] Pratap T, Patra K and Dyakonov AA 2015 Modeling cutting force in micro-milling of Ti-6Al-4V titanium alloy *Procedia Engineering* **129** 134-139.

[8] Garcia U and Ribeiro MV 2016 Ti6Al4V titanium alloy end milling with minimum quantity of fluid technique use *Materials and Manufacturing Processes* **31(7)** 905-918.

[9] Davim JP 2014 *Machining of titanium alloys* Springer, Heidelberg.

[10] Oliveira FBD, Rodrigues AR, Coelho RT and Souza AFD 2015 Size effect and minimum chip thickness in micromilling *International Journal of Machine Tools & Manufacture* **89** 39-54.

[11] Berestovskyi D, Hung WNP and Lomeli P 2014 Surface finish of ball-end milled microchannels *Transaction ASME Journal of Micro-Nano Manufacturing* **2** 041005-1-10.

[12] Arif M, Rahman M and San WY 2012 An experimental investigation into micro ball-end milling of silicon *Journal of Manufacturing Processes* **14** 52-61.

[13] Qiu Y, Gu ML, Zhang FG and Wei Z 2014 Influence of tool inclination on micro-ball-end milling of quartz glass *Materials and Manufacturing Processes* **29(11-12)** 1436-1440.

[14] Aurich JC, Dornfeld D, Arrazola PJ, Franke V, Leitz L and Min S 2009 Burrs-analysis, control and removal *CIRP Annals – Manufacturing Technology* **58** 519-542.

[15] Chen MJ, Ni HB, Wang ZJ and Jiang Y 2012 Research on the modeling of burr formation process in micro-ball end milling operation on Ti-6Al-4V *The International Journal of Advanced Manufacturing Technology* **62** 901-912.

IOP Conf. Series: Materials Science and Engineering **229** (2017) 012012 doi:10.1088/1757-899X/229/1/012012

Experimental and Numerical Investigations of Applying Tip-bottomed Tool for Bending Advanced Ultra-high Strength Steel Sheet

Pusit Mitsomwang[1,*], Rattana Borrisutthekul[1], Ken Klaiw-awoot[1] and Aran Pattalung[1]

[1]School of Metallurgical Engineering, Suranaree University of Technology, 111 University avenue, Muang, Nakhon Ratchasima, 30000, Thailand

*pusit@sut.ac.th

Abstract. This research was carried out aiming to investigate the application of a tip-bottomed tool for bending an advanced ultra-high strength steel sheet. The V-die bending experiment of a dual phase steel (DP980) sheet which had a thickness of 1.6 mm was executed using a conventional bending and a tip-bottomed punches. Experimental results revealed that the springback of the bent worksheet in the case of the tip-bottomed punch was less than that of the conventional punch case. To further discuss bending characteristics, a finite element (FE) model was developed and used to simulate the bending of the worksheet. From the FE analysis, it was found that the application of the tip-bottomed punch contributed the plastic deformation to occur at the bending region. Consequently, the springback of the worksheet reduced. In addition, the width of the punch tip was found to affect the deformation at the bending region and determined the springback of the bent worksheet. Moreover, the use of the tip-bottomed punch resulted in the apparent increase of the surface hardness of the bent worksheet, compared to the bending with the conventional punch.

1. Introduction

Due to the requirement of light weight parts and/or structures, heavy sheet materials tend to be replaced by sheets which have higher strength-to-weight ratio. Among high strength materials, advanced ultra-high strength steels are developed and increasingly used. Koo and Luton invented a high strength steel sheet having a yield strength higher than 830 MPa. The high strength was obtained from the formation of the tempered martensite, bainite and the precipitated fine carbide [1]. Vandepputte et al. reported the production of an ultra-high strength steel sheet. By controlling the chemical compositions and the rolling processes, the multiple phase steel consisting of ferrite, martensite and bainite, was produced. Its guaranteed minimum tensile strength was approximately 1000 MPa [2]. Also, Zhong et al. developed a carbon-manganese steel sheet. Its compositions allowed the transformation of austenite into ferrite and martensite during the heat treatment following the rolling. The ultimate strength of this steel reached 1150 MPa [3]. Although the ultra-high strength steels possess the advantages in the permanent deformation resistance and the high failure strength,

Content from this work may be used under the terms of the Creative Commons Attribution 3.0 licence. Any further distribution of this work must maintain attribution to the author(s) and the title of the work, journal citation and DOI.
Published under licence by IOP Publishing Ltd

their formability are typically limited, especially when compared with lower strength steels such as mild steel.

Springback, a shape error of formed parts, is one of the most severe problems in the bending and other sheet forming processes of ultra-high strength steel sheets. Our previous study showed that the dramatic springback of the ultra-high strength steel sheet mainly caused by its high yield strength and the presence of the large elastic deformation portion in the bending region of the worksheet. Based on this understanding, a key to successfully reduce the springback of the high strength steel sheet seems to be the reduction of the elastic deformation and/or increase the plastic deformation portions at the bending region. There are few techniques used for reducing the springback of high strength steel sheets during bending. Mori et al. employed an electrical resistance heater to heat up a SPFC980Y high-strength steel sheet before subjecting them to an U-die bending tool. They found that the springback of the bent worksheet decreased when the worksheet temperature increased from room temperature to 800 °C [4]. Lee et al. studied the 90° V-die bending of a DP980 steel sheet heated by a near-infrared rays. The angle of the bent sheet reduced from 95° (bent at room temperature) to 90.5° when bending at 600 °C. In these research works, the springback reduction seemed to be contributed by the lower yield strength of the steel sheets under elevated temperature [5]. Yanagimoto and Oyamada also carried out the V-die bending of HSS-1 high-strength steel sheet. To increase the plastic deformation portion during bending, they derived the benefit from high-temperature creep of the steel and detected this phenomenon when the worksheet temperature was higher than 477 °C [6]. Although, the bending under elevated temperature was effectively used for suppressing the springback of the high strength steel sheets. However, since steels are rapidly oxidized at elevated temperature and normal atmosphere, the surface quality of steel sheets after hot bending become problematic. In addition, another concern of the hot bending is that the strength of steels would be deteriorated.

Owing to the limitation mentioned above, bending techniques that can reduce the springback of ultra-high strength steels and are performed at room temperature are still necessary. Semiatin explained an attractive bending method for reducing the springback of bent metal sheets, namely the bending with a punch having a bottom [7]. Using this type of punch, a high compressive stress occurred at the bending region facilitates the plastic deformation of bending metals. Then, the springback of bent sheets would be reduced. However, until now, the author has found no reports concerning the application of a tip-bottomed tool for bending high strength steel sheets. Therefore, in this research work, the bending of an ultra-high strength steel sheet with a tip-bottomed punch was investigated, experimentally at room temperature. The applicability of the tip-bottomed punch for suppressing the springback of the ultra-high strength steel was examined. In addition, the deformation of the worksheet during bending was simulated using FE analysis, compared with that of the bending using a conventional tool and discussed. Moreover, the micro Vicker hardenss test was conducted for bent worksheets to look into the change of mechanical properties due to the tip-bottomed punch bending.

2. Experimental investigation of bending

2.1 Material and experimental procedhres

The material used in this research work was an advanced ultra-high strength dual phase steel (DP980) sheet which had a thickness of 1.6 mm. Table 1 and 2 shows its chemical compositions and mechanical properties. Before bending test, raw DP980 sheet was cut into the bending worksheets, which had a width of 20 mm and a length of 55 mm using a laser cutting machine. Then, the worksheets were cleaned ethanol and dried.

Table 1 Chemical compositions of DP980 ultra-high strength steel sheet.

Chemical compositions / wt%									
C	Mn	Si	Cr	S	P	Mo	Ni	Al	Fe
0.097	2.27	0.394	0.70	0.001	0.008	0.08	0.013	0.04	Bal.

To bend the worksheets, the bending die-set shown in figure 1 was prepared and utilized. There were four springs installed at the guide pin of the die-set. Each spring had a stiffness of 2.96 N·mm^{-1}. During bending, the upper punch attached to the punch block was moved downward by the moving cross head of a universal testing machine (UTM, Model: Instron 5582), while the lower die attached to the die block was stationary. The velocity of the moving cross head (V) was fixed to be 2.5 mm·min^{-1}. In order to measure the bending load resistance of the worksheet, a load cell, which had a capacity of ±100 kN) was installed between the moving cross head of the UTM and the punch shoe.

Table 2 Mechanical properties of DP980 ultra-high strength steel sheet (Test method: Uni-axial tension, Standard: Based on ASTM E8, Velocity of moving cross head: 5 mm·min^{-S}).

Mechanical properties of steel sheet			
Young's modulus E / GPa	0.2% offset yield strength σ_{YS} / MPa	Tensile strength σ_{TS} / MPa	Breaking strain ε_{Break}
179.7	714.4	1058.2	0.099

To examine the applicability of the bottomed-tip punch for bending the ultra-high strength steel worksheet, two bending experiments were executed. First, the worksheets were bent using a conventional 90° punch and 90° die, as shown in figure 2 (a). Second, a 90° punch which had a tip at the bottom, as illustrated in figure 2 (b), was used. In this case, the die dimension was the same as that used in the first experiment. The punch and die were made of a JIS-SKD11 cold-work tool steel which had a hardness ranging from 60 to 63 HRC. For both experiments, the maximum bending depth $d_{B(Max.)}$ was fixed as 13.275 mm. The $d_{B(Max.)}$ was defined clearly in figure 2 (a). After reaching the maximum bending depth, the punch was lifted upward by the springs. Then, the bent worksheets were removed from the die-set to measure the springback. In this work, the bending experiment was carried out five times for each punch case. Finally, to investigate the mechanical property at the bending region of the bent worksheet, the micro Vickers hardness test was conducted.

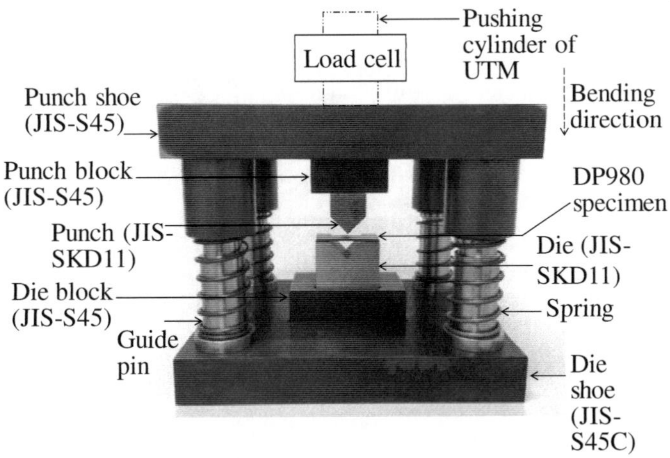

Figure 1 Die set used for bending experiments.

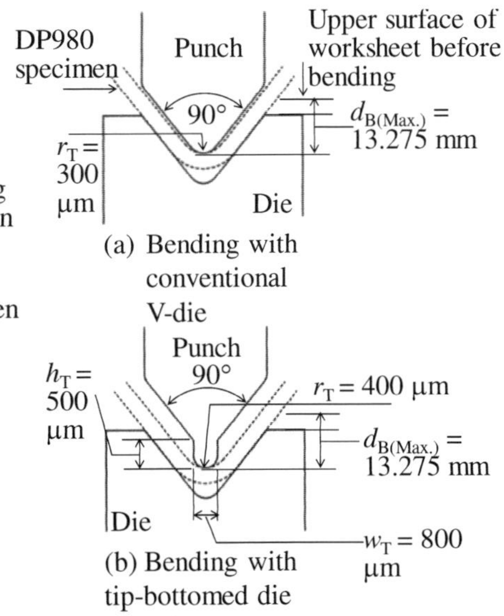

Figure 2 Bending experiments of DP980 sheet.

2.2 Experimental results of bending

Figure 3 shows the relationship between the bending load resistance of the worksheet and the indentation depth of punch for the bending with the conventional punch (C.P.) and the tip-bottomed punch (T.B.P.). In this representation, the indentation depth of zero is the position at which the tip of the punch touches the top surface of the worksheet. As shown in the figure, it was seen that the bending load resistance of the worksheet for the both punch cases was similar for $0 < d_B \leq 13.1$ mm. Then, when the indentation depth increased to $d_B \approx 13.3$ mm, the peak point of the bending load resistance ($f_{B(Peak)}$) for both tool cases was observed. Here, $f_{B(Peak)}$ in the case of the T.B.P was found to be higher than that of the C.P. case. The occurrence of higher bending load resistance at the identical indentation depth of punch seemed to indicate that higher strain hardening and/or the larger plastic deformation portion of the worksheet occurred when using the T.B.P.

After bending, the springback of the worksheets was visually examined. Figure 4 shows the representative worksheets bent by the C.P. and the T.B.P. To investigate the springback of the worksheet more precisely, the springback angle of the bent worksheets was measured by an image analyzer. Then, the springback factor (k_s), as described in figure 5, was calculated and plotted with respect to the types of the punches used, as shown in figure 6. From figures 4 and 6, it was found that the springback of the worksheet bent by the T.B.P was apparently less than that of the C.P case. The experimental results revealed that the application of the T.B.P seemed to contribute the increase of the plastic deformation portion at the bending region. Consequently, the springback of the bent worksheet decreased. To furthermore discuss the deformation and the springback of the bent worksheet, the FE analysis of the bending was carried out. Its results will be shown in the next sections.

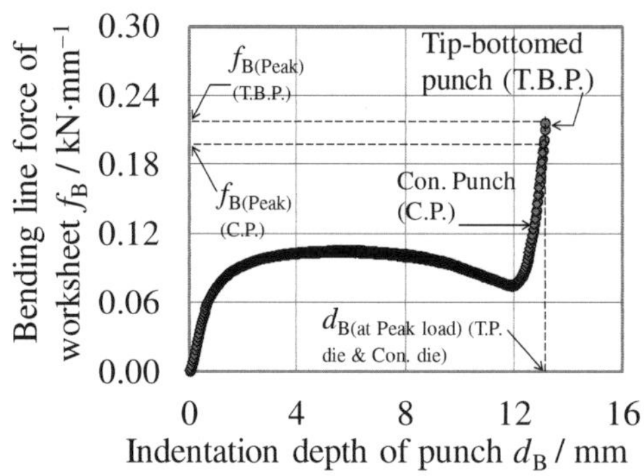

Figure 3 Bending load resistance of worksheet for both tool cases.

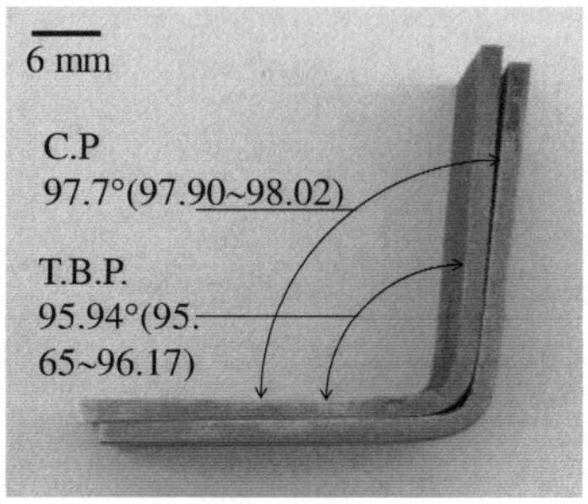

Figure 4 Springback of worksheet for both tool cases.

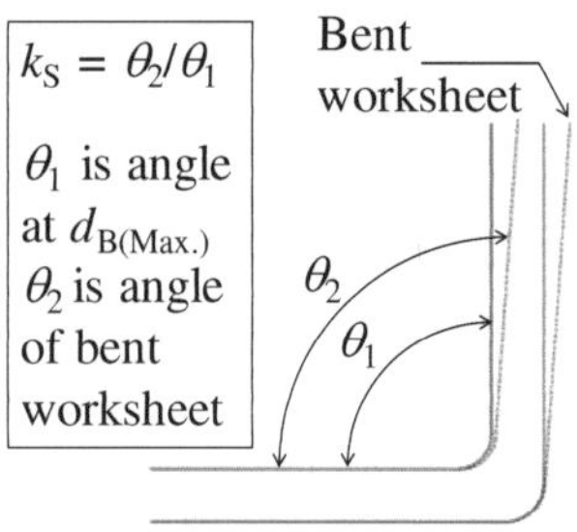

Figure 5 Definition of springback factor.

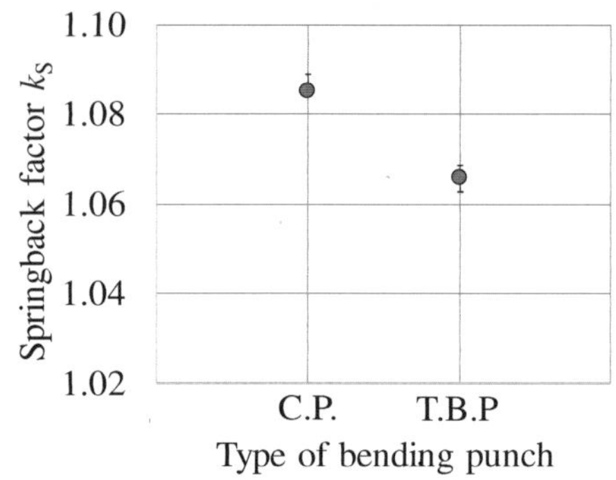

Figure 6 Springback factor for bending tools.

3. Finite element (FE) analysis of bending

3.1 Development of FE model and its verification

For the FEM simulation, a commercial software, Simufact Forming version 12.0.3 was employed to simulate the bending of the DP980 worksheet. A two-dimensional FE model shown in figure 7 was developed and used. Here, the worksheet was assumed to be deformable body and was constructed with the 4-node plane strain quadrilateral element. The thickness of the worksheet was considered as 1.6 mm. The bending zone of the worksheet was modeled using fine elements, which had a side length of 42 μm. For far zone, larger elements having a side length of 120 μm were used. The deformable worksheet was assumed to be isotropic elasto-plastic with work hardening. Its constitutive equation determined by the uni-axial tensile test, as shown in figure 8, was assumed to the analysis. The punch and die were assumed to be rigid body.

Due to the crushing of many elements at the bending region, the calculation tended to be fault. To overcome this problem, an automatic re-meshing function was utilized to re-generate new faultless elements. The re-meshing was executed when the strain change of elements reached 0.4. For the new re-meshing elements, their side length was controlled to be approximately 50 μm. The coulomb friction model was assumed to all contact interfaces. The friction coefficient for all interfaces was fixed to be 0.1. Since there are no separation or crack of the worksheet during bending, no damage model was applied to the bending simulation. After simulating, the simulated bending load resistance, the springback of the worksheet and the distribution of the equivalent stress were investigated under the variation of the punch types.

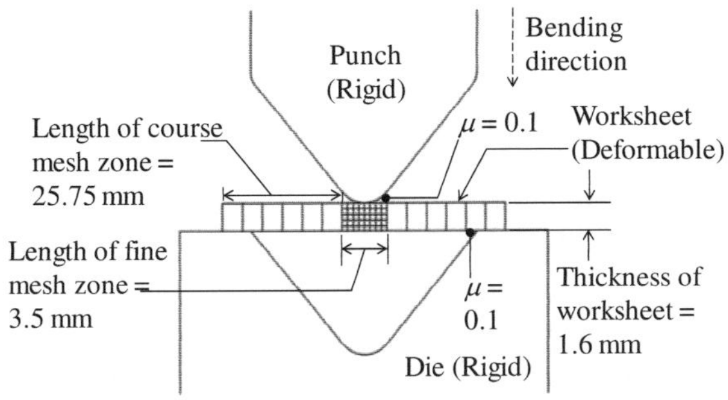

Figure 7 Schematic of two-dim. FE model for bending.

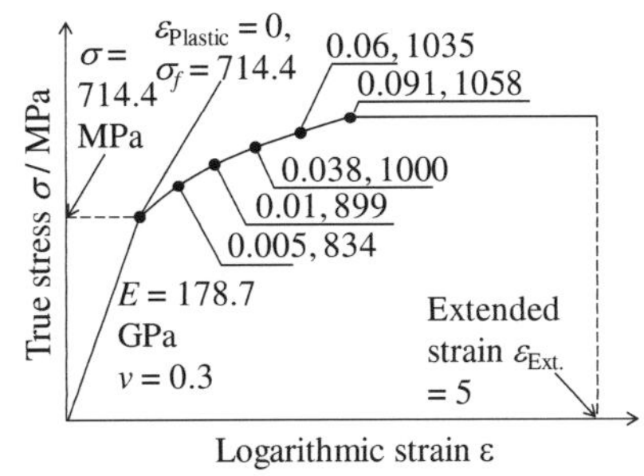

Figure 8 Material model assumed for FE analysis.

3.2 Verification of developed FE model

To ensure the applicability of the developed FE model for simulating the bending of the DP980 worksheet, the verification of the FE model was carried out. In order to do this, the bending load resistance measured in the bending experiments was plotted and compared with the bending load resistance simulated by FE analysis.

Figure 9 shows the experimental and simulated bending load resistance for the C.P. and T.B.P cases. Apart from the bending load resistance. The applicability of the developed FE model for the simulation was further checked by comparing the experimental springback factor with that simulated by FE analysis, as shown in figure 10.

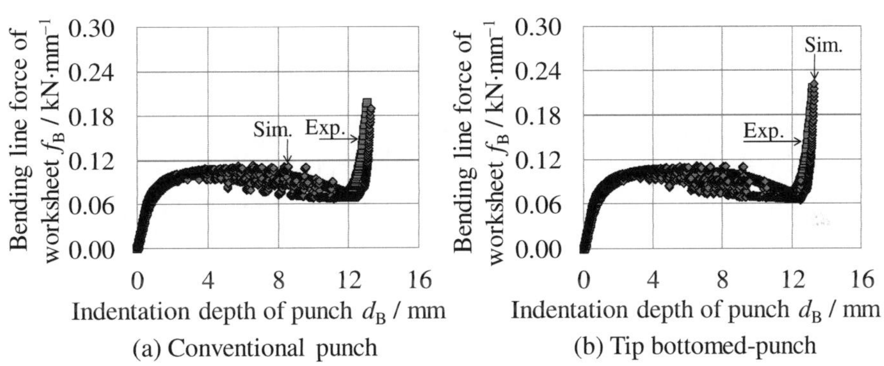

Figure 9 Experimental and simulated bending load resistance.

Seeing figure 9, the simulated bending load resistance of the worksheet was similar to that of the experiment. In addition, the similarity of the load resistance was observed for all punch cases. For the springback of the worksheet, as seen in figure 10, although the magnitude of the springback factor simulated by FEM was slightly different from that of the bending experiments, the tendency of the simulated springback factor had a good agreement with the experimental results. This comparison indicated that the developed FE model was applicable for simulating the bending of the DP980 worksheet.

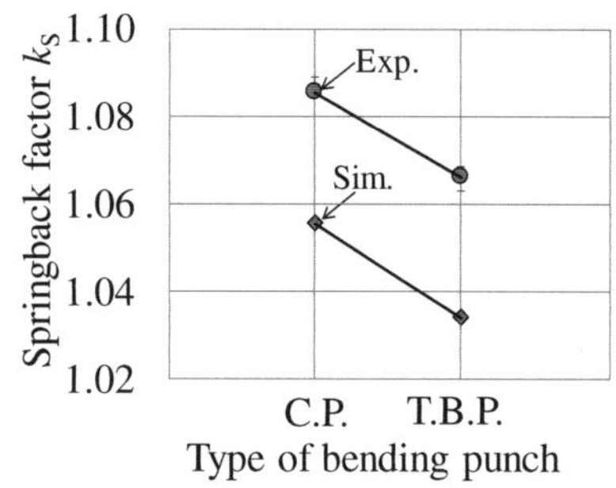

Figure 10 Experimental and simulated springback factor.

3.3 Effect of punch type on deformation of bending worksheet

After getting the FE model, the bending with the C.P. and the T.B.P. was analyzed. For discussing the deformation of the bent worksheet, the contour band diagrams of the equivalent stress ($\sigma_{Equi.}$) were plotted for the punch types used. In the diagrams, the red bands represent the area where the equivalent stress in the worksheet was higher than the yield strength of the worksheet evaluated by the uni-axial tension test. In other words, the red bands indicates the zones where the worksheet plastically deforms, while other colours indicate the zones that the worksheet elastically deforms.

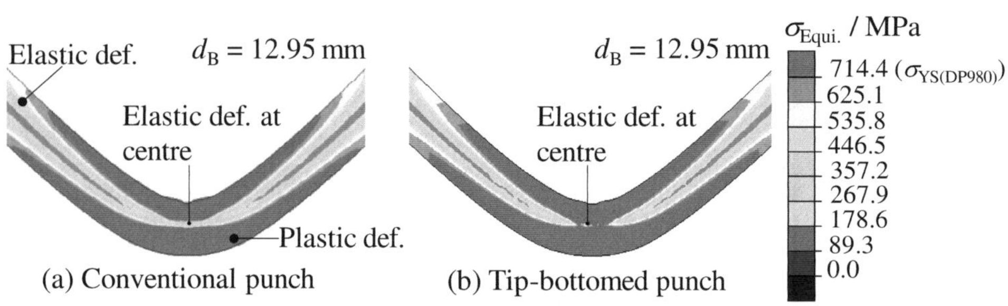

Figure 11 Contour band diagram of equivalent stress in the bent worksheet for punch types.

Figure 11 shows the representative overviews of the bent worksheets and the distribution of the equivalent stress for the C.P. and the T.B.P. cases. From this figure, the different deformation at the bending region of the worksheets was observed. Namely, in the T.B.P. case, the elastic deformation portion at the centre of the bending region seemed to be smaller than that of the C.P. case. For other zone, the stress distribution were fairly similar to each other.

To see the deformation at the bending region more clearly, the zoom figures showing the distribution of the equivalent stress at the centre of the bending were made. However, since the distribution of the equivalent stress slightly varied during the indentation of punch, the authors plotted the contour band diagrams of the equivalent stress at the indentation where an expected high equivalent stress occurred, namely at the indentation just before the $d_{B(Max.)}$ and at the $d_{B(Max.)}$. This plot is shown in figure 12. From the figure, it was found that the elastic deformation at the centre of the bending region in the case of the T.B.P. was limited, compared to the case of the C.P. This deformation characteristic was observed for all the indentation depths. Figures 11 and 12 indicated that the application of the T.B.P. contributed the plastic deformation to occur at the central region of bending. This seemed to result in the suppression of the springback when considering the T.B.P.

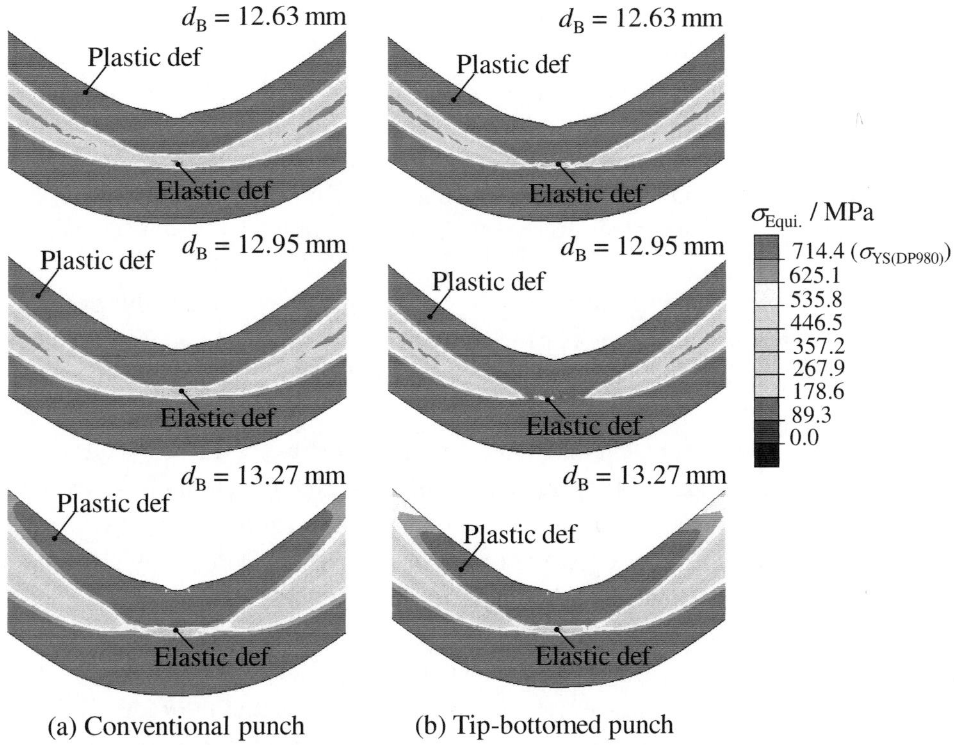

Figure 12 Contour band diagram of equivalent stress at bending zone of worksheet at various indentation depths of punch.

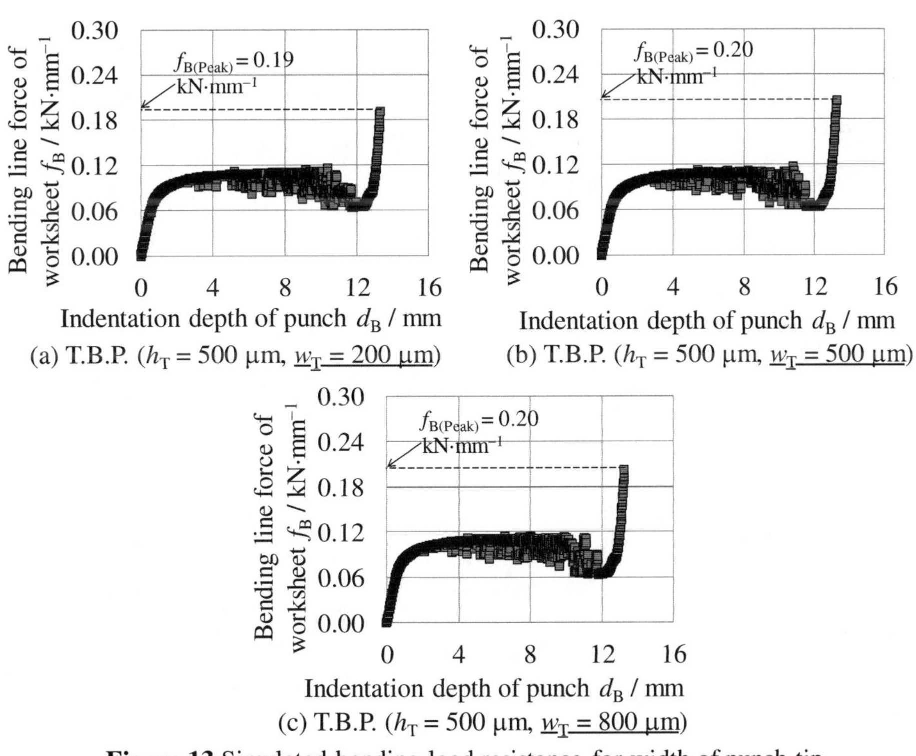

Figure 13 Simulated bending load resistance for width of punch tip.

IOP Conf. Series: Materials Science and Engineering **229** (2017) 012012 doi:10.1088/1757-899X/229/1/012012

3.4 Effect of geometry of tip-bottomed punch on deformation of bending worksheet

For the T.B.P. bending, the tip of the punch tends to indent to the worksheet during bending. Thus, the geometry of the punch tip would be an important factor affecting the deformation of the worksheet at the bending region and the springback of the bent worksheet. In this work, to investigate the effects of the punch tip geometry, i.e. the width of the punch tip, the width of the punch tip (w_T) was varied in the FE as 200, 500 and 800 μm and simulated. The angles of the punch and the die were still fixed to be 90°. Other simulation conditions which were not mentioned here were the same as those explained in section 3.1.

Figure 13 shows the plot of the bending load resistance when varying the width of the punch tip. As seen in this figure, the $f_{B(Peak)}$ in the cases of w_T = 500 and 800 μm appeared to be higher than that of the case of w_T = 200 μm, while the characteristic of the bending load resistance for $0 < d_B < d_{B(Max.)}$ was similar for all punch cases.

Figure 14 represents the contour band diagrams of equivalent stress at the bending region for w_T = 200, 500 and 800 μm. Seeing figure 14 (a), in the case of w_T = 200 μm, at all the indentation depths, the shallow-deep indentation occurred on top surface of the worksheet. In this case, a quite large elastic deformation portion was observed at the centre of the bending region. For the cases of w_T = 500 and 800 μm, at d_B = 12.63, 12.95 and 13.27 mm, on the contrary, the wider indentations occurred on the top surface of the worksheets. Seeing the elastic deformation portion in the cases of w_T = 500 and 800 μm, it appeared to be limited, compared to that of the w_T = 200 μm case.

Figure 15 illustrates the relationship between the springback factor and the width of the punch tip. As shown in this figure, the springback in the case of w_T = 200 μm tended to be larger than that of the cases of w_T = 500 and 800 μm.

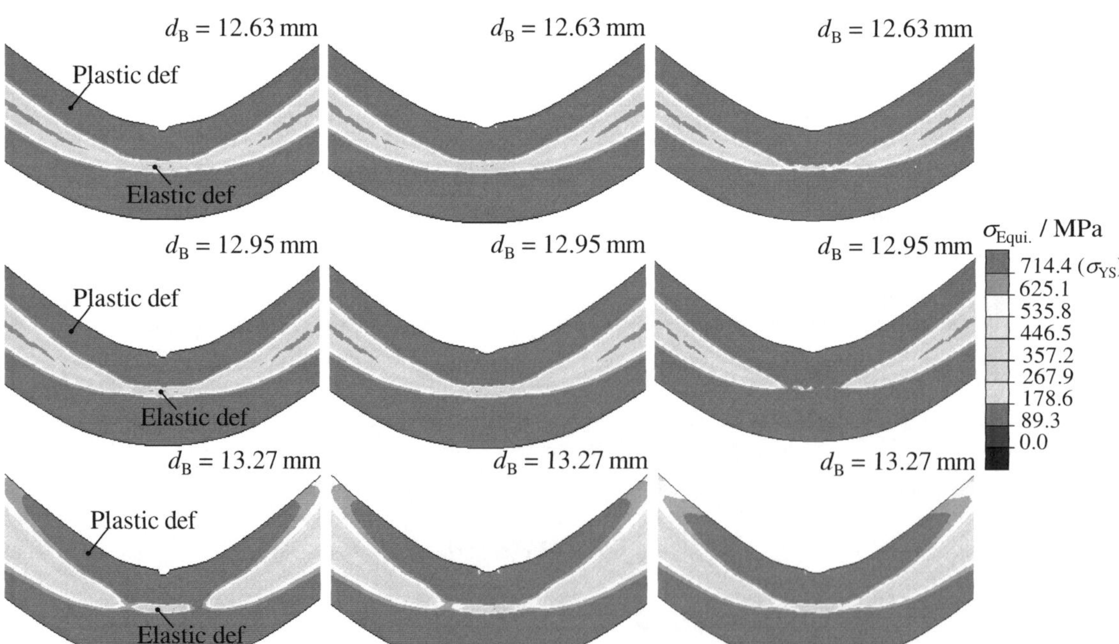

(a) T.B.P. (h_T = 500 μm, w_T = 200 μm) (b) T.B.P. (h_T = 500 μm, w_T = 500 μm) (c) T.B.P. (h_T = 500 μm, w_T = 800 μm)

Figure 14 Contour band diagram of equivalent stress at bending zone of worksheet at various indentation depths of punch for variation of width of bottom.

Considering the results shown in figures 13 – 15, it was revealed that the application of the punch, which had a narrow tip, resulted in the limited plastic deformation on the top surface of the worksheet. However, it could not contribute the plastic deformation to occur at the centre of the bending region. When wider punch tip was considered, the plastic deformation seemed to be promoted not only a limited area near the top surface of the worksheet, but also the centre of the bending retion. Consequently, the springback of the bent worksheet tended to be reduced.

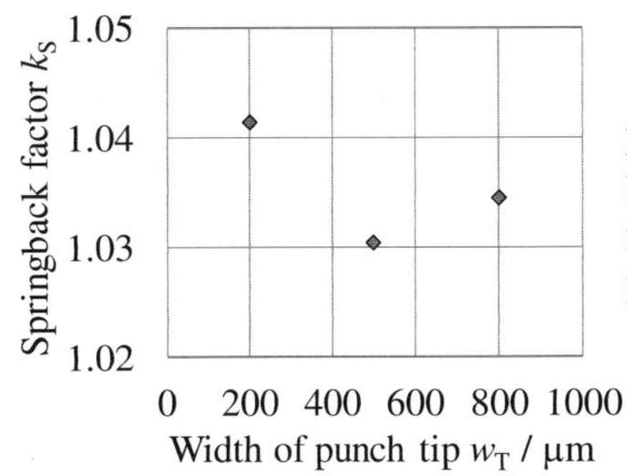

Figure 15 Retationship between springback factor and width of punch tip.

4. Mechanical property of worksheet at bending zone

As depicted in the previous sections, the application of the T.B.P could suppress the springback of the ultra-high strength steel bent worksheet. However, the mechanical property after bending of the worksheet is an important factor determining the durability of formed ultra-high strength steel parts. Thus, in this work, the mechanical property at the bending region of the worksheets bent by the T.B.P was evaluated by micro Vickers hardness test and compared with the hardness of the worksheet bent by the C.P.

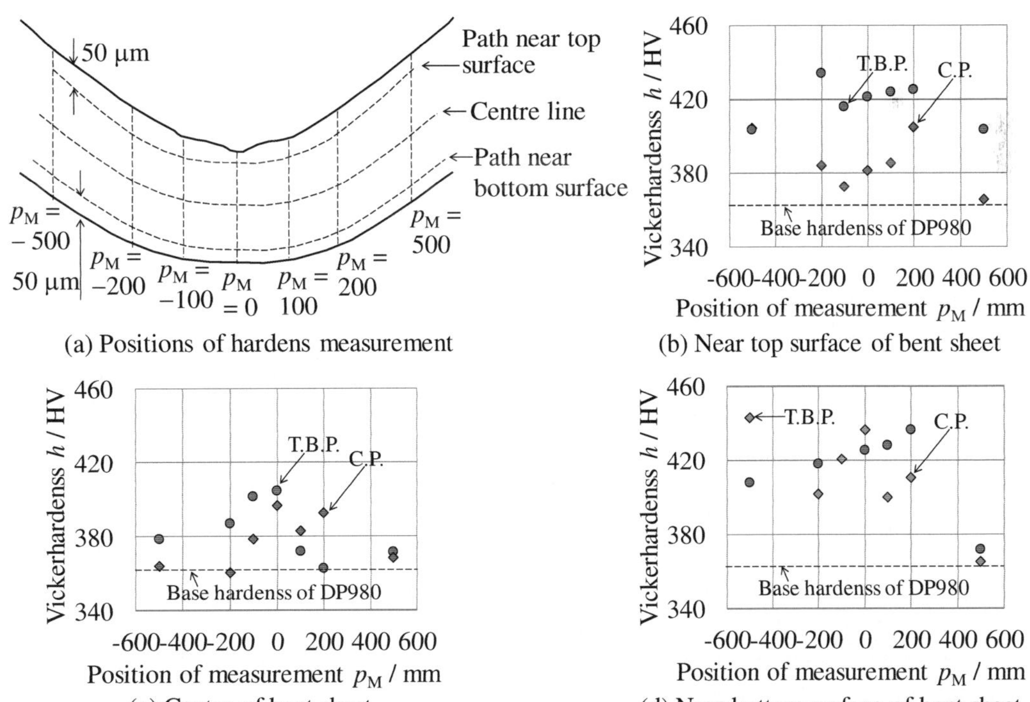

Figure 16 Hardness of the bent worksheets at the bending region and the measurement positions.

For the hardness test, the indentation load and the dwell time were set as 50 gf and 10 seconds, respectively. The measurement was performed along the three paths, i.e. near the top surface, at the centre and near the bottom surface of the worksheet, as shown in figure 16 (a). From the measurement results shown in figures 16 (b) - (d), it was revealed as follows: (i) near the top surface of the bent worksheet, for all measurement positions (p_M from –600 to 600 μm), the hardness remarkably increased from the base hardness of the DP980 (368.7 HV). This was observed in both C.P. and T.B.P. cases. (ii) the application of the T.B.P. provided significantly higher hardness near the top surface of the bent worksheets, compared to the case of C.P. Namely, the hardness in the cases of the T.B.P was higher than the base hardness of the DP980 approximately 13.9%, while the increase was only 4.6% for the C.P. case. This higher hardness seemed to be caused by higher strain hardening at the top surface of the bent worksheet when applying the T.B.P. (iii) At the centre of the bent sheet, the hardness for both punch cases tended to be higher at the position near the centre of bending (p_M between –200 to 200μm), while it appeared to be identical to the base hardness of the DP980 for the far zones ($-600 \leq p_M < -200$ μm and between ($200 < p_M \leq 600$ μm). (iv) near the bottom surface of the bent worksheet, the hardness measured in both punch cases was apparently increased from the base hardness. However, the magnitude of the hardness for both punch cases were similar to each other.

Here, it was confirmed the application of the T.B.P. was not only reduced the springback of the ultra-high strength steel worksheet but also increased the hardness, especially near the top surface of the bent worksheet.

5. Conclusions

In this research work, the application of tip-bottomed punch was applied for bending an advanced ultra-high strength steel (DP980) sheet. The bending characteristics of the worksheet such as the bending load resistance, the springback and the deformation of the worksheet, were investigated. Based on the experimental and simulation results, the following conclusions were obtained:

(i) the application of the tip-bottomed punch could reduce the springback of the DP980 bent worksheet.

(ii) The distribution of the equivalent stress appeared to be useful for explaining the deformation of the worksheet at the bending region. As found in this work, the springback of the worksheet was found to correspond to the elastic deformation portion at the bending region, especially at the centre of the bending region.

(iii) The use of the tip-bottom punch contributed the plastic deformation to occur at the centre of the bending region. Thus, it contributed to suppress the springback of the bent worksheet. However, a certain wide tip of punch was necessary.

(iv) The application of the tip-bottom punch caused the increase of the hardness, especially near the top surface of the bent worksheet. The top surface hardness was found to increase approximately 13.9% from the base hardness of the DP980 worksheet.

Acknowledgments

The authors would like to acknowledge Toyota Daihatsu Engineering & Manufacturing Co., Ltd, Thailand, for supporting ultra-high strength steel sheets used for the current research work. The financial support for the research work was the "Fund for young researcher of Suranaree University of Technology (SUT), research project entitled study of bending of high strength dual phase steel sheet in V-die bending process" Here, the authors would like to thank SUT.

IOP Conf. Series: Materials Science and Engineering **229** (2017) 012012 doi:10.1088/1757-899X/229/1/012012

References

[1] Koo J and Luton, M J 1995, European patent no. EP 0-796-352-B1
[2] Vandeputte S, Mesplont C and Jacobs S 2002, US Patent, Patent No. US 8-715-427-B2
[3] Zhong Y, Wang L, Xiong W, Zhi J and Hu G 2013, US Patent, Patent no. US 2015/0337416-A1
[4] Mori K, Maki S and Tanaka Y 2005, *CIRP Annal-Manu. Tech.* **54** 209-12
[5] Lee E, Hwang J, Lee C, Yang D and Yang W 2014, *J. of Mat. Proc. Tech.* **214** 784-93
[6] Yanagimoto J and Oyamada K 2007, *CIRP Annal-Manu. Tech.* **54** 265-68
[7] Semiatin S L 2006, *ASM Handbook Volume 14B Metalworking: Sheet Forming* (Ohio: ASM Int.)

Research on Morphology and Growth Mechanism of Xonotlite Crystal

F Liu [1,2,3], S Chen [1,3], Q Lin [1,2], X D Wang [2] and J X Cao [1,2,3,*]

[1] School of Chemistry and Chemical Engineering, Guizhou University, Guiyang, Guizhou 550025, PR China
[2] Key Laboratory of Green Chemical and Clean Energy Technology, Guiyang, Guizhou 550025, PR China
[3] Key Laboratory of Efficient Utilization of Mineral and Green Chemical Technology, Guiyang, Guizhou 550025, PR China

[*] jxcao@gzu.edu.cn

Abstract. The xonotlite crystals were synthesized via the hydrothermal synthesis manner from CaO and SiO_2 as the raw materials with their Si/Ca molar ratio of 1.0. The morphologies and their growth mechanisms of xonotlite crystal were explored by taking the various siliceous sources into account in this paper. These products were characterized by XRD and SEM techniques to investigate their crystalline phase and morphology. The results indicated that siliceous material had little influence on the crystalline phases of xonotlite crystals but posed a great impact on their morphologies. Based on the various morphologies characteristics of the products, the growth mechanism for the morphologies of spherical particles and fibers were further revealed according with the in situ transformation mechanism and dissolution-precipitation mechanism respectively.

1. Introduction

Xonotlite has been widely utilized in the fields of thermal insulation, architecture and so on, for its distinct physical and chemical properties, such as thermal stability, biological activity and environmental friendliness[1-5]. Therefore, the preparation of xonotlite crystals has become one of the research hotspots in recent years. In general, the xonotlite products used in industry were fabricated using dynamic hydrothermal synthesis method from CaO and SiO_2 as the raw materials with their Si/Ca molar ratio of 1.0[6-9]. The previous researches revealed that raw materials posed a great impact on the growth of xonotlite crystals[10-14]. Due to those factors such as raw materials and unclear growth mechanism of xonotlite crystals, the morphologies of the products were quite different and uncontrollable. Accordingly, based on the previous studies in our group[15-16], the morphologies and their growth mechanism of xonotlite crystals were particularly explored by taking H_2SiO_3 and K_2SiO_3 as siliceous materials respectively into account, aiming to make the morphologies of the xonotlite crystals be controllable.

2. Materials and methods

The xonotlite crystals were synthesized via the hydrothermal synthesis method.

H_2SiO_3 and K_2SiO_3 solutions were used as siliceous materials respectively. $Ca(OH)_2$ suspension as calcareous material was prepared by mixing $CaCl_2$ with KOH $100 g \cdot L^{-1}$ at the molar ratio of 1:2. The

Content from this work may be used under the terms of the Creative Commons Attribution 3.0 licence. Any further distribution of this work must maintain attribution to the author(s) and the title of the work, journal citation and DOI.
Published under licence by IOP Publishing Ltd

calcareous material, distilled water, and siliceous material (H_2SiO_3 or K_2SiO_3 solution) were homogeneously mixed, and then the mixture was transferred into a reaction chamber with a stirrer at a heating rate of 1.5 °C·min^{-1} and a stirring rate of 200 rpm. The hydrothermal treatment proceeded under the following conditions: the Ca/Si molar ratio of 1.0, the reaction temperature of 225 °C, the reaction time of 15 h. After the hydrothermal treatment, the suspensions were filtrated and washed three times by distilled water. The resultant powders were dried at 75 °C for 12 h. Technological flow diagram of hydrothermally synthesizing xonotlite products was shown in figure 1.

The crystalline phase composition and microstructure characteristics of the samples were characterized by X-ray diffraction (XRD, Model D/max-2200, Japan). A scanning electron microscope (SEM, Model JSM-6490LV, operating at 25 KV, Japan) was used to observe the morphology of xonotlite crystals.

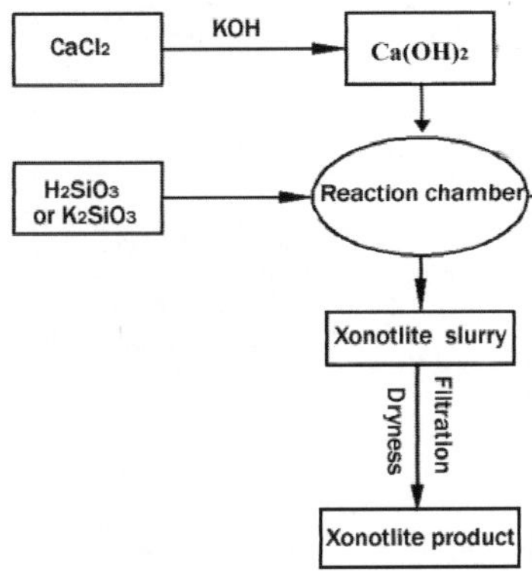

Figure 1. Technological flow diagram of hydrothermally synthesizing xonotlite products

3. Results and discussion

Figure 2 shows the XRD patterns of the products synthesized hydrothermally at 225 °C for 15 h from various siliceous materials. Clearly, the main crystalline phases prepared by hydrothermal synthesis method with various siliceous materials were xonotlite (corresponding to $Ca_6Si_6O_{17}(OH)_2$, standard JCPDS card No. 23-0125).

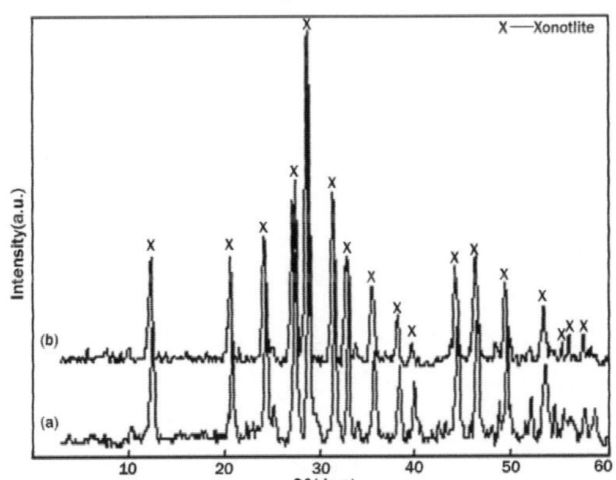

Figure 2. X-ray diffraction patterns of the products hydrothermally synthesized from various siliceous materials
(a) H_2SiO_3 as siliceous material; (b) K_2SiO_3 as siliceous material

From the analysis result of XRD patterns, the existence of a series of strong diffraction peaks showed that the xonotlite crystal from K_2SiO_3 solution as siliceous material was well-crystallized and almost no impurity. Meanwhile, the intensity of diffraction peak was also strong when using H_2SiO_3 solution as siliceous material, indicating that the pure xonotlite crystalline phase was completely obtained in the synthesis system.

(a)

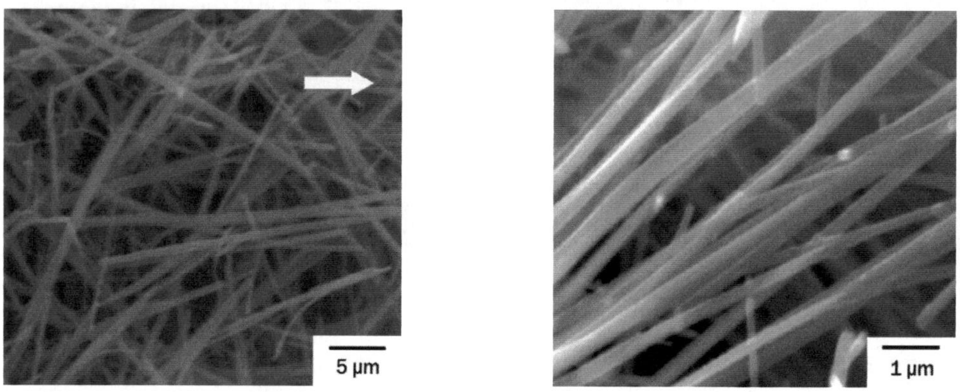

(b)

Figure 3. SEM images of the products hydrothermally synthesized from various siliceous materials
(a) H_2SiO_3 as siliceous material; (b) K_2SiO_3 as siliceous material

Figure 3 displays the SEM images of the products synthesized from various siliceous materials. Using H_2SiO_3 as siliceous material, the products obtained by hydrothermal synthesis were spherical particles with diameters of 20-50 μm, which were formed by a large number of short xonotlite fibers disordered intertwined. While using K_2SiO_3 as the siliceous material, the morphology of xonotlite crystal was typical fiber and its aspect ratio was approximately 50-100, with the length of 20-40 μm and the diameter of approximately 400 nm.

The dissociation constant values of H_2SiO_3 were extremely low ($K_{a(1)}$=2.5×10^{-10}, $K_{a(2)}$=1.6×10^{-12}, $K_{a(3)}$=1.0×10^{-12}), then H_2SiO_3 was extremely difficult to dissolve in the water at room temperature when used H_2SiO_3 as siliceous material. The growth mechanism of xonotlite spherical particles synthesized by the hydrothermal synthesis method could be explained by in situ transformation mechanism. It means that these reactions took place on the surface of precursor particles without dissolution to obtain crystal particles of products. The schematic diagram of in situ transformation mechanism was shown in figure 4.

In the process of hydrothermal synthesis for xonotlite crystals, the solubility of the siliceous precursor increased with the elevation of the reaction temperature and pH value, so that the siliceous precursor had been partially ionized into $HSiO_3^-$ and SiO_3^{2-} ions, which initially reacted with Ca^{2+} and OH^- ions introduced from $Ca(OH)_2$ ionization to synthesize poorly crystalline C-S-H (hydrated calcium silicate) gels. Then the C-S-H gels wrapped on the surface of spherical aggregates of siliceous material to form spherical colloidal particles. At this time, the remaining Ca^{2+} and OH^- spread on the thin layers of spherical aggregate to participate in reaction until siliceous material completely depleted. As a result, the spherical colloidal particles turned into spherical gel particles with distinct hollow structure. As extension of the reaction time, a series of reactions from C-S-H into the acicular xonotlite crystals took place on the surface of the hollow spherical gel.

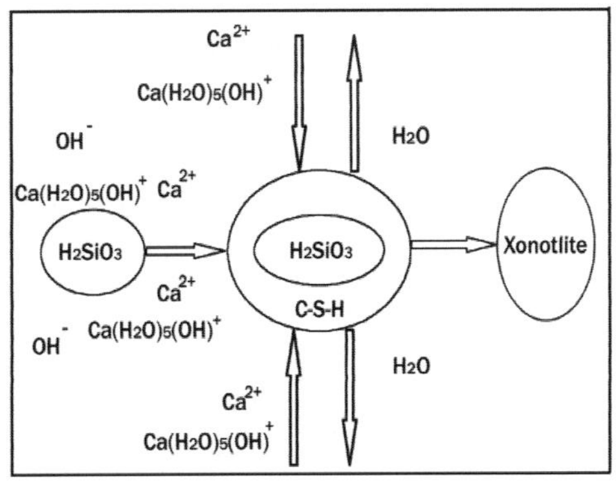

Figure 4. The schematic diagram of in situ transformation mechanism

The aforementioned growth mechanism of xonotlite spherical aggregates conformed to the in situ transformation mechanism that the growth of the xonotlite crystal existed on the surface of the undissolved siliceous precursor particles. After a period of time, the siliceous materials surrounded by a special thin layers of C-S-H gel gradually dissolved, and ultimately turned into the spherical particles with distinct hollow structure.

Using K_2SiO_3 as siliceous material, the growth mechanism of xonotlite fibers synthesized via the hydrothermal synthesis method could be interpreted by the dissolution-precipitation mechanism. It means that the precursor expeditiously dissolved to form hydrated ions or other complex ions, resulting in the formation of xonotlite crystal nucleus through transition phase. Meanwhile, remaining precursors provided ions for the growth of crystal nucleus or the formation of new crystal nucleus.

The dissolution-precipitation mechanism for the growth of xonotlite fibers involved several steps. At room temperature, $Ca(OH)_2$ exhibits obviously solubility, and the $Ca(H_2O)_5(OH)^+$ hydrated ions and SiO_3^{2-} anionic groups are considerably formed by the dissolution of $Ca(OH)_2$ in aqueous solution. The collinear reactions, taking place between $Ca(H_2O)_5(OH)^+$ hydrated ions and SiO_3^{2-} anionic groups, formed the Ca-O bonds which had been squeezed and finally developed into CaO_2 chips with the characteristics of tobermorite. Both sides of the CaO_2 chips combined with $[SiO_4]$ triple chains and OH^- ions respectively to synthesize semi-crystalline C-S-H gels. With the reaction time increasing, C-S-H gels crystallized into C-S-H(II) firstly, then generated into C-S-H(II), which was similar to the tobermorite. But C-S-H(II) was unstable at the reaction temperature (above 220 °C), resulting in the interlayer water molecules of C-S-H(II) strongly rotated and activated to the extent of transiting to adjacent lattices.

The gap between the CaO_2 chips and $[SiO_4]$ triple chains was too narrow, so the interlayer water molecules could only dissociate into OH^- ions attracting free Ca^{2+} ions. The triple chains adjacent to each other condensed into double triple chains of $[Si_6O_{17}]^{10-}$, which was paralleled to the b-axis. The Ca^{2+} ions from the chains of $[Si_6O_{17}]$ had adopted six-coordinated to form $Ca\text{-}O_6$ octahedrons connecting by edges, and those octahedrons had played a role of connector between dual chains. The octahedrons formed a crystal structure with double chains when the characteristics of CaO_2 chips gradually disappeared. The xonotlite crystal was mainly grown along the b-axis direction, and the growth rate of the crystal in the b-axis direction was much faster than that in other directions, ultimately leading to the formation of acicular or fibrous shape.

From the above analysis, at the initial stage, a large amount of SiO_3^{2-} ions introduced from siliceous material was an essential prerequisite to the formation of xonotlite fibers. In the experimental process of hydrothermal synthesis, the nucleation and growth of the xonotlite had generally finished during the high temperature stage. However, $Ca(OH)_2$ had a unique dissolution characteristic at constant temperature, and its solubility decreased with hydrothermal temperature increasing. Thus, the solubility of $Ca(OH)_2$ was extremely low at reaction temperature of 225 °C so that the rate of Ca^{2+} ions releasing into the solution was also slow and controllable, contributing to the nucleation and growth of the xonotlite fiber without the appearance of explosive nucleation section.

4. Conclusions

Siliceous material had little influence on the crystalline phases of xonotlite crystals but posed a great impact on their morphologies. The morphologies of xonotlite crystal via hydrothermal synthesis route were significantly determined by the various precursor states of siliceous materials. The xonotlite crystal from H_2SiO_3 and K_2SiO_3 as siliceous materials could be grown to form the various morphologies of spherical particles and fibers respectively. The growth mechanism for the morphology of spherical particles was accorded with the in situ transformation mechanism due to the reactions on the surface of the precursor particles without dissolution. The growth mechanism for the morphology of fibers was conformed to the dissolution-precipitation mechanism, and a large amount of SiO_3^{2-} ions introduced from siliceous material was an essential prerequisite to the formation for xonotlite fibers.

Acknowledgments

We gratefully acknowledge the supports of this work by the National Natural Science Foundation of China (Grant No. 21666007), Science Technology Foundation of Guizhou (Grant No. 2014-2007) and Program for "Hundred" High-level Talents in Guizhou (Grant No. 2016-5655).

References

[1] Zou J J, Guo C B, Wei C D and Jiang Y S 2016 *Res. Chem. Intermed* **42** 519-30
[2] Hsiang H, Chen W S and Huang W C 2016 *Mater. Struct* **49** 905-15
[3] Liu L, Yue Y and Cao J X 2014 *Mater. Sci. Forum* **809-810** 672-5
[4] Zhang S G, Xie H B and Li G Z 2006 *Energy-Saving. Mater* **34** 28-30
[5] Liu F, Wang X D and Cao J X 2012 *Chin. J* **57** 3323-28
[6] Zou J, Guo C, Wei C, Jiang Y and Li F 2016 *Mater. Chem. Phys* **172** 121-8
[7] Hartmann A, Schulenberg D and Buhl J C 2015 *J. Mater. Sci. Technol. Chem. Eng* **3** 39-55
[8] Liu F, Zhu B and Cao J X 2011 *Adv. Mater. Res* **148-149** 1755-8
[9] Yang J, Zhang X, Ma H W, Wang M W and Wu H 2014 *Key Eng. Mater* **633** 7-10
[10] Black L, Garbev K and Stumm A 2009 *Adv. Appl. Ceram* **108** 137-44
[11] Guo X Y, Ma S H, Lü S Q, Zheng S L and Zou X 2015 *Chin. J. Nonferrous. Metals* **25** 534-44
[12] Spudulis E, Šavareika V and Špokauskas A 2013 *Mater. Sci* **19** 190-6
[13] Baltakys K and Prichockiene E 2010 *Mater. Sci-Poland* **28** 295-304
[14] Bao Y W, Jiang D Y and Gong J H 2017 *Key Eng. Mater* **726** 569-75
[15] Liu F, Wang X D and Cao J X 2013 *Int. J. Min. Met. Mater* **20** 88-93
[16] Liu F, Zeng L K Cao J X and Li J 2010 *J. Wuhan. Univ. Technol* **25** 295-7

IOP Conf. Series: Materials Science and Engineering **229** (2017) 012014 doi:10.1088/1757-899X/229/1/012014

Effects of Cryogenic Treatment on the Strength Properties of Heat Resistant Stainless Steel (07X16H6)

D S Nadig*, M R Bhat, Pavan V K and Chandan Mahishi

Indian Institute of Science, Bangalore 560012, INDIA

*nadig@ccf.iisc.ernet.in

Abstract. Cryogenic treatment on metals is a well known technology where the materials are exposed to cryogenic temperature for prolonged time duration. The process involves three stages viz. slow cooling, holding at cryogenic temperature and warming to room temperature. During this process, hard and micro sized carbide particles are released within the steel material. In addition, soft and unconverted austenite of steel changes to strong martensite structure. These combined effects increase the strength and hardness of the cryotreated steel. In this experimental study, the effects of cryogenic treatment, austenitising and tempering on the mechanical properties of stainless steel (07X16H6) have been carried. After determining the strength properties of the original material, the specimens were cryotreated at 98K for 24 hours in a specially developed cryotreatment system. The effects of austenitising prior to cryogenic treatment and tempering post cryotreatment on the mechanical properties of steel samples have been experimentally determined and analysed.

1. Introduction

The subzero treatment on metals has been extensively employed since many decades for various applications. However this treatment was restricted to about 193K (dry ice temperature). In the recent years extensive research has been carried out in studying the effect of the cold treatment down to cryogenic temperatures by varying operating parameters [1].Even though major application of cryogenic treatment has been in the field of tooling industry in improving the tool life of the cutting tools, it has other important applications like reduction of internal stresses in metals, imparting dimensional stability to gauges and precision machined parts, enhancement of strength properties, etc.
Cryogenic treatment follows a procedure where the test samples are cooled down to cryogenic temperature at a definite rate, held for longer time duration and warmed to room temperature at a definite rate. This cryo treatment cycle can be operated for various conditions of temperature, duration, rate of cooling and warming. A typical cryotreatment cycle is shown in figure 1.

Content from this work may be used under the terms of the Creative Commons Attribution 3.0 licence. Any further distribution of this work must maintain attribution to the author(s) and the title of the work, journal citation and DOI.
Published under licence by IOP Publishing Ltd

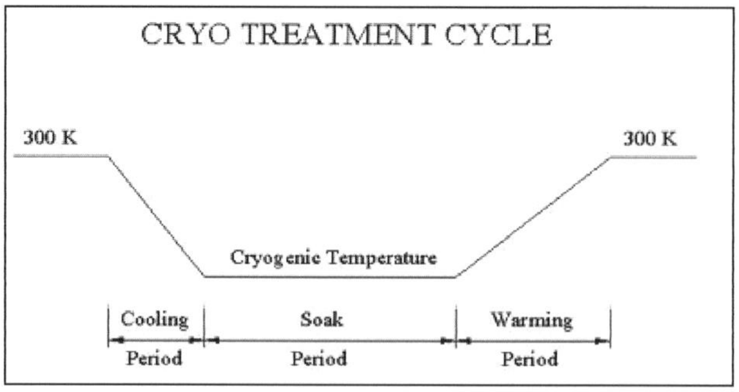

Figure 1. Cryotreatment cycle

During the cryotreatment process on steels, hard and micro sized carbide particles are released within the material and they are evenly spread over the entire volume. In addition, soft and unconverted austenite of steel changes to strong martensite structure [2] [3]. These combined effects tend to increase the strength and hardness of the cryotreated steel [4] [5]. In this experimental study, the effects of cryotreatment on the mechanical properties viz. strength (tensile and impact) and hardness of stainless steel (07X16H6) are studied.

2. Development of cryotreatment unit

In order to cryotreat the steel samples, a cryotreatment unit was designed and developed with capability to cryotreat for various rates of cooling, soak and warming periods. This developed cryotreatment unit has the unique feature that the test samples are be cooled to cryogenic temperature by indirect cooling without causing thermal shocks by spray of LN2 droplets.

The cryotreatment system incorporates mainly the cryotreatment chamber and an auxiliary LN2 supply system to supply controlled quantity of LN2 to the chamber to maintain the rate of cooling, soaking and warm up. The controlled LN2 supply to the chamber is carried out by using a solenoid valve activated by a PID controller. The Schematic diagram of the cryotreatment system is as shown in the figure 2.

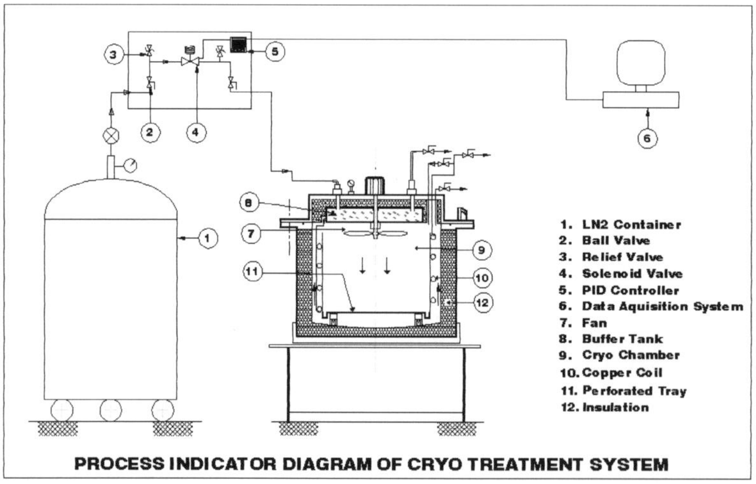

Figure 2. Process indicator diagram of the cryotreatment system

The cryotreatment unit is made of stainless steel with polyurethane foam (PUF) insulation. It has a fan-motor assembly mounted centrally in the top cover. The samples housed in the meshed tray located at the bottom are gradually cooled to cryogenic temperature by the cold nitrogen gas forced convection continuously produced by the rotating fan housed below the buffer tank containing LN2 and the cold exchange from LN2 circulating in the copper tube heat exchanger brazed on the outer wall of the metallic shroud housed around the meshed tray. The photograph of the developed unit is shown in figure 3.

Figure 3. Photograph of the cryotreatment unit.

The LN2 supply is regulated by a solenoid valve operated by a PID controller with predetermined set points. Various cryotreatment cycles can be programmed to suit the requirements. The temperatures of the specimens are measured using Platinum Resistance Temperature Detectors (RTD).

3. Cryotreatment trials and tests

In order to study the effects of cryotreatment on mechanical properties, trials were conducted on stainless steel (07X16H6).This is a corrosion resistant stainless steel used for general engineering applications. The valve material for LH2 and LOX in cryogenic engine are produced by using this selected grade of steel. The material was machined to produce test samples for the Tensile, Hardness and Impact tests as per ASTM standards E8, E10 and E23 respectively [6] [7] [8].

The machined test samples were categorized into four different groups in order to try various combinations of cryotreatment and heat treatment processes. It was ensured that each batch has a minimum of 6 samples in order to obtain authentic and reliable results. The samples were divided into 4 batches as under:

Batch A: Raw material (in annealed) condition.

Batch B: Cryotreatment alone at 98K for 24 hours.

Batch C: Cryotreatment at 98K for 24 hours followed by tempering at 400 Celsius for 1 hour.

Batch D: Austenitisation at 990 Celsius for 1 hour followed by cryotreatment at 98K for 24 hours and tempering at 400 Celsius for 1 hour.

In order to get the reference values of the mechanical properties of the selected material, tests were conducted to determine the strength (both tensile and impact strength) and hardness values [7] [8]. At the same time, microscopic structures were obtained using high resolution optical metallurgical microscope. (Model: NIKON-Epiphot 200)The test specimens were machined as per the specifications of ASTM E8M to carry out tensile strength properties. The tests were conducted on servo hydraulic dynamic test system (Make: Instron; Model: 1341). Properties viz. Ultimate tensile strength, Yield strength and percentage elongations were determined using the test system. Test samples for impact tests were machined in accordance with as per ASTM E23.Charpy impact strength were carried on FIT 300(EN) system. Hardness values (HRC) were also determined as per ASTM E 10.

After obtaining the initial data, the test samples of batch D were austenitised at 990 Celsius for 1 hour. All the samples (batches B, C and D) were successfully cryotreated in the newly developed cryotreatment unit at 98K for 24 hours with 4 hours of cool down and 9 hours of warm up time to ambient condition. The entire cycle consumed around 260 litres of liquid nitrogen which was supplied to the unit at 22 psi from pressurized container. The temperatures of the samples were monitored throughout the run. After the cryotreatment run the samples of batch A and C were tempered at 400 Celsius for 1 hour.

All the samples of batches B, C and D were again subjected to tensile, impact and hardness tests (including microstructure study). These readings were compared with the original test results and comparative study was carried out.

4. Results and analysis

Comparative studies of test results were carried out. The results of the study are as under:

4.1. *Tensile strength*

The test results are listed in the table 1.

Table 1. Strength and ductility properties.

Batch	A	B	C	D
UTS (MPa)	1954	1902	1987	2509
YS (MPa)	1901	1834	1949	2404
% Elongation	16.13	17.89	23.7	9.29

The results clearly indicate significant increase in tensile strength of the samples of batch D compared to the raw material samples (batch A). This is due to the conversion of soft and ductile retained austenite into hard and strong martensite. At the same time, the micro-sized carbide particles are precipitated during cryotreatment and fill the voids between atoms. During tempering still finer nano sized carbide particles known as η particles further fill up the voids and make the structure much more dense and strong [9]. Not much significant changes were observed in strength properties of batch B and C.

4.2. Hardness

The test results are listed in table 2.

Table 2. Rockwell hardness (HRC).

Batch	A	B	C	D
Hardness (HRC)	35	37	38	41

The hardness values were recorded using the hardness tester with scale C, using diamond cone indenter with load of 150Kg. Due to precipitation of hard carbide particles both during cryotreatment and tempering, samples of batch A showed marked increase in hardness values. However, improvements in batches B and C were marginal.

4.3. Impact strength

The test results are listed in table 3.

Table 3. Charpy Impact strength (J).

Batch	A	B	C	D
Impact strength	50.4	47	27	23.4

Extending the concept of formation of hard martensite and precipitation of hard carbide fillers it is expected that the material becomes harder and absorbs less energy during impact loading (Charpy). In the case of batches A and C (subjected to tempering operation in addition to cryotreatment), the results indicate proportional changes with regard to energy absorbed before failure [10].

4.4 Microstructure study

The photographic images of the microstructures were obtained using the metallurgical microscopes. The images are shown in the figure 4.

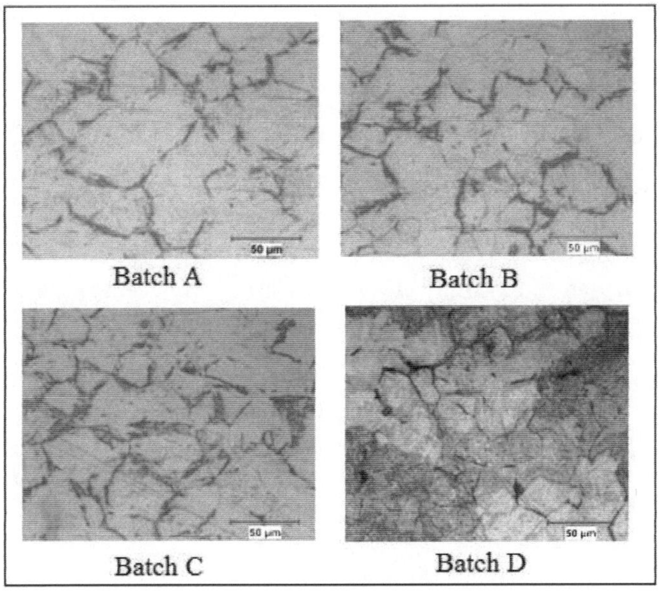

Figure 4. Photographs of microstructure

Comparing the microstructures of all the samples the samples of batch D indicated marked presence of martensite plates nucleating at the grain boundaries of austenite region. No major observable changes were found in batches B and C compared with samples of batch A.

5. Conclusions

The entire experimental study was carried out on all the feasible combinations of austenitising, cryotreatment and tempering. It is clearly evident that the process of cryotreatment can render significant changes when the samples are subjected to austenitising before cryotreatment and tempering post cryotreatment.

References

[1] Barron R F 1982 Cryogenic treatment of metals to improve wear resistance *Cryogenics* vol **22** pp 409-413

[2] A. Bensely et al 2007 Effect of Cryogenic treatment on tensile behaviour of case carburised steel 815M7 *Material Characterization* vol **58** pp 485-491

[3] Yi He et al 2002 Strengthening and toughening of a 2800-MPa grade Maraging steel *Materials letters* vol **56** pp 763-769

[4] Paolo Baldissera 2010 Deep cryogenic treatment of AISI 302 stainless steel: Part I Hardness and tensile properties *Materials and Design*, vol **31** pp 4725-4730

[5] A Molinari et al 2001 Effect of deep cryogenic treatment on the mechanical properties of tool steels *Journal of Materials processing technology* vol **118** pp 350-355

[6] ASTM E8M *Standard Test Methods for Tension Testing of Metallic Materials*

[7] ASTM E18 *Standard test methods for Rockwell hardness of metallic materials*

[8] ASTM E23 *Standard Test Methods for Notched Bar Impact Testing of Metallic Materials*

[9] Pen-Li-Yen and Dennis J Kamody 1997 Formation of Eta carbide in special Cryogenic and Tempering Process *Journal of Industrial heating*

[10] S. Zhirafar, A Rezaeian and M Pugh 2007 Effect of cryogenic treatment on the mechanical properties of 4340 steel *Journal of Materials processing technology* vol **186** pp 298-303

Effect of Crystallizable Solvent on Phase Separation and Charge Transport in Polymer-fullerene Films

A Kaewprajak[1,2], K Lohawet[1], T Wutikhun[1], B Meemuk[1], P Kumnorkaew[1] and T Sagawa[2,*]

[1] National Nanotechnology Center, National Science and Technology Development Agency, 111 Thailand Science Park, Phahonyothin Rd., Khlongnueng, Khlongluang, Pathumthani 12120, Thailand.
[2] Graduate School of Energy Science, Kyoto University, Yoshida-honmachi, Sakyo-ku, Kyoto 606-8501, Japan.

*sagawa@energy.kyoto-u.ac.jp

Abstract. The effect of 1,3,5-trichlorobenzene (TCB) as crystallizable solvent on poly[N-9′-heptadecanyl-2,7-carbazole-alt-5,5-(4′,7′-di-2-thienyl-2′,1′,3′-benzothiadiazole)] (PCDTBT) and [6,6]-phenyl C_{71} butyric acid methyl ester ($PC_{71}BM$) bulk heterojunction (BHJ) was investigated. We found that phase separation of PCDTBT and $PC_{71}BM$ and formation of the condensed network of polymers were appropriately regulated by addition of TCB in the BHJ films, which were confirmed by optical microscopic, AFM, and TEM observations in addition to current-voltage analyses. Through the formation of a good continuous pathway for carrier transport by the addition of TCB, 2.5 times enhancement of the hole mobility in the BHJ film was attained from 5.82×10^{-5} $cm^2 V^{-1} s^{-1}$ without TCB to 1.48×10^{-4} $cm^2 V^{-1} s^{-1}$ with 20 mg ml^{-1} of TCB.

1. Introduction

Organic photovoltaics (OPVs) have been attractive because they can be prepared onto flexible substrates by using simple printing techniques. [1-3] To improve the photovoltaic performance of OPVs, morphology and charge transport in bulk heterojunction (BHJ) solar cells are important. Many methods were used to optimize the morphology of polymers and fullerenes in BHJ such as thermal annealing [4, 5], solvent annealing [6, 7] and addition of solvent additives [8, 9]. One effective method to control the morphology of polymer-fullerene is adding high boiling point additive in polymer blend such as 1,8-diiodooctane (DIO) [10], 1,8-dibromooctane [11] and chloronaphthalene [12]. The solvent additive can selectively dissolve $PC_{71}BM$, which aggregates to smaller domains and promote a better donor-acceptor network into smaller domains.

In this work, we investigated the effect of 1,3,5-trichlorobenzene (TCB) as the crystallizable solvent on phase separation and formation of the highly oriented network of polymers in terms of the improvement of charge transport in poly[N-9′-heptadecanyl-2,7-carbazole-alt-5,5-(4′,7′-di-2-thienyl-2′,1′,3′-benzothiadiazole)] (PCDTBT) and [6,6]-phenyl C_{71} butyric acid methyl ester ($PC_{71}BM$) BHJ films. The concentration of TCB in PCDTBT:$PC_{71}BM$ were ranged from 0 to 50 mg ml^{-1} in dichlorobenzene (DCB). We found that the appropriate amount of TCB in the PCDTBT:$PC_{71}BM$

Content from this work may be used under the terms of the Creative Commons Attribution 3.0 licence. Any further distribution of this work must maintain attribution to the author(s) and the title of the work, journal citation and DOI.
Published under licence by IOP Publishing Ltd

effectively improved the morphology and formation of the condensed network of polymers, which resulted in the enhancement of the hole mobility in the BHJ films.

2. Experimental

The hole-only devices were prepared with a device structure of indium-tin-oxide (ITO) coated glass/PEDOT:PSS/polymer-fullerene/Au. The polymer and fullerene are poly[N-9′-heptadecanyl-2,7-carbazole-alt-5,5-(4′,7′-di-2-thienyl-2′,1′,3′-benzothiadiazole)] (PCDTBT) (1-Material) and [6,6]-phenyl C_{71} butyric acid methyl ester ($PC_{71}BM$) (American Dye Source, Inc.), respectively. The glass-ITO substrates (Geomatec, 10 Ω sq^{-1}) patterned were ultrasonically cleaned with deionized water, acetone, tetramethylammonium hydroxide (Semico clean 56), deionized water and isopropyl alcohol for every 15 min. The glass-ITO substrates were dried with heat gun and treated with UV–ozone treatment for 15 min. PEDOT:PSS (Clevios™ P VP AI 4083) after passing through a 0.45 μm filter was coated by convective deposition onto the cleaned glass-ITO substrate at 3000 μm s^{-1} solution volume at 20 μL with two time coating. The sample was annealed on hot plate at 120 °C for 20 min under atmospheric conditions. The solvent mixtures were prepared by dichlorobenzene (DCB) (Sigma-Aldrich) and 1,3,5-trichlorobenzene (TCB) (Tokyo chemical industry co. ltd.) at a concentration of 0, 10, 20, 30, 40, and 50 mg ml^{-1} stirred at 45 °C for 3 h. The mixed solvents with different concentration of TCB were added with polymer blend at same volume. The solution of polymer-fullerene from a 1:4 (w/w) of PCDTBT and $PC_{71}BM$ were dissolved with the prepared solvents of TCB and DCB at a concentration of 40 mg ml^{-1}. The solution was stirred at 45 °C for overnight until PCDTBT and $PC_{71}BM$ were dissolved completely. The polymer blend was coated on top of the PEDOT:PSS layer by convective deposition at coating speed 1250 μm s^{-1} with volume at 10 μL one time coating. The bulk-heterojunction films were kept under vacuum for 1 h. Each sample was transferred to the vacuum chamber of the thermal evaporator. The vacuum chamber was pumped down to a base pressure of 2.4 x 10^{-3} Pa. The Au electrodes with a thickness of 100 nm were deposited onto the polymer layer via thermal evaporation.

3. Results and Discussion

Figure 1 shows the preparation process of $PCDTBT:PC_{71}BM$ film by convective deposition technique. Coating process starts from injection of a microliter droplet of $PCDTBT:PC_{71}BM:DCB:TCB$ mixture between the deposition blade and the glass substrate. When the substrate was moved from meniscus of polymer blend solution with constant velocity at a deposition angle (α) of 45°, a thin film of polymer blend was coated horizontal deposition on the glass substrate. TCB in the thin film of $PCDTBT:PC_{71}BM:TCB$ was removed by evaporation in a vacuum oven at room temperature.

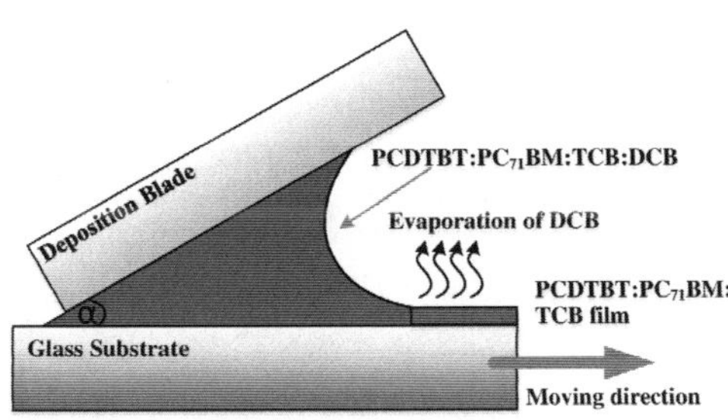

Figure 1. Schematic illustration of coating process of $PCDTBT:PC_{71}BM:DCB:TCB$ solution on the glass substrate by convective deposition.

The $PCDTBT:PC_{71}BM$ films with different concentrations of TCB were observed by an optical microscope with polarized light and high dynamic range (HDR) mode as displayed in Figure 2. In Figure 2a, phase-separated fullerene-rich clusters as shown in the dark areas and polymer-rich part in the bright areas were observed. [13,14] When the amount of TCB in polymer blend film was increased to10 and 20 mg ml^{-1}, the domains of PCDTBT-rich and $PC_{71}BM$-rich (bright and dark areas respectively) were observed as an uniformed phase separation and smaller size than those of the film without TCB as indicated in Figure 2b and 2c, respectively. When TCB concentration increased more than 20 mg ml^{-1}, the domain size of the PCDTBT-rich part and the $PC_{71}BM$-rich one gradually increased as displayed in Figure 2d to Figure 2f. The phase separated states of donor and acceptor were improved in terms of homogeneously and uniformity by the addition of TCB as a crystallizable solvent. In addition, when TCB concentration was increased more than 20 mg ml^{-1}, the domain size of donor and acceptor became larger because of aggregation of TCB at higher nucleation density.

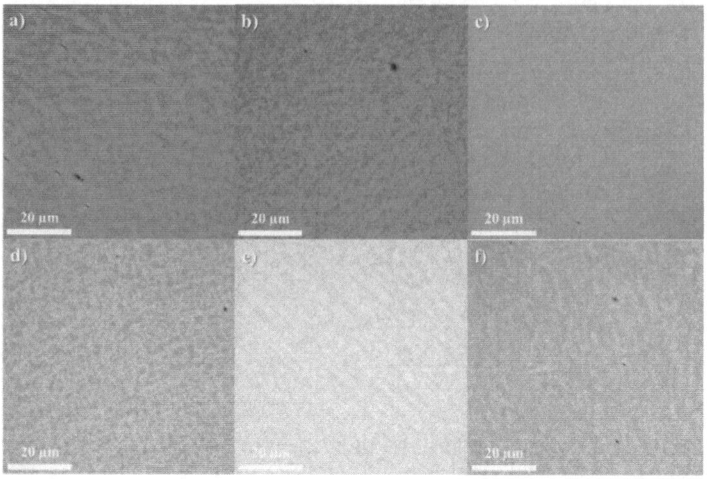

Figure 2. Optical microscopic images of $PCDTBT:PC_{71}BM$ films prepared by convective deposition with different concentrations of TCB of (a) 0, (b) 10, (c) 20, (d) 30, (e) 40, and (f) 50 mg ml^{-1}, respectively. Films were observed under the conditions of polarized light with high dynamic range (HDR) function. Scale bar: 20 μm.

AFM topographic images (5 μm x 5 μm) of $PCDTBT:PC_{71}BM$ films with different concentration of TCB are indicated in Figure 3. Figure 3a shows the topographic image of the film without TCB with large grain size of $PCDTBT:PC_{71}BM$. The root mean square (RMS) roughness of polymer film without TCB was 0.51 nm. When the TCB was added from 10 to 50 mg ml^{-1}, the RMS roughness were slightly increased between 0.57 and 0.65 nm. As a result of the process for elimination of TCB in the polymer blend films under the vacuum conditions, vacancy and crevice on the surface of the films might be formed.

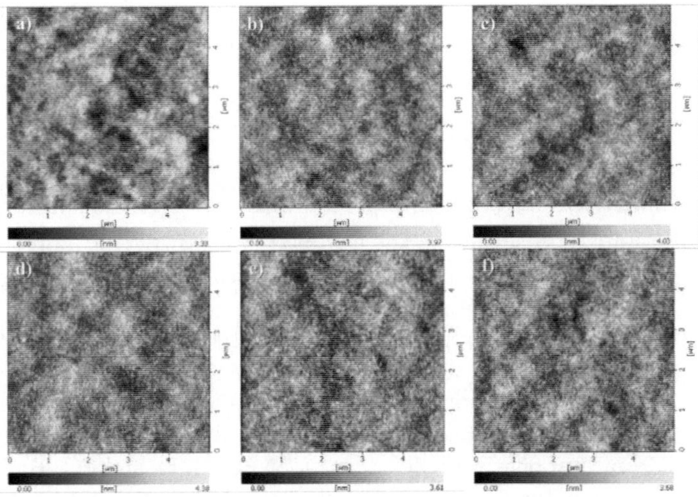

Figure 3. AFM images as tapping-mode of topography and phase mode of PCDTBT:$PC_{71}BM$ films prepared by convective deposition with different concentrations of TCB of (a) 0, (b) 10, (c) 20, (d) 30, (e) 40, and (f) 50 mg ml^{-1}, respectively.

Figure 4 displays the bright-field TEM images and electron diffraction (ED) patterns (insets) of PCDTBT:$PC_{71}BM$ films with different concentration of TCB. Small amount of fibrillar PCDTBT in the polymer blend films with TCB[15] was observed in Figure 4b-4f. As a result, the films with TCB had highly condensed PCDTBT network and better connected pathways for charge transport. The ED pattern of the film without TCB in Figure 4a presents a blur electron diffraction ring of its amorphous nature of PCDTBT. [16] Contrarily, crystallization of PCDTBT:$PC_{71}BM$ film with TCB was confirmed by the outer electron diffraction ring, which was indicated by white arrows right hand side of diffraction rings in Figure 4c-4f. [16]

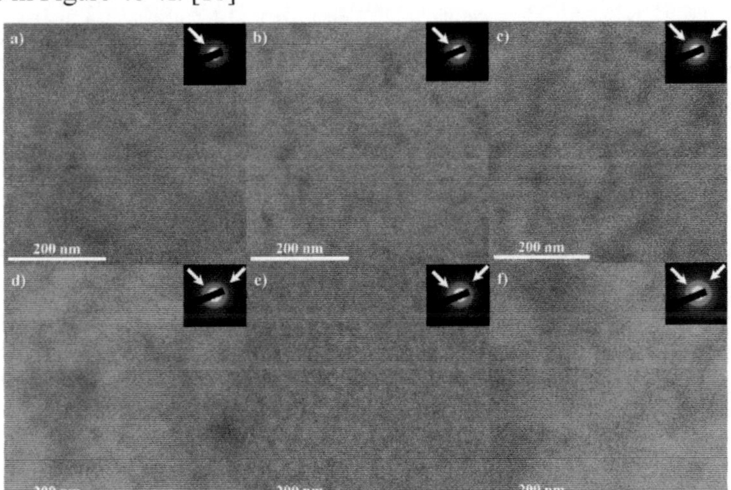

Figure 4. Bright-field TEM images of PCDTBT:$PC_{71}BM$ films prepared by convective deposition with different concentrations of TCB of (a) 0, (b) 10, (c) 20, (d) 30, (e) 40, and (f) 50 mg ml^{-1}. The insets are the electron diffraction patterns indicating the relative orientations to those of the bright field image.

To study the charge transport properties in the PCDTBT:$PC_{71}BM$ films, which were prepared with different TCB concentration, the hole-only devices in a configuration of glass-ITO/PEDOT:PSS/PCDTBT:$PC_{71}BM$/Au were prepared and characterized. The dark *J-V* characteristics of the hole-only devices are displayed in Figure 5. The hole mobilities were estimated from the dark current voltage curves in Figure 5 by applying the space-charge limited current (SCLC) theory as shown in Equation (1). [17, 18]

$$J = \frac{9}{8}\varepsilon_0 \varepsilon_r \mu_h \frac{V^2}{L^3} \qquad (1)$$

Where ε_0 is the permittivity of free space, ε_r is the dielectric constant of the polymer, μ_h is the hole mobility, V is the applied voltage, and L is the thickness of the polymer film. The field independent mobility in the PCDTBT:PC_{71}BM film without TCB was calculated as a reference with the positive charge mobility of 5.8 x 10^{-5} cm^2 V^{-1} s^{-1}. With the concentration of TCB at 10, 20, 30, 40 and 50 mg ml^{-1}, the positive charge mobilities of the devices are 8.9×10^{-5}, 1.5×10^{-4}, 1.4×10^{-4}, 1.2×10^{-4} and 8.1×10^{-5} cm^2 V^{-1} s^{-1}, respectively. The maximum positive charge mobility of 1.5×10^{-4} cm^2 V^{-1} s^{-1} was observed when TCB concentration was 20 mg ml^{-1}. The increment of the hole mobility presumably dues to the improved crystallinity of PCDTBT, which was confirmed by the ED patterns as shown in Figure 5 and uniform phase separated states of PCDTBT and PC_{71}BM in Figures 3 and 4 resulted in the formation of continuous charge pathways. [6]

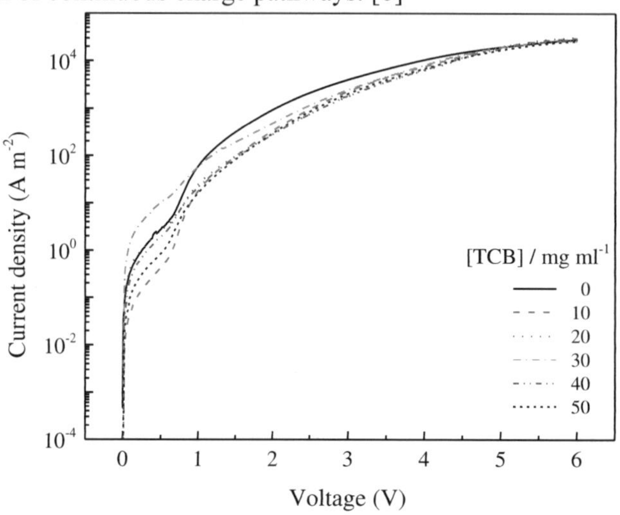

Figure 5. Dark current density (J)-voltage (V) curves of hole-only devices, consisted of glass-ITO/PEDOT:PSS/PCDTBT:PC_{71}BM/Au, at room temperature. The PCDTBT:PC_{71}BM layers ware prepared with different concentrations of TCB from 0 to 50 mg ml^{-1}.

4. Conclusion

In conclusion, we have demonstrated the effect of TCB crystallizable solvent on the improvement of the morphology and charge transport in PCDTBT:PC_{71}BM films. The films with TCB can improve of phase separated states and the domain sizes of PCDTBT and PC_{71}BM bulk heterojunction. The improvement of the hole mobilities by using TCB is presumably derived from the improvement of phase separation of PCDTBT and PC_{71}BM, which made a formation of good continuous pathway for carrier transport. The highest hole mobility of 1.5 x 10^{-4} cm^2 V^{-1} s^{-1} in this series of the TCB addition was obtained when TCB was 20 mg ml^{-1}. That is 2.54 times improvement of hole mobility as compared to the device without TCB (5.8 x 10^{-5} cm^2 V^{-1} s^{-1}). Therefore, addition of TCB as crystallizable solvent can be effective method for improvement of phase separation of polymer-fullerene and formation of the condensed network of polymers in the BHJ films for organic electronic applications.

References

IOP Conf. Series: Materials Science and Engineering **229** (2017) 012015 doi:10.1088/1757-899X/229/1/012015

[1] Wan Q, Guo X, Wang Z, Li W, Guo B, Ma W, Zhang M, and Li Y 2016 *Adv. Funct. Mater.* **26** 6635-6640.
[2] Krebs FC 2009 *Solar Energy Mater. Solar Cells.* **93** 465-475.
[3] Gu KL, Zhou Y, Gu X, Yan H, Diao Y, Kurosawa T, Ganapathysubramanian B, Toney MF and Bao Z 2017 *Org. Electronics.* **40** 79-87.
[4] Verploegen E, Mondal R, Bettinger CJ, Sok S, Toney MF and Bao Z 2010 *Adv. Funct. Mater.* **20** 3519-3529.
[5] Kadem B, Hassan A and Cranton WJ 2016 *Mater. Sci.: Mater. Electron.* **27** 7038-7048.
[6] Liu J, Liang Q, Wang H, Li M, Han Y, Xie Z and Wang L 2014 *J. Phys. Chem. C.* **118** 4585-4595.
[7] He Z, Liu F, Wang C, Chen J, He L, Nordlund D, Wu H, Russell TP and Cao Y 2015 *Mater. Horizons.* **2** 592-597.
[8] Etzold F, Howard IA, Forler N, Cho DM, Meister M, Mangold H, Shu J, Hansen MR, Mullen K, and Laquai F 2012 *J. Am. Chem. Soc.* **134** 10569-10583.
[9] Liao H-C, Ho C-C, Chang C-Y, Jao M-H, Darling SB and Su W-F 2013 *Mater. Today.* **16** 326-336.
[10] Liang Y, Xu Z, Xia J, Tsai S-T, Wu Y, Li G, Ray C and Yu L 2010 *Adv. Mater.* **22** E135-E138.
[11] Huang D, Li Y, Xu Z, Zhao S, Zhao L and Zhao J 2015 *Phys. Chem. Chem. Phys.* **17** 8053-8060.
[12] Yi C, Hu X, Liu HC, Hu R, Hsu C-H, Zheng J and Gong X 2014 *J. Mater. Chem. C.* **3** 26-32.
[13] Piersimoni F, Degutis G, Bertho S, Vandewal K, Spoltore D, Vangerven T, Drijkoningen J, Van Bael M K, Hardy A, D'Haen J, Maes W, Vanderzande D, Nesladek M and Manca J 2013 *J. Polym. Sci. Part B: Polym. Phys.* **51** 1209-1214.
[14] Liu H-W, Chang D-Y, Chiu W-Y, Rwei S-P and Wang L. 2012 *J. Mater. Chem.* **22** 15586-15591.
[15] Park SH, Roy A, Beaupré S, Cho S, Coates N, Moon JS, Moses D, Leclerc M, Lee K and Heeger AJ 2009 *Nat. Photon.* **3** 297-302.
[16] Kim T, Jae Yang S, Kyun Kim S, Soo Choi H and Rae Park C 2014 *Nanoscale.* **6** 2847-2854.
[17] Goh C, Kline RJ, McGehee MD, Kadnikova EN and Fréchet JMJ 2005 *Appl. Phys. Lett.* **86** 122110.
[18] Wang ZB, Helander MG, Greiner MT, Qiu J and Lu ZH 2010 *J. Appl. Phys.* **107** 034506.

IOP Conf. Series: Materials Science and Engineering **229** (2017) 012016 doi:10.1088/1757-899X/229/1/012016

Castor Oil Based Polyurethanes: Synthesis and Characterization

AD Macalino[1], VA Salen[1] and LQ Reyes[1,*]

[1]School of Chemical Engineering and Chemistry, Mapua Universtiy, Manila, Philippines

*lqreyes03@gmail.com

Abstract. In this study, polyurethanes based on castor oil and 1,6-hexamethylene diisocyanate (HMDI) were synthesized with varying weight ratio of the castor oil and HMDI. The formation of urethane linkages was verified through the use of a fourier transform infrared spectroscopy (FTIR). The hydrophilicity of the films was evaluated through the use of a contact angle meter and it was found that the contact angle of all the films were below 90 degrees which confirms their hydrophilicity. The thermal stability of the PU films were studies through the use of a thermal gravimetric analyzer and found that all of the polyurethane films exhibited two weight loss events at elevated temperatures wherein the first weight loss event was observed to occur at 285°C to 384°C while the second weight loss event was observed at around 521°C to 551°C. The hardness, elastic modulus, and tensile elongation of the PU films were determined by using a universal testing machine (UTM) where it was found out that the hardness and the elastic modulus of the film is directly proportional with HMDI loading while the tensile elongation is inversely proportional to it. Lastly, it was known through the swelling studies of the PU films that it does not swell, this is due to the presence of unreacted triglycerides in the material, which prevents water from permeating to the films.

1. Introduction

Bio-based materials are now receiving considerable attention in the promotion of sustainable chemistry in the production of materials by its replacement of petroleum-derived raw materials in the production of different products needed in various industries [1]. The shift to a more green way of chemistry is due to the increasing concerns about the environment and a shift to a more sustainable engineering are pushing scientist and engineers to research new ways to better understand renewable resources and how can they be used efficiently in the conversion of raw materials to products [2]. Polyurethanes, which are usually produced using diisocyanate is one of the materials receiving intensive studies today due to their potential in being used in different fields such as biomedical engineering and tissue engineering. These new polyurethanes which uses materials derived from vegetable oils like canola oil, castor oil, etc. can be used as a scaffold in which isolated cells may be seeded into which allows the cells to grow and proliferate until it forms a tissue construct which can then be implanted back to the patient [3]. These new polyurethanes may exhibit the ability to biodegrade, has excellent mechanical, chemical, and thermal properties, and has high wear resistance

Content from this work may be used under the terms of the Creative Commons Attribution 3.0 licence. Any further distribution of this work must maintain attribution to the author(s) and the title of the work, journal citation and DOI.
Published under licence by IOP Publishing Ltd

and good blood and tissue compatibility [4]. Thus, the goal for the next generation development of bio-based polyurethanes is to develop a material which minimizes the side effect of diisocyanate which is toxic and to enhance traits which can improve the efficiency of the produced product. Another goal for this is to lessen the consumption of petroleum derived polymers which may contribute to the promotion of green chemistry and sustainable engineering. Polyurethanes mainly consist of a flexible soft segment and a hard segment derived from diisocyanate [1]. The effects of modifying the chemical structure of polyurethane is also not widely known, it may exhibit specific traits like it being rigid or flexible, or they can exhibit biodegradation or biostability when its chemical structure is modified which can be exploited to fit the traits of polyurethane to the needs of a specific industry. The main objective of this study is to synthesize polyurethanes using castor oil (with ricinoleic acid) and 1,6-hexamethylene diisocyanate. The second objective of this study is to characterize and compare the chemical, mechanical, and thermal properties of the synthesized polymer with different weight ratios of the hard and soft segments using different apparatuses. Its swelling study is also studied. This study will focus on synthesizing bio-based polymer and characterizing its properties that can be used in polymer production as an alternative source of raw material.

2.Methodology

2.1. Synthesis of castor oil based polyurethane films

Different weight ratios of castor oil and HMDI (95:5, 90:10, 85:15, 80:20, 70:30, 60:40, and 50:50) is added in a 50 mL beaker. Then, the mixtures are heated at 55 degrees Celsius and stirred at 220 rpm until the mixture is deemed to be viscous and also free from air bubbles. The mixture is then poured into glass slides and allowed to settle for 3 to 4 days until the reaction is complete after which the resulting polymer film is removed from the glass slide and then stored at a cabinet for characterization.

2.2. Characterization of the polyurethane films

The formation of urethane linkage for the PUs will be determined using Shimadzu IR Prestige 21 Fourier-Transform infrared spectrophotometer (FTIR). The thermal stability of the sample was determined by thermal gravimetric analysis using a Mettler Toledo TGA/SDTA 851e. The heating range set was from 30°C to 1000°C at a heating rate of 10°C per minute. The sample will be scanned for 32 times with a scan rate of 8 cm^{-1} from 4000 to 600 cm^{-1}. Culture plates with bacteria were then incubated at 35°C for 24 hours. Surface topography of the polymers was determined by shell electron microscopy using JEOL JSM5310. The polarity of a polyurethane film t was tested by using a Theta Life Optical Tensiometer. First, the sample was affixed in the glass slide to make sure that the sample was flat for better results. Then, the sample was placed on the instrument. The volume of water drop used was 5±0.1 microliter for every trial and allowed to settle for 1 minute. The effect of varying the weight ratio of castor oil and HMDI on the mechanical properties of the PUs will be evaluated using an Instron Single Column Universal Testing Machine 3343 (UTM) equipped with an Instron 2519-103 Force Transducer with a 100N capacity. The samples were cut into 40 mm x 10 mm strips and were mounted onto the UTM. The mechanical properties of the PU samples such as elastic modulus, tensile strength and maximum tensile extension will then be determined. Stress-strain curves will be collected at a strain rate of 10 mm/min.

3.Results and Discussion

3.1. Synthesis of PU based on castor oil and and 1, 6-hexamethylene diisocyanate

Polyurethanes (PU) films were successfully prepared by the coupling of 1,6-hexamethylene diisocyanate (HMDI) with castor oil, which is a polyhydroxylated triglyceride. The coupling reaction may facilitate the covalent linking between of triglyceride molecules. The proposed molecular structures of PUs with varying degrees of cross-linking are shown in Figure 1. Adding excess amounts of HMDI facilitates the formation of urethane linkages between triglyceride molecules and then inter and intrachain crosslinking bridges are formed.

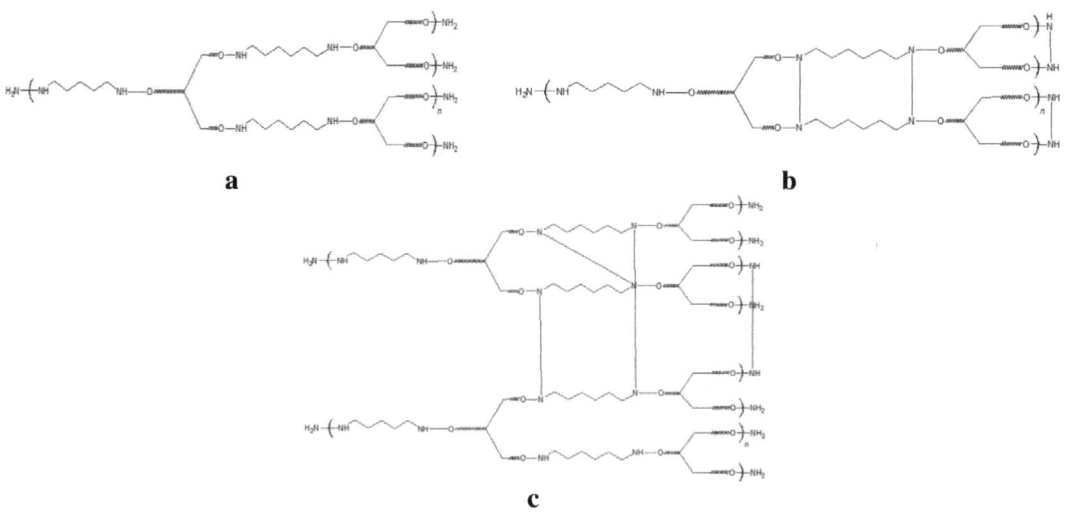

Figure 1. Proposed structure of castor oil and HMDI based PU: (a) uncrosslinked PU, (b) lightly crosslinked and (c) highly crosslinked PU structures

3.2. Molecular structure of the PU materials

The occurrence of urethane bond between HMDI and castor was determined by Fourier Transform Infrared (FTIR) Spectroscopy. The IR spectra of the of the pristine HMDI, castor oil and PU films are shown in Figure 2. The peaks at the wavenumber around 1535 cm^{-1} is indicative of urethane linkage (C-N-H) formation in all of the synthesized PU samples. In addition, vibration peak found at 3342 cm^{-1} corresponds to the amide N-H stretching in urethane. These two peaks are absent to the IR spectra of HMDI; while, pristine HMDI exhibited peaks at 2254 cm^{-1} to 2310 cm^{-1}, a unique peak which is attributed to the unreacted isocyanate group O=C=N. The peak at 1743 cm^{-1} which is attributed to the ester carbonyl (C=O) stretching of castor oil and urethane carbonyl. The bands between 2854 cm^{-1} and 2927 cm^{-1} are attributed to the methylene (CH_2) asymmetric and symmetric stretching vibration of the hydrocarbon chain of castor oil and HMDI. The peaks at 3008 cm^{-1} to 3014 cm^{-1} in the spectra of the PU films and castor oil is attributed to the methyl group (CH_3) stretching.

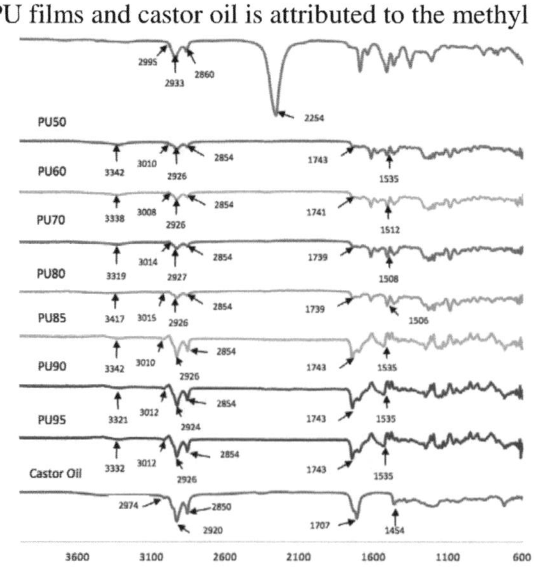

Figure 2. FTIR spectra of the PU materials

IOP Conf. Series: Materials Science and Engineering **229** (2017) 012016 doi:10.1088/1757-899X/229/1/012016

3.3. Thermal stability of PU materials

The thermograms of PU films and their derivatives are shown in Figures 3a and 3b respectively. All PUs exhibited two weight loss events at elevated temperatures. The first and the major weight loss event was observed to occur at 285°C to 384°C, which might be attributed to the thermal degradation or thermolysis of the ester linkage between the ricinoleic fatty acid to the glycerol backbone of castor oil. In addition, the first stage of decomposition is known to be dominated by urethane bond decomposition. The correlation of the amount of residue with the amount of unreacted isocyanate in polymers have found that even though the first stage of decomposition is subjected to urethane bond decomposition, the weight loss at higher conversion was possibly affected by polymer component. In addition, it can be observed that as the HMDI content is increasing, a shoulder peak becomes more prominent and is shifting from higher temperature 325°C (PU95) to 285°C (PU50). These shoulder peaks are quite near to the degradation temperature of pristine HMDI, thus it can be inferred that this shoulder peak is in fact the degradation temperature of HMDI. A second weight loss event is observed at around 521°C to 551°C and this peak is observed to shift to higher temperature with increasing HMDI loading. This weight loss event can be attributed to the decomposition of the hard segments of the PU films. The high amount of HMDI during the reaction promotes formation of allophanate bonds between urethane linkages. In the presence of excess of HMDI, almost all of the hydroxyl groups of castor reacts with HMDI; however, the same condition promotes the side reaction were excess HMDI reacts with the urethane linkage forming an allophanate. PU50 and PU60 are highly crosslinked because of allophanate bridges resulting to a more apparent second degradation event occurring at higher temperature. PU with lower HMDI content exhibited this second weight loss event but it occurs at a lower temperature relative to crosslinked PUs.

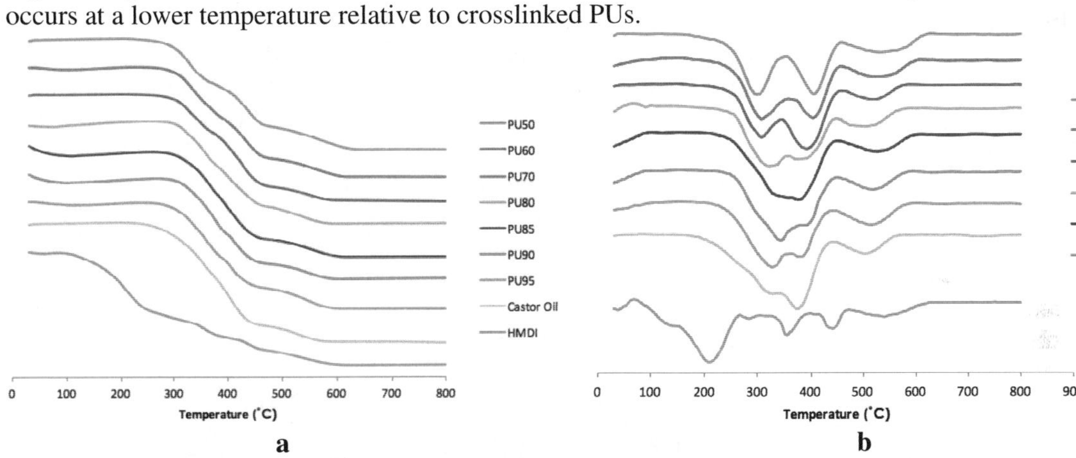

Figure 3. (a) Thermograms and (b) derivative of thermograms of the PU materials

3.4. Tensile properties of the PU materials

Stress strain curves of the polyurethanes (PU50, PU60, PU70, PU80) are separately shown in Figure 4 and the corresponding tensile properties of the PU materials are shown in Table 1. All of the polyurethanes have been found to be flexible. Since castor oil represents the soft segment part of the PU, the flexibility of the PU increases as the castor oil content increases. Based on stress-strain curve of all of the PU's, it was found that PU50 is least flexible because it has the least amount of castor oil content among the PU's. As the PU50 is the least flexible, it followed by PU60, PU70, PU80 to PU95 as PU95 is the most flexible among the PU due to its castor oil content. Regarding with the tensile strength, it was found the PU50 has the highest value of tensile strength among the PU's. Tensile strength of the PU was found directly proportional to the HMDI content of the PU since HMDI represent the hard segment part of the PU which is responsible for its hardness. Following the trend, PU50 has the highest value of tensile strength followed by PU60, PU70, PU80 to PU95 as PU95 has the least value of the tensile strength due to its HMDI content. In addition, the elasticity of PU's varies significantly as the castor oil to HMDI weight ratio change.

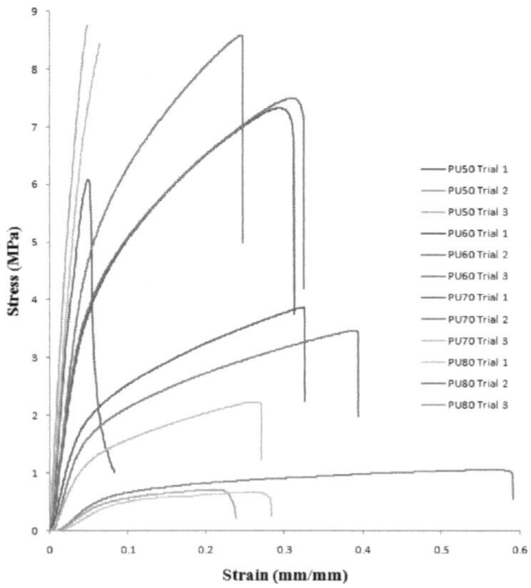

Figure 4. Stress-strain curves of the PU materials

Table 1. Elastic Modulus and Tensile elongation of Polyurethane

Polyurethane Code	Elastic Modulus (MPa)	Tensile Elongation (mm)
PU50	208.37804	2.57917
PU60	88.55964	11.78859
PU70	30.78883	13.20859
PU80	7 .877735	14.83083
PU85	2.64017	5.58287
PU90	2.05394	5.77198
PU95	0.19691	4.58656

3.5. Surface properties of the PU materials

The surface morphology of the PU films was observed and recorded by scanning electron microscopy and the micrographs are shown in Figure 5. The surface of the films with high HMDI content, PU50 and PU60, showed the highest surface roughness at high magnification (15,000X), with the presence of peaks and valleys. These features were not so apparent at lower magnification. The roughness of the PU's surface might be due to the shrinkage of the polymer chains after covalent crosslinking. The surface morphology of the PUs with intermediate amount of HMDI exhibited a smoother surface morphology than those that have high HMDI content and this might be due to the less extensive crosslinking of the polymer matrices. PUs with the lowest HMDI content, PU85, PU90 and PU95, did not completely dry up allowing the samples to react for 4 days which makes it unsuitable for electron microscopy. It might be due to presence of completely or partially unreacted triglycerides.

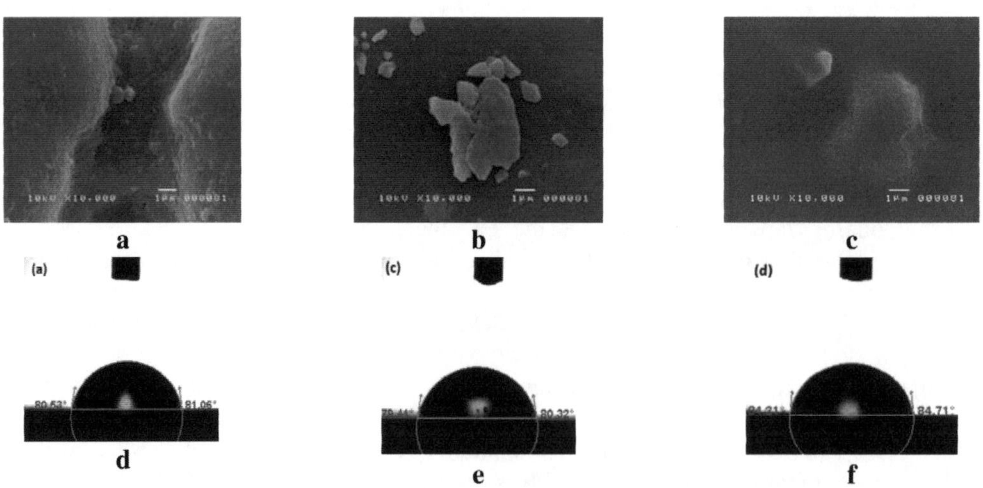

Figure 5. Electron micrographs of (a) PU50, (b) PU70 and (c) PU80. Water contact angle of (d) PU50, (e) PU70 and (f) PU80

The hydrophilicity of the polyester urethane films is evaluated through the measurement of the contact angle formed as water is dropped onto the surface of the films via a contact angle meter. Coatings can be described by their measured contact angle, a coating with contact angle below 90 degrees is considered to be hydrophilic, between 90 to 150 degrees is considered to be hydrophobic, and above 150 degrees is considered as ultra-hydrophobic. As seen on the figures above, the contact angle in all of the PEU film samples is lower than 90 degrees, which indicates that the film is hydrophilic in nature. Although vegetable oils are hydrophobic in nature due to the hydrophobicity of triglycerides, the use of a hydrophilic isocyanate such as HMDI together with changes in the surface roughness of the PEU films should make a PEU film hydrophilic. As such, this type of PEU can be used as biomaterials, filter membranes, biosensors, water permeable materials, etc. [5], [6].

4.Conclusion

Polyurethanes films were prepared by the coupling reaction of castor oil and HMDI. The polyurethane films exhibited the formation urethane linkage (C-N-H) formation, amide N-H stretching, ester carbonyl (C=O) stretching between castor oil and urethane carbonyl based on FTIR spectroscopy. The PU films were found to be thermally stabile by thermogravimetry with degradation temperature greater than 300°C. The surface morphology of the PU films was observed to increase in surface roughness with increasing HMDI content due to the formation of highly crosslinked polymer network. The PU films were found to be slightly non polar by water contact angle meter. The tensile properties of the PU films varied with the HMDI content; where PUs with high HDMI content were rigid and inelastic while PUs with low HMDI content were soft and pliable. The present study demonstrated a facile method of preparing PUs with a wide range of physical properties.

References

[1] M. A. Corcuera, L. Rueda, B. Fernandez d'Arlas, A. Arbelaiz, C. Marieta, I. Mondragon, and A. Eceiza, "Microstructure and properties of polyurethanes derived from castor oil," *Polym. Degrad. Stab.*, vol. 95, no. 11, pp. 2175–2184, 2010.

[2] S. S. Narine, X. Kong, L. Bouzidi, and P. Sporns, "Physical properties of polyurethanes produced from polyols from seed oils: I. Elastomers," *JAOCS, J. Am. Oil Chem. Soc.*, vol. 84, no. 1, pp. 55–63, 2007.

[3] M. W. Laschke, a. Strohe, C. Scheuer, D. Eglin, S. Verrier, M. Alini, T. Pohlemann, and M. D. Menger, "In vivo biocompatibility and vascularization of biodegradable porous polyurethane scaffolds for tissue engineering," *Acta Biomater.*, vol. 5, no. 6, pp. 1991–2001, 2009.

[4] H. Madra, S. B. Tantekin-Ersolmaz, and F. S. Guner, "Monitoring of oil-based polyurethane synthesis by FTIR-ATR," *Polym. Test.*, vol. 28, no. 7, pp. 773–779, 2009.

[5] K. Kojio, Y. Mitsui, and M. Furukawa, "Synthesis and properties of highly hydrophilic polyurethane based on diisocyanate with ether group," *Polymer (Guildf).*, vol. 50, no. 15, pp. 3693–3697, 2009.

[6] M. Furukawa, Y. Mitsui, T. Fukumaru, and K. Kojio, "Microphase-separated structure and mechanical properties of novel polyurethane elastomers prepared with ether based diisocyanate," *Polymer (Guildf).*, vol. 46, no. 24, pp. 10817–10822, 2005.

IOP Conf. Series: Materials Science and Engineering **229** (2017) 012017 doi:10.1088/1757-899X/229/1/012017

Effect of DT4 Interlayer on Properties of Hot-roll Bonding TA2/Q235B Plate

Chao Yu[1], Hong Xiao[1, a], Na Li[2], Zichen Qi[1] and Zhongkai Ren[1]

[1] National Engineering Research Center for Equipment and Technology of Cold Strip Rolling, Yanshan University, Qinhuangdao, Hebei, China
[2] College of Mechanical Engineering, Yanshan University, Qinhuangdao, Hebei,China.

[a] xhh@ysu.edu.cn

Abstract. In this paper, a Q235B/TA2/DT4/Q235B plate was bonded by hot-rolling in a vacuum, and the effect of including a DT4 interlayer within the TA2/Q235B plate was studied. The microstructure and properties of the composite plate at different reduction ratios were investigated by scanning electron microscopy, as well as tensile-shear, bending and tensile tests. The results show that when the reduction ratio is below 18%, the shear strength of the interface is higher with the DT4 interlayer than without it. At 35% reduction, the shear strength is similar in both cases. At a reduction ratio of 68%, with the DT4 interlayer, fracture of the bonding interface occurs on the TA2 side, whereas without the DT4 interlayer, fracture occurs on both the TA2 side and the compound layer on the interface. Including the DT4 interlayer improves the bending and tensile properties of the TA2/Q235B plate appreciably.

1. Introduction

Titanium clad steel plates are not only have the corrosion resistant of titanium but also have the weldability, formability and thermal conductivity of carbon steel, and it also saves precious titanium metals. Due to these characteristics, titanium clad steel plates are widely used in the petrochemical, shipbuilding, and marine industries, among others [1-2]. Currently, several methods have been used to manufacture titanium clad steel plates, such as explosive welding, explosive-rolling bonding and rolling bonding. Each method has its own advantages, compared with the explosive cladding and explosive-rolling bonding methods, the rolling bonding method can produce large-sized and thin titanium clad steel plates with high production efficiency, low pollution and low energy consumption [3].

While producing dissimilar metals composite plates, intermetallic compounds are easily formed on the interface, and they have a great influence on the properties of the composite plates. Many studies have improved the mechanical properties of composite plates by using a intermediate layer. In the production of titanium clad steel plates, the formation of brittle compounds such as TiC, FeTi and Fe_2Ti on the interface may cause the degradation of mechanical properties [4]. To prevent the formation of brittle compounds that reduce the bonding strength, some studies have used niobium, copper, silver and nickel as the interlayer [4-7], however, because of high costs, these materials are not widely used in the production of titanium clad steel plates. C. Yu [8] used DT4 interlayer to hot-roll bonding TA2/Q235B plate with the single pass reduction ratio of 25%, the TA2 and Q235B could be bonded well at heating temperature of 850 °C. However, in the industrial production, in order to obtain a large wide composite plate, it often requires multi pass rolling, therefore, on the basis of study on

Content from this work may be used under the terms of the Creative Commons Attribution 3.0 licence. Any further distribution of this work must maintain attribution to the author(s) and the title of the work, journal citation and DOI.
Published under licence by IOP Publishing Ltd

IOP Conf. Series: Materials Science and Engineering **229** (2017) 012017 doi:10.1088/1757-899X/229/1/012017

literature [8], this paper uses DT4 as interlayer, to study the effect of DT4 on properties of hot-roll bonding TA2/Q235B plate after multi pass rolling.

2. Experimental

The assembly pattern is shown in Figure 1, in which the thicknesses from top to bottom are 6 mm, 3.7 mm, 1 mm and 5 mm. Because the phase transition temperature of titanium is 882 °C, the mechanical properties of the titanium change radically near this temperature. Above 882 °C, Fe readily dissolves in titanium to form Fe_2Ti and FeTi compounds. Therefore, in order to stabilize the mechanical properties of TA2 during rolling, the heating temperature of the billet was chosen to be 850 °C to limit both the deformation resistance during rolling and the formation of Fe_2Ti and FeTi compounds. The billets were heated in a furnace for 2 h, rolled at a speed of 50 mm/s, and cooled to room temperature in air. The rolling reduction ratios were 8%, 18%, 35%, 54% and 68%, the first 3 of which were single pass rolling reductions, and the last 2 were cumulative reduction ratios based on the prior reduction. The surface temperature of the plate after 68% rolling reduction was approximately 553.3 °C.

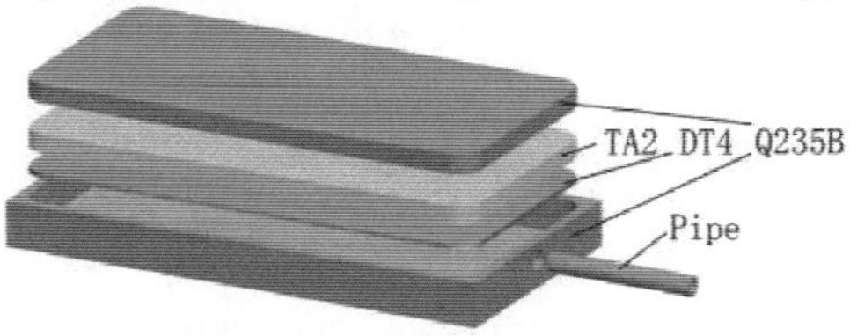

Figure 1. Assemble pattern.

The tensile-shear, bending and tensile tests were carried out using an Inspekt Table 100 kN instrument. The speed was 1 mm/min. The tension-shear test samples are shown schematically in Figure 2. The morphology of the fracture surface was investigated using scanning electron microscopy (SEM, Zeiss Sigma 500). The distribution of elements across the interface was examined by SEM equipped with an energy dispersive spectrometer (EDS).

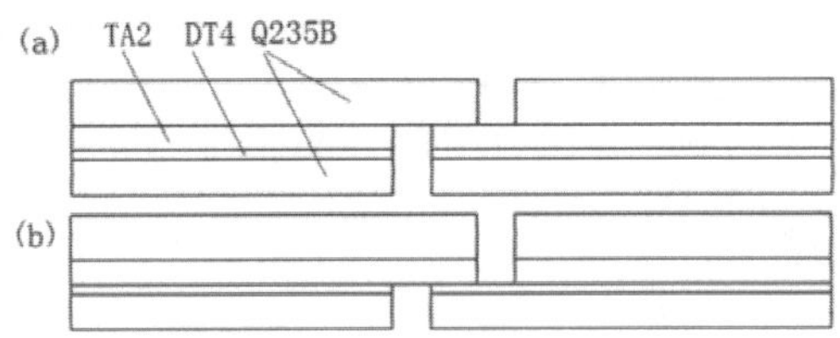

Figure 2. Schematic of the tension-shear samples: (a) measuring the tension-shear strength of the side without the DT4, (b) measuring the tension-shear strength of the side with the DT4.

3. Results and discussion

3.1. Macroscopic bonding property test of the clad plates

The effect of the rolling reduction ratio on the interfacial shear strength of titanium clad steel plate with and without the DT4 interlayer is shown in Figure 3. When the reduction ratio is between 8%~68%, the shear strength of the interface increases with increasing reduction ratio: without a DT4 interlayer, the shear strength increased from 66.9 MPa to 271.2 MPa, and with a DT4 interlayer, the shear strength increased from 144.8 MPa to 262.8 MPa. When the reduction ratio is less than 18%, the shear strength is much higher with the DT4 interlayer than without it. When the reduction ratio is greater than 35%, the shear strengths on both sides are similar. The shear strength of the interface reaches the standard value of 140 MPa from the GB/T 8547-2006 type 1 titanium clad steel plate when the reduction ratio is 18% without the DT4 interlayer and 8% with it. The shear strength reaches the standard value of 196 MPa from the type 0 titanium clad steel plate when the reduction ratio is 35% without the DT4 interlayer and 18% with it. It can be concluded that the addition of the DT4 interlayer improves the shear strength of the composite plate at small reduction ratios. When the reduction ratio is large, the shear strength of the interface on both sides is close to the shear strength of the titanium matrix, meaning the effect of the DT4 interlayer on the shear strength is not as obvious.

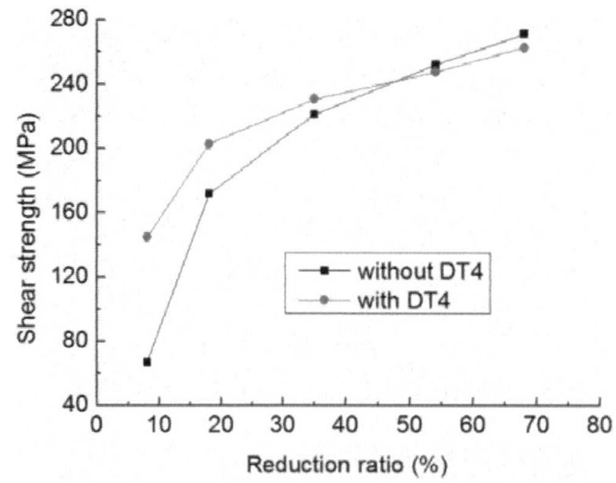

Figure 3. Effect of reduction ratio on the shear strength of titanium clad steel plates.

3.2. Interfacial structure of TA2/Q235 clad plates

Figure 4 and Figure 5 show secondary electron images and element surface scanning analysis images of the fracture surface on the Q235 side and the DT4 side, respectively, at a 68% reduction. As shown in Figure 4 (b) and (c), there is a large amount of Ti and a smaller amount of Fe distributed on the fracture surface. The microstructures of these two main element distribution regions are respectively shown in Figure 4(d) and (e). Figure 4 (d) reveals brittle fracture characteristics on the surface, indicating that brittle intermetallic compounds have been formed there. Figure 4 (e) presents obvious ductile fracture features with many dimples. Figure 6 shows that fractures occur mainly on the Ti side, though some occur in the brittle region at the interface. From Figure 5 (a) and Figure 5 (b), the fracture surface is full of Ti, and as shown in Figure 5 (c), there is a large number of dimples distributed along the fracture surface, which is characteristic of plastic fracture. This indicates that the fracture absolutely occurs on the TA2 side, where the bonding strength between DT4 and TA2 is high.

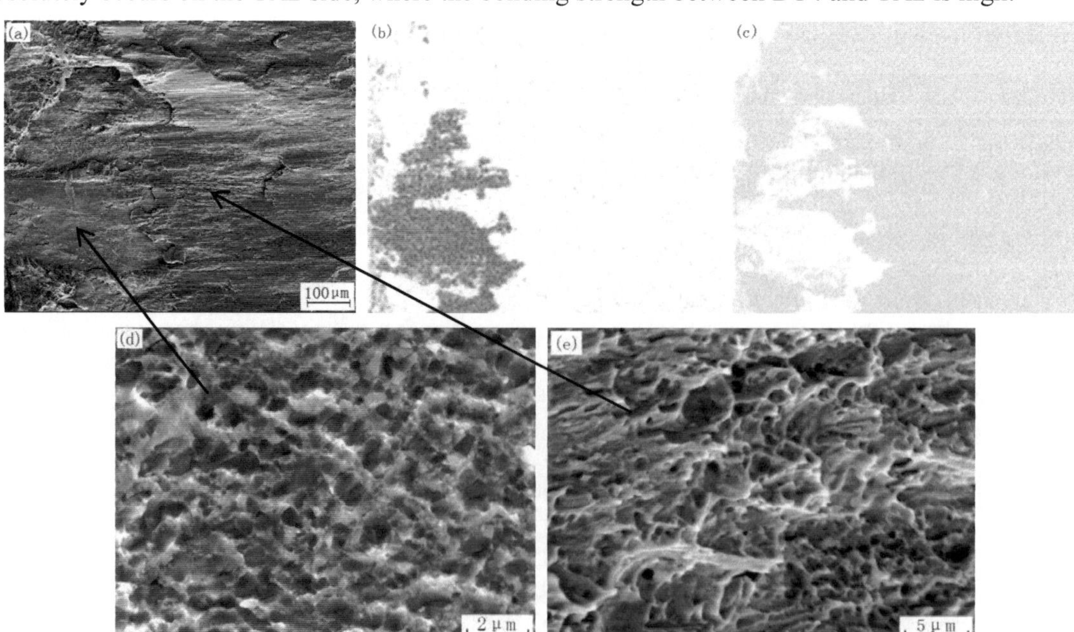

Figure 4. SEM and element surface scanning analysis of the fracture surface on the Q235 side at a reduction ratio of 68%: (a) overall morphology, (b) Fe distribution on Fig. 4 (a), (c) Ti distribution on fig. 4 (a), (d) and (e) microstructure of areas selected from Fig. 4 (a).

Figure 5. SEM and element surface scanning analysis of the fracture surface on the DT4 side at a reduction ratio of 68%: (a) overall morphology, (b) Fe distribution on Fig. 5 (a), (c) microstructure of the fracture surface.

Figure 6 shows the element distribution profiles across the interface at a bonding reduction ratio of 68%. The Ti and Fe diffusion distances are similar for both interfaces and are approximately 1.5 μm. Comparing the intensity of C across the interface between Figure 6 (a) and (b), it can be seen that, without the DT4, the C is concentrated near the interface. Conversely, with the DT4, such concentration is not as obvious. This shows that adding the DT4 interlayer can effectively reduce C diffusion to the interface from Q235, thus reducing the formation of brittle compounds such as TiC on the interface.

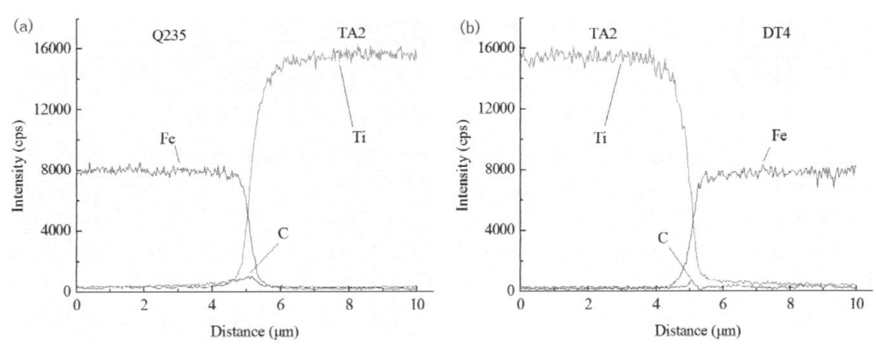

Figure 6. Element distribution profiles across the interface (a) between Q235 and TA2, (b) between TA2 and DT4.

3.3. Mechanical properties of TA2/Q235 clad plates

Figure 7 shows macroscale photographs of a bent specimen at a reduction ratio of 68%. Whether the DT4 is on the interior or the exterior, there are no cracks after bending. An optical specimen was taken from the bent specimen to observe the microstructure, which is shown in Figure 8. Without the DT4, there are some microcracks on the TA2 side whether the DT4 is on the interior or the exterior. Because the amount of deformation is greater on the outside than the inside, there are more cracks on the outside. With the DT4, no microcracks were found on the interface. Therefore, when TA2 and Q235 are composited by rolling, brittle compounds are easily formed on the interface. The poor plasticity of such compounds results in the formation of microcracks during bonding. This suggest that the addition of a DT4 interlayer can effectively inhibit the generation of compounds, thereby improving the bending performance of the composite plate.

IOP Conf. Series: Materials Science and Engineering **229** (2017) 012017 doi:10.1088/1757-899X/229/1/012017

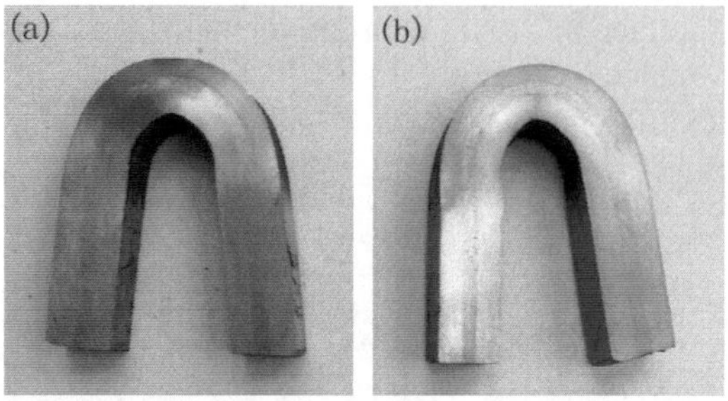

Figure 7. Macro photographs of bent specimen: (a) DT4 at interior, (b) DT4 at outside.

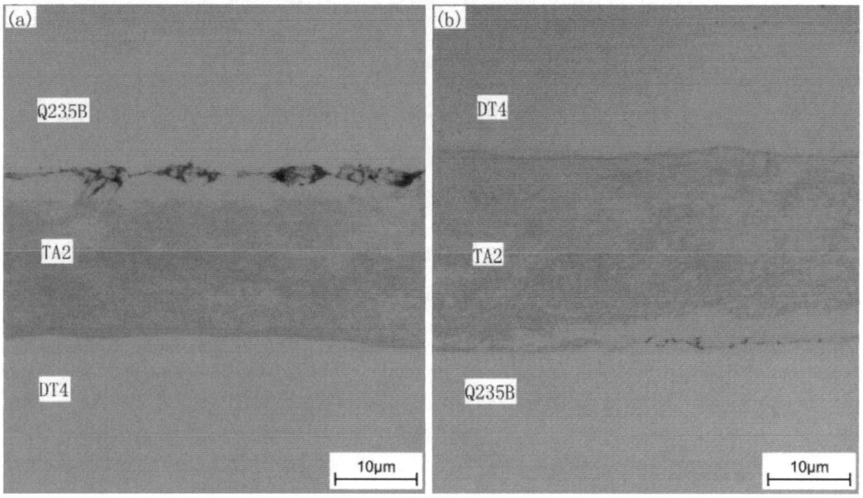

Figure 8. Optical micrograph of bent specimen: (a) DT4 at interior, (b) DT4 at outside.

Tensile specimens were taken along the rolling direction of the composite plate with 68% reduction ratio, and the tensile test results are shown in Figure 9. Without the DT4, when the strain was 0.356, Q235 and TA2 became separated, resulting in the invalid. With the DT4, when the strain was 0.386, the DT4 and TA2 became separated, and when the strain was 0.397, the composite plate underwent overall fracture. This indicates that since adding a DT4 interlayer can effectively reduce the formation of brittle compounds, crack sources are reduced on the interface during tensile deformation. Therefore, with the DT4 interlayer, the tensile properties are improved, and the fracture invalid is relieved.

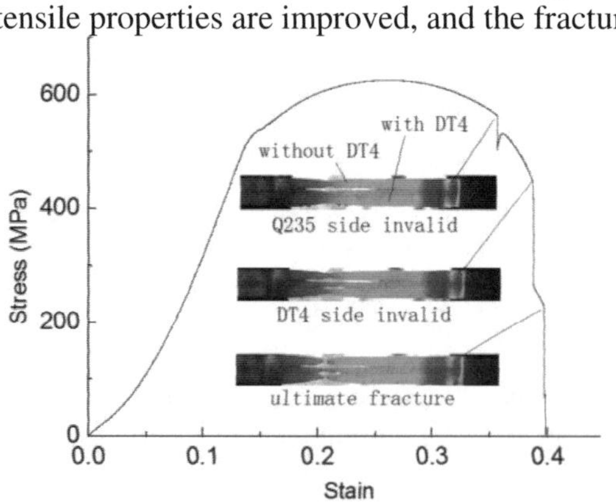

Figure 9. Tensile test results.

4. Conclusions

(1) A Q235B/TA2/DT4/Q235B plate was bonded by hot-rolling in a vacuum to compare the effect of the DT4 interlayer on the TA2/Q235B plate. When the heating temperature is 850 °C and the rolling speed is 50 mm/s, the shear strength of the interface increases with reduction ratio on both sides with or without DT4. When the reduction ratio is below 18%, with the DT4 interlayer, the shear strength of the interface is higher than without it. After 35% reduction, the shear strengths of the interfaces on both sides are similar.
(2) When the reduction ratio is 68%, with the DT4 interlayer, shear fracture occurs on the TA2 side. Without the DT4 interlayer, the fracture occurs mainly on the Ti side, though some also occur at the brittle region near the interface.
(3) With a DT4 interlayer, the bending and tensile properties of the TA2/Q235B plate can be improved.

Acknowledgments
The project was supported by the Natural Science Foundation of HeBei Province (E2015203311) and the National Natural Science Foundation of China (No. 51474190).

References
[1] Ha J S and Hong S I 2016 Deformation and fracture of Ti/439 stainless steel clad composite at intermediate temperatures Materials Science and Engineering: A 651 805-9.
[2] Su H, Luo X B, Chai F, Shen J C, Sun X J and Lu F 2015 Manufacturing technology and application trends of titanium clad steel plates Journal of Iron and Steel Research 22 977-82.
[3] Hwang Y M, Hsu H H and Hwang Y L 2000 Analytical and experimental study on bonding behavior at the roll gap during complex rolling of sandwich sheets International Journal of Mechanical Sciences 42 2417-37.
[4] Kundu S and Chatterjee S 2010 Evolution of interface microstructure and mechanical properties of titanium/304 stainless steel diffusion bonded joint using Nb interlayer ISIJ International 50 1460–5.
[5] Ha J S and Hong S I 2013Design of high strength Cu alloy interlayer for mechanical bonding Ti to steel and characterization of their tri-layered clad Materials & Design 51 293-9.
[6] Laik A, Shirzadi A A, Sharma G, Tewari R, Jayakumar T and Dey GK, Microstructure and interfacial reactions during vacuum brazing of stainless steel to titanium using ag-28 pct cu alloy, Metallurgical and Materials Transactions A 46 771-82.
[7] Kundu S and Chatterjee S 2006 Effect of bonding temperature on interface microstructure and properties of titanium–304 stainless steel diffusion bonded joints with Ni interlayer Materials Science and Technology 22 1201-7.
[8] Yu C, Xiao H, Yu H, Qi Z C and Xu C 2017 Mechanical properties and interfacial structure of hot-roll bonding TA2/Q235B plate using DT4 interlayer Materials Science and Engineering: A 695 120-5.

IOP Conf. Series: Materials Science and Engineering **229** (2017) 012018 doi:10.1088/1757-899X/229/1/012018

Interfacial and Electrical Properties of Ge MOS Capacitor by ZrLaON Passivation Layer and Fluorine Incorporation

Yong Huang[1, 2], Jing-Ping Xu[1,a], Lu Liu[1], Zhi-Xiang Cheng[1], Pui-To Lai[3,b] and Wing-Man Tang[4]

[1] School of Optical and Electronic Information, Huazhong University of Science and Technology, Wuhan 430074, China
[2] School of Information and Engineering, Hubei University for Nationalities, Enshi,Hubei, 445000, China
[3] Department of Electrical and Electronic Engineering, The University of Hong Kong, Pokfulam Road, Hong Kong
[4] Department of Applied Physics, The Hong Kong Polytechnic University, Hung Hom, Kowloon, Hong Kong

[a]jpxu@hust.edu.cn; [b]laip@eee.hku.hk

Abstract. Ge Metal-Oxide-Semiconductor (MOS) capacitor with HfTiON/ZrLaON stacked gate dielectric and fluorine-plasma treatment is fabricated, and its interfacial and electrical properties are compared with its counterparts without the ZrLaON passivation layer or the fluorine-plasma treatment. Experimental results show that the sample exhibits excellent performances: low interface-state density (3.7×10^{11} $cm^{-2}eV^{-1}$), small flatband voltage (0.21 V), good capacitance-voltage behavior, small frequency dispersion and low gate leakage current (4.41×10^{-5} A/cm^2 at $V_g = V_{fb} + 1V$). These should be attributed to the suppressed growth of unstable Ge oxides on the Ge surface during gate-dielectric annealing by the ZrLaON interlayer and fluorine incorporation, thus greatly reducing the defective states at/near the ZrLaON/Ge interface and improving the electrical properties of the device.

1. Introduction

Ge-based MOSFET with high-k gate dielectric has been widely investigated due to its higher carrier mobility [1, 2] and easier integration of Ge on Si than III-V semiconductors on Si. However, the development of Ge MOSFET has been hampered due to the absence of a good thermally-grown oxide on the Ge substrate. In order to overcome the drawback of the unstable and volatile Ge native oxide, a high-quality interface between high-k dielectric and Ge is needed to establish the Ge MOS technology [3]. So introducing an interfacial passivation layer (IPL) becomes very important for improving the performance of Ge MOS devices. But the dielectric constant (k) of many IPL materials, e.g. SiO_2 (~ 3.9) [4], GeON (~ 7) [5], Al_2O_3 (~ 8) [6], AlON (~ 10) [7], is usually not very large, thus limiting further scaling of the devices.

It is well known that Zirconium dioxide (ZrO_2) has a high k value (~ 20), a large bandgap (5.8 eV) and has been used as high-k gate dielectric for Si [8], SiGe [9] and GaAs [10] MOS devices with good

Content from this work may be used under the terms of the Creative Commons Attribution 3.0 licence. Any further distribution of this work must maintain attribution to the author(s) and the title of the work, journal citation and DOI.
Published under licence by IOP Publishing Ltd

interfacial and electrical properties. However, ZrO_2 may easily crystallize due to its lower crystallization temperature (400 - 500 ℃) [11], thus leading to a large gate leakage current [12]. Luckily, it has been demonstrated that incorporating lanthanum (La) into ZrO_2 can increase the crystallization temperature, and also can suppress the growth of the interfacial layers, thus it can reduce the defects in the dielectric and improve the interface quality [13, 14]. So in this work, La-doped ZrON (ZrLaON) would be used as IPL to improve the interfacial properties of the Ge MOS devices.

As one of the most promising high-k materials, Hf-based oxide/silicate/aluminate are often employed as gate dielectric. However, their moderate dielectric constants limit further device scaling [15]. While it is well known that Ti oxide has much higher k value than Hf oxide, Ti-incorporated Hf oxynitride (HfTiON) can achieve higher k value [16]. So, in this work, the Ge MOS device with ZrLaON as passivation interlayer and HfTiON as high-k layer is proposed and prepared. On the other hand, fluorine incorporation was reported to be another good way for passivating the Ge surface because it is capable of passivating the oxygen vacancies in high-k materials [17, 18]. Therefore, the fluorine-plasma is also used to treat the Ge MOS capacitor with the stacked gate dielectric of HfTiON/ZrLaON in this work, and its electrical and interfacial characteristics are compared with its counterparts without ZrLaON passivation layer or fluorine-plasma treatment.

2. Experiments

Ge MOS capacitors were fabricated on n-type (100) Ge wafers with a resistivity of 0.02 ~ 0.10Ωcm. The wafers were cleaned using ethanol, acetone and trichloroethylene, followed by dipping in diluted HF (1:50) for 30 s, and then rinsed in deionized water for several times to remove the native oxide. After N_2 drying, the wafers were immediately transferred to the vacuum chamber of the Denton Vacuum Discovery Deposition System. A ~ 2-nm ZrLaON film was deposited by co-sputtering Zr and La_2O_3 targets (as IPL) at room temperature in Ar/N_2 (= 24/12) ambient, followed by the in-situ deposition of a ~ 6-nm HfTiN gate dielectric by co-sputtering Hf and Ti targets in the same ambient. Next, a sample received a fluorine-plasma treatment at a flow rate of CHF_3/O_2 = 10/1 for 5 min.

For comparison, a control sample without the ZrLaON interlayer (i.e. HfTiON was directly deposited on Ge) was prepared. So, the samples can be divided into three groups: the sample with ZrLaON passivation layer and fluorine-plasma treatment (denoted as ZrLaON+F), the sample with ZrLaON passivation layer but no fluorine-plasma treatment (denoted as ZrLaON) and the control sample without any passivation layer. Subsequently, post-deposition annealing (PDA) was carried out at 500 ℃ for 300 s in N_2 (500 sccm) + O_2 (50 sccm) to transform HfTiN into HfTiON, respectively. Finally, Al was thermally evaporated and patterned as gate electrode with an area of 7.85×10^{-5} cm^2, followed by a forming-gas (95% N_2 + 5% H_2) annealing at 300 ℃ for 20 min to reduce contact resistance.

High-frequency (HF) capacitance-voltage (C-V) and gate leakage current-voltage (J_g-V_g) measurements were performed using HP4284A precision LCR meter and HP4156A semiconductor parameter analyzer, respectively. Physical thickness of the gate dielectrics was determined by spectroscopic ellipsometry. The chemical structures at/near the high-k/Ge interface were analyzed by X-ray photoelectron spectroscopy (XPS).

3. Results and Discussion

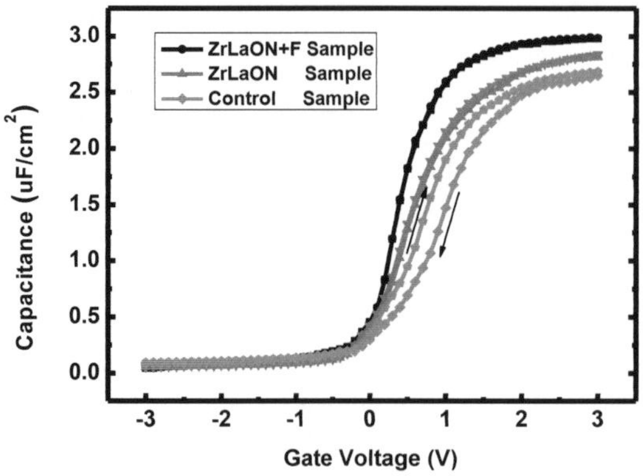

Figure 1. HF (1 MHz) C-V curve of all the samples.

Fig.1 shows the typical HF C-V curves of the three samples. Obviously, the control sample has a much smaller accumulation capacitance than the ZrLaON+F and ZrLaON samples, implying that a low-k GeO_x interlayer exists in the former and is greatly suppressed in the latter due to a blocking role of the ZrLaON interlayer against the in-diffusion of oxygen and out-diffusion of germanium. As a result, a stretch out of the C-V curve as well as a small kink in the depletion region are observed for the control sample in Fig.1, indicating a high interface-state density. While for the ZrLaON+F and ZrLaON samples, no such distortion occurs and a large slope of the C-V curve from depletion to accumulation is obtained, implying a high-quality ZrLaON/Ge interface. Furthermore, a smaller C-V hysteresis for the ZrLaON+F sample (10 mV) and ZrLaON sample (40 mV) than the control sample (220 mV) is observed in Fig. 1, indicating less slow states or deep-level traps in the former (especially for the ZrLaON+F sample) than in the latter. The flatband voltage (V_{fb}) of the samples is determined from their flatband capacitance [19], and equivalent oxide-charge density (Q_{ox}) is calculated as $-C_{ox}(V_{fb} - \varphi_{ms})/q$, with φ_{ms} as the work-function difference between Al and Ge, and C_{ox} as the accumulation capacitance (oxide capacitance) per unit area. The interface-state density near midgap (D_{it}) is estimated from the 1-MHz C-V curve using the Terman's method [20] for the purpose of comparison, as listed in Table 1.

Table 1. Electrical and Physical Parameters Extracted from HF C-V curves

Sample	ZrLaON+F	ZrLaON	Control
C_{ox} (μF/cm^2)	2.97	2.81	2.63
V_{fb} (V)	0.21	0.29	0.59
D_{it} (cm^{-2}eV^{-1})	3.7×10^{11}	6.3×10^{11}	7.9×10^{12}
Q_{ox} (cm^{-2})	-3.39×10^{12}	-4.75×10^{12}	-9.87×10^{12}
t_{ox} (nm)	6.9	7.1	7.2
CET (nm)	1.16	1.23	1.31
k	23.2	22.5	21.4

Obviously, the ZrLaON+F and ZrLaON samples have smaller V_{fb} and D_{it} than the control sample, with the smallest for the ZrLaON+F sample (V_{fb} = 0.21 V and $D_{it} = 3.7\times10^{11}$ eV^{-1}cm^{-2}), Which should be attributed to the fact that the ZrLaON interlayer can block the out-diffusion of Ge and in-diffusion of oxygen at the interface, and the F incorporated in the HfTiON/ZrLaON stack dielectric by the F plasma treatment can effectively occupy the oxygen vacancies in the ZrLaON passivation layer and passivate the traps at/near the interface [21, 22], resulting in a reduction of defect traps in the dielectric bulk and at/near the ZrLaON/Ge interface.

This is further supported by the small frequency dispersion of its C-V curves and small distortion in the depletion region even for 50-kHz C-V curve, as shown in Fig. 2. The frequency dispersion in the accumulation region should be attributed to border traps with much larger trapping time constant than the interface traps [23, 24] because they can exchange charges with the conduction band of the semiconductor through tunneling, thus contributing to the total gate capacitance by capturing and emitting charges. As the measurement frequency increases, they are too slow to respond to the applied voltage, thus resulting in lower total gate capacitance [23]. The larger accumulation capacitance of the ZrLaON+F sample than the ZrLaON sample is also attributed to the F incorporation at/near the HfTiON/ZrLaON and ZrLaON/Ge interfaces, which can enhance the suppression of growth of the low-k interlayer and give high-quality ZrLaON/Ge interface.

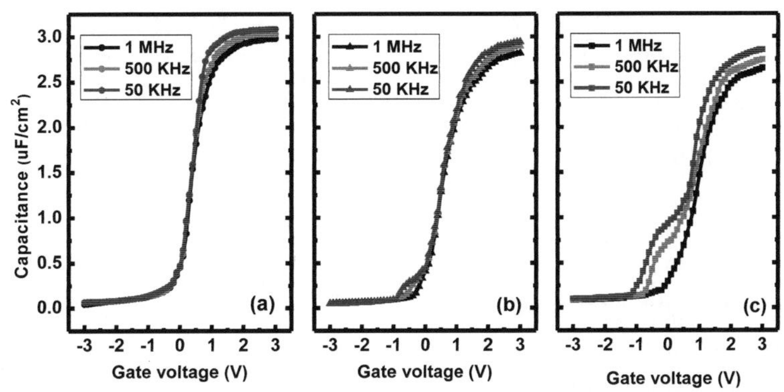

Figure 2. Frequency dispersion of C-V curves at room temperature. (a) ZrLaON+F sample, (b) ZrLaON sample, and (c) control sample.

The capacitance equivalent thickness (CET) is calculated as CET = k_0k_{SiO2}/C_{ox}, where k_0 and k_{SiO2} are the vacuum permittivity and relative permittivity of SiO_2, and thus the equivalent k value of the gate dielectric can be calculated as k = $k_{SiO2}t_{ox}$/CET, as also listed in Table I. Obviously, the ZrLaON+F sample has the smallest CET (1.16 nm) and largest k value (23.2) due to the enhanced suppression of the low-k GeO_x growth, as mentioned above. Also, the smallest Q_{ox} is obtained for the ZrLaON+F sample, followed by the ZrLaON sample, indicating an effective blocking role of the ZrLaON interlayer (especially with Fluorine incorporation) against the inter-diffusion of elements [25].

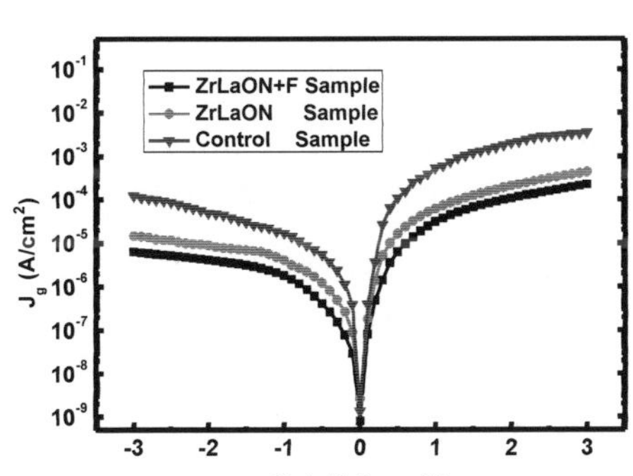

Figure 3. J_g-V_g Characteristics of all the samples.

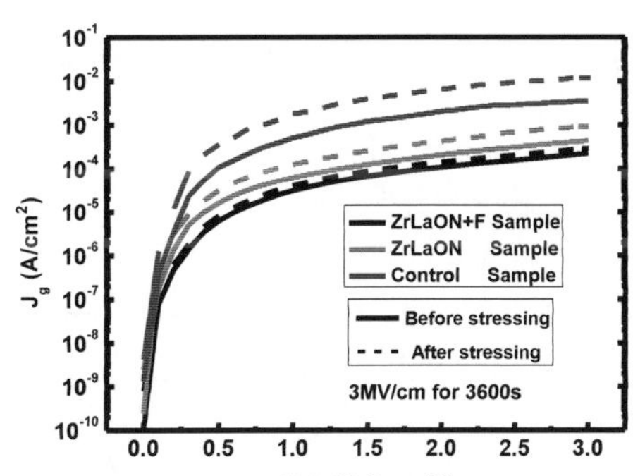

Figure 4. Gate leakage current density of all the samples before and after a high-field stressing at 3 MV/cm for 3600 s.

The gate leakage properties of the three samples are shown in Fig. 3. Obviously, the ZrLaON+F sample has the lowest gate-leakage current among the three samples (e.g. at $V_g = V_{fb} + 1$ V, $J_g = 4.41 \times 10^{-5}$ A/cm^2, 9.69×10^{-5} A/cm^2 and 1.31×10^{-3}A/cm^2 for the ZrLaON+F, ZrLaON and control samples respectively), which is closely related to its smallest D_{it} and Q_{ox} due to the absence of the undesirable GeO_x interlayer and the suppression of intermixing between Ge and high-k dielectric [26, 27]. Furthermore, a high-field stress at 3 MV/cm [= $(V_g - V_{fb})/t_{ox}$] for 3600 s is used to examine the reliability of the devices. The J_g-V_g properties are measured before and after the stressing, as shown in Fig. 4. The increase in gate leakage current after the stressing for the three samples is due to the trap-assisted tunneling of electrons from the substrate to the gate via newly-generated interface and near-interface traps. Obviously, the post-stressing increase of the leakage current is smallest for the ZrLaON+F sample, which can be associated with less generation of the interface and near-interface traps during the constant-voltage stressing due to very little GeO_x at/near the ZrLaON /Ge interface [28], as confirmed by XPS result below.

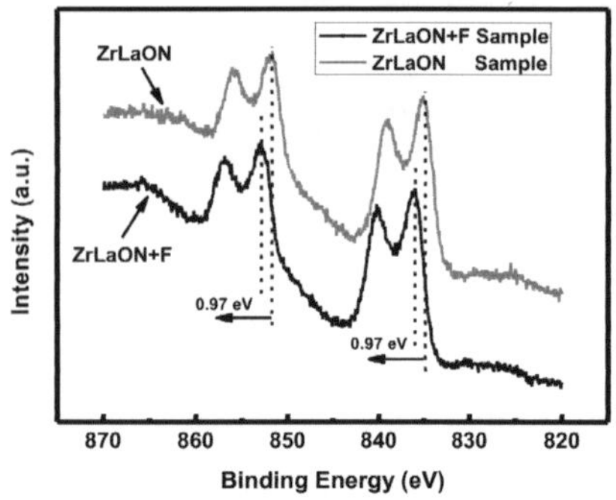

Figure 5. La 3d XPS spectrum at the ZrLaON interlayer of the ZrLaON+F and ZrLaON samples.

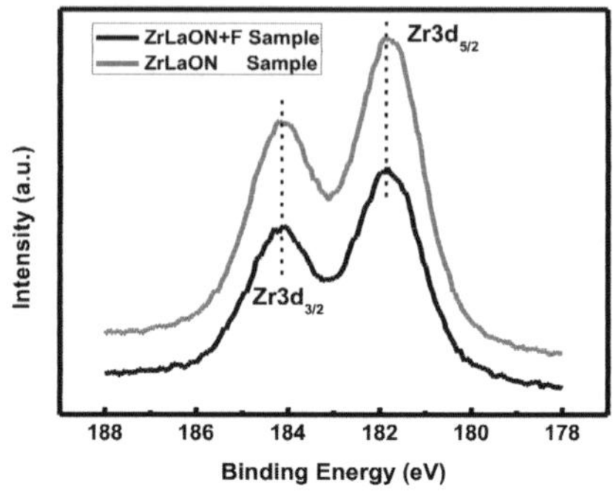

Figure 6. Zr 3d XPS spectrum at the LaTaON interlayer of the ZrLaON+F and ZrLaON samples.

To identify the composition and chemical status of the ZrLaON interlayer, and further analyze their effects on the chemical states of the interface between HfTiON and Ge, the HfTiON film is etched to a distance of ~ 5 nm from the Ge surface using an in-situ Ar+ ion beam in the XPS chamber. Fig. 5 shows the La 3d XPS spectrum of the ZrLaON+F and ZrLaON samples. As compared to the ZrLaON sample, the La 3d peaks of the ZrLaON+F sample exhibit a shift of 0.97 eV towards higher energy, which should be contributed to the higher electronegativity of F (4.0) than that of O (3.5).

Figure 6 shows the Zr 3d XPS spectra of the ZrLaON+F and ZrLaON samples, Two strong peaks at 181.9 eV (Zr 3d 5/2) and 184.1 eV (Zr 3d 3/2) are detected and the spin-orbit splitting energy of 2.2 eV is consistent with the Zr hump [29]. Also, the peak located at 181.9 eV should be from the mixture of major Zr-O bonds (182.0 eV [29]) and minor Zr-N bonds (180.1 eV [30]) (or Zr-O-N bonds at 180.0-182.5 eV [29]).

To further analyze the Ge oxides or oxynitride at/near the interface, the XPS spectrum of Ge 2p for the two samples are shown in Fig. 7, and the energy position and atomic percentage of the species are listed in Table 2.

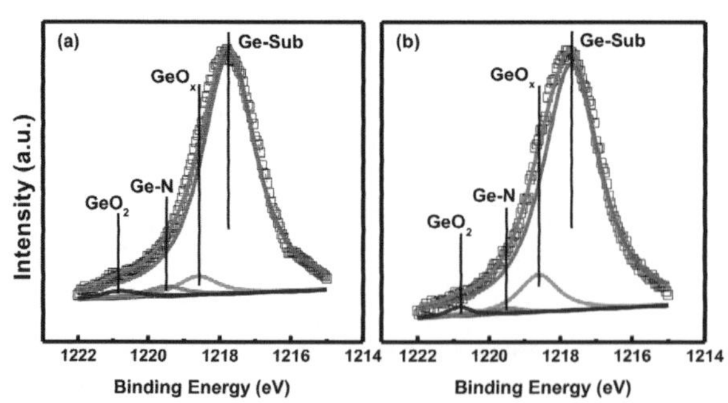

Figure 7. Ge 2p spectrum of (a) ZrLaON+F sample, (b) ZrLaON sample.

Table 2. Energy Position and Atomic Percentage of the Species In Fig. 7

Sample		ZrLaON+F sample	ZrLaON sample
	Energy Position (eV)	Atomic percentage (%)	Atomic percentage (%)
Ge-Sub	1217.7	90.3	86.4
GeO_x	1218.6	5.2	9.3
GeON	1219.5	2.6	1.1
GeO_2	1220.8	1.9	3.2

As can be seen, the GeO_2, GeON, GeO_x and Ge-Sub are found at 1220.8 eV, 1219.5 eV, 1218.6 eV and 1217.7 eV respectively [31, 32] for the two samples, demonstrating that there exist GeO_2, GeON and GeO_x at/near the ZrLaON/Ge interface. However, it is worth noting that the content of GeO_x is obviously lower for the ZrLaON+F sample (5.2% from the GeO_x/Ge2p peak-area ratio) than the ZrLaON sample (9.3%), and similarly, the content of GeO_2 is also lower for the ZrLaON+F sample (1.9% from the GeO_2/Ge3d peak-area ratio) than the ZrLaON sample (3.2%), implying fluorine incorporation can effectively occupy the oxygen vacancies in the ZrLaON passivation layer. On the contrary, the content of GeON is higher for the ZrLaON+F sample (2.6% from the GeON/Ge3d peak-area ratio) than the ZrLaON sample (1.1%). These results indicate that fluorine incorporation can further reduce the oxide traps, beneficial for formation of GeON and block the inter-diffusion of elements near the interface, resulting in the best interface quality and thus electrical properties, as shown for the ZrLaON+F sample above.

4. Conclusion

The effects of ZrLaON passivation interlayer and fluorine-plasma treatment on the interfacial and electrical properties of Ge MOS capacitor with HfTiON gate dielectric are investigated. Experimental results show that the Ge MOS devices with ZrLaON interlayer have less interface states and oxide charges, which can be further reduced by fluorine incorporation. XPS analyses indicate that the ZrLaON passivation layer can effectively suppress the growth of unstable native oxides at the Ge surface, and also fluorine incorporation can suppress the inter-diffusion of elements and reduce the oxide traps at/near the high-k/Ge interface, thus giving good electrical properties of the Ge MOS device. Therefore, the ZrLaON interlayer combined with the fluorine-plasma treatment is a promising way for improving the interface quality and thus obtaining high-performance Ge MOS devices.

Acknowledgments

This work was supported mainly by the National Natural Science Foundation of China under Grant 61274112, Grant 61176100, and Grant 61404055, in part by the University Development Fund of the University of Hong Kong under Grant 00600009, and in part by the University Development Fund of the Hong Kong Polytechnic University under Grant 1-ZVB1.

References

[1] C. O. Chui, H. Kim, D. Chi, P. C. McIntyre, and K. C. Saraswat, IEEE Trans. Electron Devices **53**, 1509 (2006).
[2] Y. Kamata. Mater. Today, **11**, 30 (2008).
[3] M. Caymax, M. Houssa, G. Pourtois, F. Bellenger, K. Martens, A. Delabie, and S. Van Elshocht, Appl. Surf. Sci. **254**, 6094 (2008).
[4] T. Das, C. Mahata, C. K. Maiti, G. K. Dalapati, C. K. Chia, D. Z. Chi, S. Y. Chiam, H. L. Seng, C. C. Tan, H. K. Hui, G. Sutradhar, and P. K. Bose, J. Electrochem. Society, **159**, G15 (2012).
[5] C. O. Chui, F. Ito, and K. C. Saraswat, IEEE Electron Device Lett. **25**, 613(2004).
[6] M. K. Lee, and C.-F. Yen, Appl. Phys. A Mater. Sci. Process. **116**, 2051 (2014).
[7] L. S. Wang, L. Liu, J. P. Xu, S. Y. Zhu, Y. Huang, and P. T. Lai, IEEE Trans. Electron. Devices, **61**, 742 (2014).

[8] B. O. Cho, J. Wang, L. Sha, and J. P. Chang, Appl. Phys. Lett. **80**, 1052 (2002).
[9] T. Ngai, W. J. Qi, R. Sharma, J. Fretwell, X. Chen, J. C. Lee, and S. Banerjee, Appl. Phys. Lett. **76**, 502 (2000).
[10] S. Kundu, S. Roy, P. Banerji, S. Chakraborty, and T. Shripathi, J. Vac. Sci. Tech. B, **29**, 031203 (2011).
[11] J. M. Gaskell, A. C. Jones, H. C. Aspinall, S. Taylor, P. Taechakumput, P. R. Chalker, P. N. Heys, and R. Odedra, Appl. Phys. Lett. **91**, 112912 (2007).
[12] S. Jeon, C.-J. Choi, T.-Y. Seong, and H. Hwang, Appl. Phys. Lett. **79**, 245 (2001).
[13] C. Wiemer, A. Debernardi, A. Lamperti, Molle, O. Salicio, L. Lamagna, and M. Fanciulli, Appl. Phys. Lett. **99**, 232907 (2011).
[14] C. Y. Han, W. M. Tang, C. H. Leung, C.-M. Che, and P. T. Lai, IEEE Trans. Electron Devices, **62**, 2313 (2015).
[15] G. He, Z. Q. Sun, M. Liu, and L. D. Zhang, Appl. Phys. Lett. **97**, 192902 (2010).
[16] Li-Sheng Wang, Lu Liu, Jing-Ping Xu, Shu-Yan Zhu, Yuan Huang, and Pui-To Lai, IEEE Trans. Electron Devices, **61**, 742 (2014).
[17] C. X. Li, C. H. Leung, P. T. Lai, J. P. Xu. Solid-State Electronics, **54**, 675(2010).
[18] K. Tse, and J. Robertson, Appl. Phys. Lett. **89**, 142914 (2006).
[19] J. P. Xu, P. T. Lai, C. X. Li, X. Zou, and C. L. Chan, IEEE Electron Device Lett. **27**, 439 (2006).
[20] L. M. Terman, Solid-State Electron. **5**, 285 (1962).
[21] K. Tse and J. Robertson, Appl. Phys. Lett. **89**, 142914 (2006).
[22] L. N. Liu, H. W. Choi, J. P. Xu, and P. T. Lai， Appl. Phys. Lett. **107**, 213501 (2015).
[23] D. Lin, A. Alian, S. Gupta, B. Yang, E. Bury, S. Sioncke, R. Degraeve1, M. L. Toledano, R. Krom, P. Favia, H. Bender, M. Caymax, K. C. Saraswat, N. Collaert, and A. Thean, Tech. Dig. - Int. Electron Devices Meet. 2002, 28.3.1.
[24] Y. Yuan, L. Wang, B. Yu, B. Shin, J. Ahn, P. C. McIntyre, P. M. Asbeck, M. J. W. Rodwell, and Y. Taur, IEEE Electron Device Lett., 32, 485 (2011).
[25] L. K. Chu, W. C. Lee, M. L. Huang, Y. H. Chang, L. T. Tung, C. C. Chang, Y. J. Lee, J. Kwo, and M. Hong, J. Cryst. Growth, **311**, 2195 (2009).
[26] A. Dimoulas, D. P. Brunco, S. Ferrari, J. W. Seo, Y. Panayiotatos, A. Sotiropoulos, T. Conard, M. Caymax, S. Spiga, M. Fanciulli, C. Dieker, E. K. Evangelou, S. Galata, M. Houssa, M. M. Heyns, Thin Solid Films. **515**, 6337 (2007).
[27] D. P. Brunco, A. Dimoulas, N. Boukos, M. Houssa, T. Conard, K. Martens, C. Zhao, F. Bellenger, M. Caymax, M. Meuris, and M. M. Heyns, J. Appl. Phys. **102**, 024104 (2007).
[28] J. P. Xu, X. F. Zhang, C. X. Li, C. L. Chan, P. T. Lai, Appl. Phys. A. **99**, 177 (2010).
[29] Y. H. Wong and K. Y. Cheong, Thin Solid Films, **520**, 6822 (2012).
[30] H. S. Jung, J. M. Park, H. K. Kim, J. H. Kim, S. J. Won, J. Lee, S. Y. Lee, C. S. Hwang, W. H. Kim, M. W. Song, N. I. Lee, D. Y. Cho, Electrochem Solid-State Lett, **13**, G71 (2010).
[31] N. Wu, Q. C. Zhang, C. X. Zhu, C. C. Yeo, S. J. Whang, D. S. H. Chan, M. F. Li, B. J. Cho, A. Chin, D. L. Kwong, A. Y. Du, C. H. Tung, and N. Balasubramanian, Appl Phys Lett, **84**, 3741 (2004).
[32] K. Prabhakaran, T. Ogina. Surf Sci, **325**, 263 (1995).

IOP Conf. Series: Materials Science and Engineering **229** (2017) 012019 doi:10.1088/1757-899X/229/1/012019

Efficiency of Nb-Doped ZnO Nanoparticles Electrode for Dye-Sensitized Solar Cells Application

Jirapat Anuntahirunrat[1], Youl-Moon Sung[1,*] and Pattarapon Pooyodying[1]

[1]Electrical Engineering, Kyungsung University, Busan 608-736, Korea

[*]ymsung@ks.ac.kr

Abstract. The technological of Dye-sensitized solar cells (DSSCs) had been improved for several years. Due to its simplicity and low cost materials with belonging to the part of thin films solar cells. DSSCs have numerous advantages and benefits among the other types of solar cells. Many of the DSSC devices had use organic chemical that produce by specific method to use as thin film electrodes. The organic chemical that widely use to establish thin film electrodes are Zinc Oxide (ZnO), Titanium Dioxide (TiO_2) and many other chemical substances. Zinc oxide (ZnO) nanoparticles had been used in DSSCs applications as thin film electrodes. Nanoparticles are a part of nanomaterials that are defined as a single particles 1-100 nm in diameter. From a few year ZnO widely used in DSSC applications because of its optical, electrical and many others properties. In particular, the unique properties and utility of ZnO structure. However the efficiency of ZnO nanoparticles based solar cells can be improved by doped various foreign impurity to change the structures and properties. Niobium (Nb) had been use as a dopant of metal oxide thin films. Using specification method to doped the ZnO nanoparticles thin film can improved the efficiencies of DSSCs. The efficiencies of Nb-doped ZnO can be compared by doping 0 at wt% to 5 at wt% in ZnO nanoparticles thin films that prepared by the spin coating method. The thin film electrodes doped with 3 at wt% represent a maximum efficiencies with the lowest resistivity of $8.95\times10^{-4}\ \Omega\cdot cm$.

1. Introduction

Solar energy is one of the most essential energy source for renewable energy. This solar energy is the clean energy with a substantial renewable resources. Dye-sensitized solar cells (DSSCs) seem to be fond of commonly, this is because of the cost value to prepared the materials and invent the device is less than any other solar cells application. DSSCs were first studied and researched vastly after B. O'Regan et al. [1] reported the efficiency of DSSCs in 1991. The technology that attractive for established DSSCs is the synthesis of thin films electrode. There are so many preparation method to prepare compounds materials and many method for electrode fabrications. The usage of ZnO nanoparticles is attractive for thin film electrodes to prepared the thin films electrodes [2],[3],[4],[5],[6] and [7] because of its good characteristics[8],[9] and [10]. Even though ZnO nanoparticles is a part of nanomaterials that has a crystal structure of wurtzite (B4) at ambient conditions. The ZnO wurtzite structure has a hexagonal unit cell of two lattice parameters. However, hexagonal structure has lo efficiencies in such a way of one-dimentional (1D) nanostructures had less surface area. Therefore, the used of ZnO nanoparticles based solar cells can be improved by optimize the crystal srtuture of active layer. Zno is the II-VI semiconductor group with a wide bandgap about 3.2 - 3.4 eV and the wurtzite structure by lattice parameters of a = b = 0.325 nm and c = 0.520 nm.

Content from this work may be used under the terms of the Creative Commons Attribution 3.0 licence. Any further distribution of this work must maintain attribution to the author(s) and the title of the work, journal citation and DOI.
Published under licence by IOP Publishing Ltd

Various foreign impurity had doped into ZnO nanoparticles thin films to changed its characteristics and properties. Many of compound materials had been use for doping to improve efficiencies of thin films electrode. Especially, an elements from Group IIIA (B, Al, Ga, In, etc.), IVA (Si, Ge, Sn,etc.), IIIB (Sc, Y, etc.) and IVB (Ti, Zr, etc.). Furthermore, the transition metals has been researched for using as dopant material for instance Mn, T, La, Fe, etc. Meanwhile, using Niobium (Nb) as dopant material is very attractive by Nb^{5+} and Zn^{2+} there is only 3 valence difference, thus, Nb atom can be grant more than 1 electron to the electrical conductivity. Also Nb have a smaller ionic radius of 0.64 Å and higher covalent radius of 1.34 Å compared to Zn values that have an ionic radius of 0.74 Å and covalent radius of 1.25 Å.

In this article, we reported the effect and efficiencies of Nb-doped ZnO nanoparticles (Nb:ZNP) electrode based on DSSCs. In the usage as thin film electrode in DSSCs, we focusing on the scope of structural and electrical properties with a good efficiencies to using for DSSCs applications. Although the amount of Nb dopant is different at wt% of percentages from 1 at wt% to 5 at wt%.

2. Experimental details

2.1 Electrode and Counter electrode preparation

In order to fabricated the thin film electrode for DSSCs. A ZnO nanoparticles solution were prepared by dissolved 2 g of zinc acetate, anhydrous [$Zn(CH_3COO)_2$, Wako] in 50 mL of 1 M methanolic solution and keep under stirring for 24 hours. Then add a few drops of sodium hydroxide 1 mol/L solution (NaOH, Junsei) while keep stirring the solution and mix it up well into a white suspension solution. Then keep it under stirring for 24 hours at room temperature. Then drop wired with niobium (V) ehthoxide in a different weight percent of solution from 0 at wt% to 5 at wt% and mix it well. The Nb:ZNP solution will get into a viscous suspension paste.

For the fabrication of thin film counter electrode using Platinum paste. By dissolved 3 g tetraammine platinum(II) chloride [$Pt(NH_3)_4Cl_2.H_2O$, Kaida] in 50 mL of 0.1 M Hydrochloric acid solution [HCl, Sigma-Aldrich] and stirred for 2 hours at room temperature until it get into sticky paste.

Nb:ZNP paste and Pt paste were adopted on 5 x 4 cm^2 conducting fluorine-doped SnO_2-(FTO)-coated glass substrate (resistance sheet: ~7 Ω/sq, Pilkington). FTO glass sheets were first washed with DI water then sonicate in ultrasonic cleaner (JAC4020, Ultrasonic) in acetone solution (CH_3COCH_3, Duksan) for 1 hour. Then drying under dry air jet. The FTO were coated with Nb:ZNP paste and Pt paste. Using spin coating technique to figure 2 x 2 cm^2 layers. Then the layers were heating at 450°C for 1 hour.

2.2 Fabrication of DSSCs

Nb:ZNP coated on FTO glasses with difference wt%, used as an electrode were fixed with a distance of 50 μm by spacers (Himilian). For counter electrode using Pt coated on FTO glass as a counter electrode. The Nb:ZNP electrode and Pt counter electrode are put together into DSSCs in the structure of FTO glass/ Nb:ZNP electrode/ Electrolyte/ Pt counter electrode/ FTO glass. The schematic of DSSCs can be shown in figure 1.

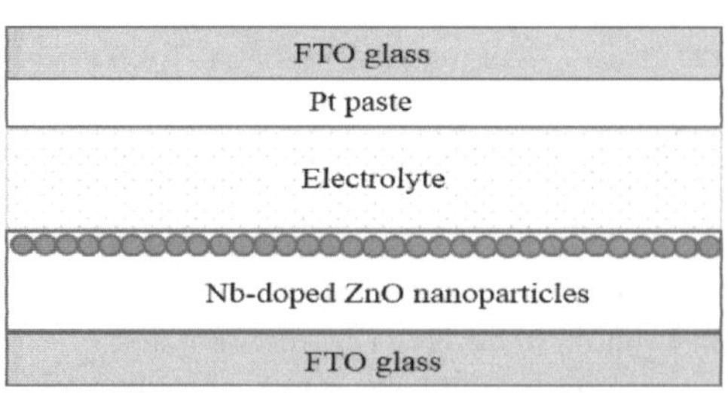

Figure 1. The schematic of DSSCs

IOP Conf. Series: Materials Science and Engineering **229** (2017) 012019 doi:10.1088/1757-899X/229/1/012019

3. Result and Discussion

The chemistry analytical to analyze the crystalline structure of the thin film electrode can be analysed by XRD analysis. The XRD pattern is shown in figure 2. The XRD patterns are exposes that it belong to the crystal structural of the hexagonal structure of wurtzite (B4) structure which growth through (002) and (101) orientation plan. The normally ZnO films that prepared by annealing at the temperature about 400 °C to 500°C normally growth through (002) orientation plan. The strongest intensity of Nb:ZNP films have the peek at an angle of 36.21° (2θ) corresponding to the ZNP (002) and (101) plan and observed corresponding to the Nb (010) plan.

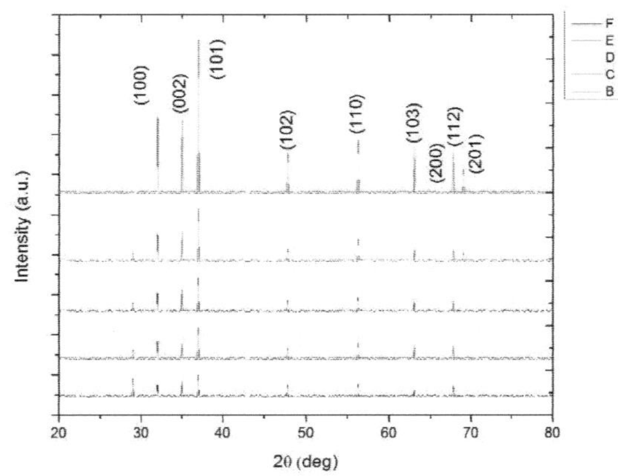

Figure 2. The XRD pattern of Nb-doped ZnO nanoparticles at difference wt%

In figure 3 is shown the SEM images of Nb:ZNP thin film electrodes that coated on FTO glasses. The morphology of Nb:ZNP were observed as clear spherical particles. The figures of SEM micrographs show the morphologies of Nb:ZNP in different doping wt%. All the result show as nanoparticles in the uniform spherical morphology with the size distribute at diameter about 25 nm.

(a)

(b)

(c)

(d)

(e)

Figure 3. SEM images of Nb-doped ZnO nanoparticles at difference wt%

The electrical properties of Nb:ZNP thin film electrodes can be analyse by the I-V curve of Nb:ZNP in different doping wt% from 1 at wt% to 5 at wt%. in figure 4.

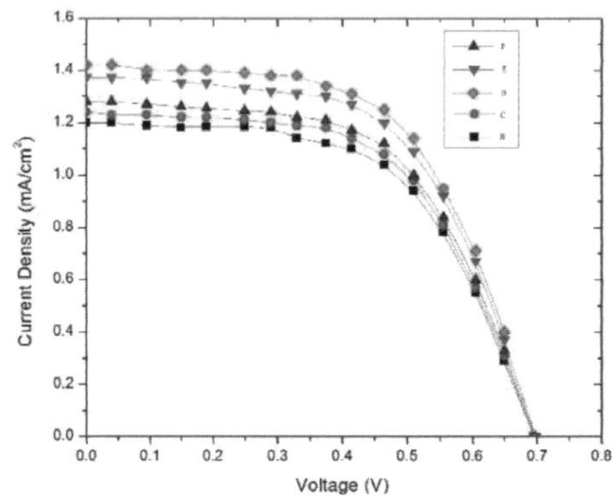

Figure 4. I-V curve of Nb-doped ZnO nanoparticles at difference wt%

All the DSSCs which was tested under a solar simulation AM 1.5 light illumination at 200 mW/cm^2. The open circuit voltage (VOC), current density (JSC), fill factor (FF) and efficiency (η) can summarize in table 1. The conversion efficiency of DSCs with Nb:ZNP and ZNP film electrode can calculated in equation (1),

$$\eta = \frac{V_{OC} I_{SC} FF}{P_{in}} \tag{1}$$

Table 1. Solar cells parameter characteristics of DSSCs using Nb:ZNP thin film electrodes

Sample	V_{OC} (V)	J_{SC} (mA/cm^2)	FF (%)	Efficiency (%)
B	0.452	11.4	19.32	8.18
C	0.536	13.6	22.82	8.48
D	0.627	16.3	38.26	9.02
E	0.582	15.1	26.84	8.68
F	0.524	14.3	24.22	8.52

The DSSCs using Nb:ZNP in sample D (3 wt%) was highest energy conversion efficiency of 9.02%, ISC of 16.3 mA/cm^2, VOC of 0.627 V and FF of 38.26% which more than that other electrodes with different doping wt%.

4. Conclusion

The synthesis and operation of DSSCs with Nb-doped ZnO nanoparticles using as thin film electrode can be improved the electrical properties and performance of ZnO nanoparticles. It can be concluded that Nb did not affect the ZnO structure. It was observed that the absorption intensity of ZnO slightly increased with increasing Niobium content. Which can improved the advantages of organic semiconductors (flexibility, solutions processing) with those of inorganic semiconductors (stability, high dielectric constant, which facilitates carrier generation processes, a high carrier mobility and thermal morphological stability of the blended materials).

Acknowledgements
This research was supported by Basic Science Research Program through the National Research Foundation of Korea(NRF) funded by the Ministry of Education(No. 2016R1D1A3B03935372).

References

[1] B. O'Regan, M. Grätzel, "A low-cost, high-efficiency solar cell based on dye sensitized colloidal TiO_2 films." Nature, vol. 353, pp.737–740, 1991.
[2] Zhifeng Liu, Chengcheng Liu, Jing Ya, Lei E, "Preparation of ZnO nanoparticles and characteristics of dye-sentisitized solar cells based on nanoparticles film", Solid State Sciences, vol. 12 Issue 1, pp. 111-114, 2010.
[3] Q. Zheng, G. Cao, "Nanostructure photoelectrodes for dye-sensitized solar cells" Nanotoday., vol. 6, pp. 91-109, 2011.
[4] Ali Elkhidir Suliman, , Yiwen Tang, Liang Xu, "Preparation of ZnO nanoparticles and nanosheets and their application to dye-sensitized solar cells", Solar Energy Materials and Solar Cells, vol. 91, Issue 18, pp. 1658-1662, 2007.
[5] S.M. Reda, "Synthesis of ZnO and Fe_2O_3 nanoparticles by sol–gel method and their application in dye-sensitized solar cells", Materials Science in Semiconductor Processing, vol. 13, Issue5-6, pp. 417-425, 2010.
[6] Lin Ke, Surani Bin Dolmanan, Lu Shen, Pramoda Kumari Pallathadk, Zheng Zhang, Doreen Mei Ying Lai, Hong Liu, "Degradation mechanism of ZnO-based dye-sensitized solar cells", Solar Energy Materials and Solar Cells, vol. 94, Issue 2, pp. 323-326, 2010.
[7] P. Wang, C. Klein, R. Humphry-Baker, S. Zakeeruddin, M. Grätzel, A High Molar Extinction Coefficient Sensitizer for Stable Dye-Sensitized Solar Cells", J. Am. Chem. Soc., vol. 127, pp" 808-809, 2005.
[8] Taro Hitosugi, Naoomi Yamada, Shoichiro Nakao, Yasushi Hirose and Tetsuya Hasegawa, "Properties of TiO_2-based transparent conducting oxides" Applications and materials science., vol. 207, Issue 7, pp. 1529-1537, 2010.
[9] Y.M. Sung, H.J. Kim, "Sputter deposition and surface treatment of TiO_2 films for dye-sensitized solar cells using reactive RF plasma", Thin Solid Films, vol. 515, pp. 4996 – 4999, 2007.
[10] D.J. Kwak, B.H. Moon, D.K. Lee, C.S. Park, Y.M. Sung, "Comparison of transparent conductive indium tin oxide, titanium-doped indium oxide, and fluorine-doped tin oxide films for dye-sensitized solar cell application", J. Elec. Eng. Technol., vol. 6 no. 5, pp. 684-687, 2011.

IOP Conf. Series: Materials Science and Engineering **229** (2017) 012020 doi:10.1088/1757-899X/229/1/012020

Synthesis of TiO_2 Nanotubes Electrode for Photo Electrochemical cells Application

Pattarapon Pooyodying[1], Youl-Moon Sung[1,*] and Jirapat Anuntahirunrat[1]

[1]Electrical Engineering, Kyungsung University, Busan 608-736, Korea

[*]ymsung@ks.ac.kr

Abstract. In this research propose a synthesis and study characteristics of electrode for photoelectrochemical cells using TiO_2 nanotubes electrode. The fabrication of electrode for photoelectrochemical cells using TiO_2 nanotubes which made from anodization process at size area 2x2 cm^2 and counter electrode used platinum pastes coated on fluoride-doped tin oxide (FTO) glass at size area 2x2 cm^2. The solar simulation was used as the light source for the photoelectrochemical cells at 100mW/cm^2. The morphological properties of TiO_2 electrode was confirmed by X-ray diffraction (XRD) and scanning electron microscope (SEM). We measure the sheet resistance, open circuit voltage and short circuit current for analyse the photoconversion efficiency. The photoconversion efficiency was 5.78%, open circuit voltage was 0.72V and short circuit current was 2.87mA.

1. Introduction

The Photo Electrochemical Cell (PEC) is the part of thin film solar cell technology that can convert the solar energy to electricity energy. Electricity energy is the importance energy to develop the technology. The solar cell is an interested application due to it has enormous advantage such as clean energy during electricity generation with solar cell panels there are no harmful greenhouse gas, neutral energy it is thus free and abundant, the solar cell panels are silent, no motion part, easy to install on roof or on ground and generate directly electricity energy. The PEC has attracted much attention in the academic and commercial due to a trend that is produced in commercially available, the price is lower than the P-N junction type of silicon semiconductor solar cell (at currently manufactured), and can be produced in large quantities [1-6]. In present, the mostly material that used for made the solar cells device were silicon which used in electronic device. In addition, the materials used for made solar cells device were gallium arsenide and cadmium telluride which those materials have high price. The PEC is mainly composed of working electrode, electrolyte and counter electrode. The working electrode consists of n-type semiconductor metal oxide films such as titanium dioxide (TiO_2), zinc oxide (ZnO), and tin oxide (SnO). In the counter electrode material is used platinum (Pt), graphene and carbon (C) electrode, can show previous a reported [7-10]

TiO_2 is the interested material science and most studied. It is used for dye-sensitized solar cells and biomedical device. In the recent research was used titanium powder or Tio_2 for dye-sensitized solar cell

Content from this work may be used under the terms of the Creative Commons Attribution 3.0 licence. Any further distribution of this work must maintain attribution to the author(s) and the title of the work, journal citation and DOI.
Published under licence by IOP Publishing Ltd

device [11]. But in this research used TiO_2 nanotubes by anodization process due to the nanoparticle has affect to efficiency of photocatalyst process.

In this work, we focused to study the characteristic of Photo Electrochemical Cell with TiO_2 nanotubes and TiO_2 electrode. The TiO_2 nanotubes was confirmed the morphological by X-ray diffraction (XRD) and scanning electron microscope (SEM). The working electrode of Photo Electrochemical Cell used TiO_2 nanotubes which made from titanium foil under size 2x2 cm^2 by anodization process. The counter electrode of Photo Electrochemical Cell used FTO glass size 2x2 cm^2 coated with platinum pastes. Working electrode and counter electrode were sealed with sealing material then injected the electrolyte. The Photo Electrochemical cell with TiO_2 nanotubes and platinum electrode was tasted by applied the light source which used solar simulator 100mW/cm^2.

2. Experimental setup

2.1. Synthesis TiO_2 nanotubes

The synthesis of the titanium nanotubes was prepared by anodization process. The titanium foil (thickness 0.25 mm, 99.7% trace metals basis, Sigma-Aldrich) at 2x2 cm^2 area size was used as the substrate. The substrate was degreased by acetone and ethanol in ultrasonic cleaner (JAC4020, Ultrasonic) for 30 min then dried. The electrolyte for anodization process was prepared by ethylene glycol contained 0.3wt% ammonium fluoride (NH_4F) and deionized water. The titanium foil was connected to the positive of power supply and the Pt plate (gauze, 100 mesh, 99.9% trace metals basis, Sigma-Aldrich) was connected to negative. The voltage of power supply for anodization process was constant at 30 V for 240 min then annealed at 450°C for 180min. The TiO_2 nanotubes electrode was investigated by X-ray diffraction (XRD; Rigaku D/max 2100H), the morphological propertie was measured by field emission scanning electron microscope (FE-SEM; Hitachi S-3000H)[12].

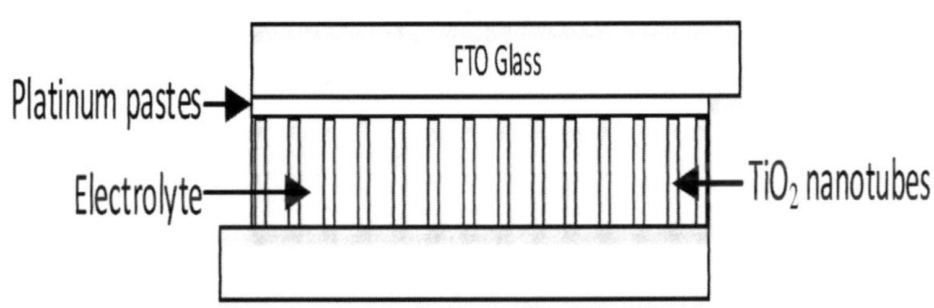

Figure 1. The structure of Photo Electrochemical Cell with TiO_2 nanotubes electrode

2.2. Preparation of platinum electrode

FTO glass (resistance sheet: ~7 Ω/sq, Pilkington) at size area 2×2 cm^2 was degreased by acetone and ethanol in ultrasonic cleaner for 30 min then dried. After the FTP glass dried, the platinum pastes (Platisol T/SP, Sloaronix) was coated on FTO glass by doctor-blade technique then annealed at 450°C for 30min.

2.3. Fabrication of Photo Electrochemical Cell with TiO_2 nanotubes electrode and Platinum electrode

The TiO_2 nanotubes was set as the photoanode of the Photo Electrochemical Cell and the platinum electrode was set as the cathode. The two electrodes were overlap and sealed with sealing film thickness 60 μm (Meltonix 1170-60, Solaronix) the structure of photo electrochemical cell is show in figure1. Then annealed at 125°C for 15min. After annealed, The Photo Electrochemical Cell was injected the electrolyte into the gap between electrodes. The Photo Electrochemical Cell was applied the light source by solar simulator at 100mW/cm^2 then measure the voltage and current for analyze. The process of fabrication of photo electrochemical cell is show in figure 2.

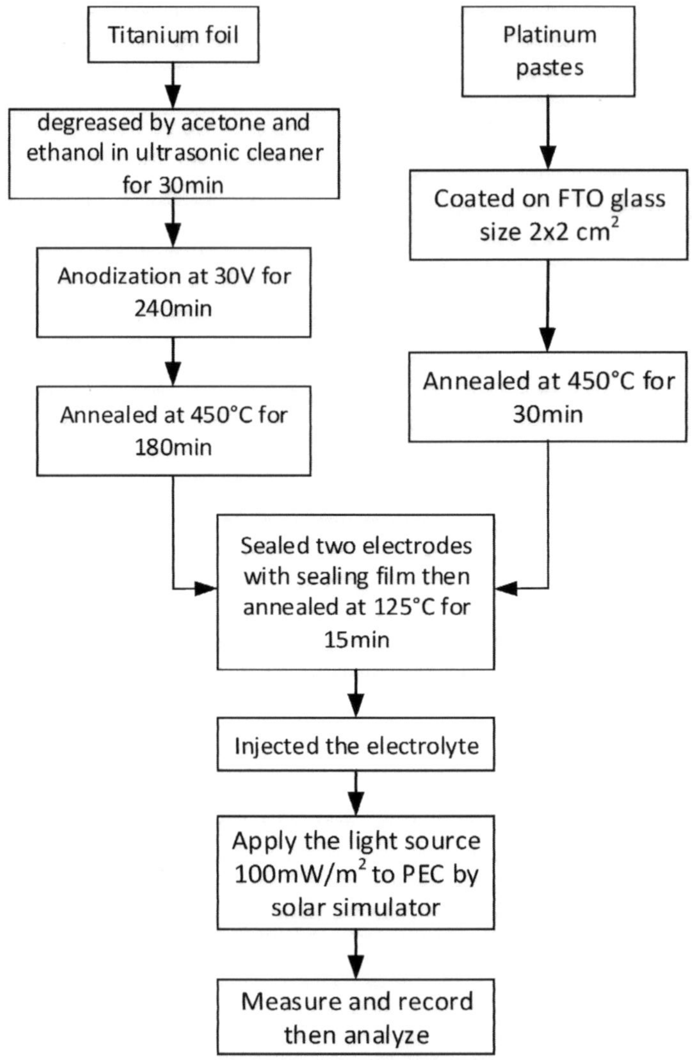

Figure 2. The block diagram for fabrication of Photo Electrochemical Cell and experimental

3. Results and Discussion

Figure 3(a) shows the XRD pattern of TiO_2 and TiO_2 nanotubes. The strong peaks of TiO_2 nanotubes shown at (100), (002), (101), (102), (110), (103), (112) and anatase peaks shown at (101) and (200). The strong peaks of TiO_2 shown at (100), (002), (101), (102), (110), (103), (112). We found that the strong peaks of TiO_2 didn't shown the anatase peaks different from the TiO2 nanotubes. Figure 3(b) shows the SEM image of TiO_2 nanotubes.

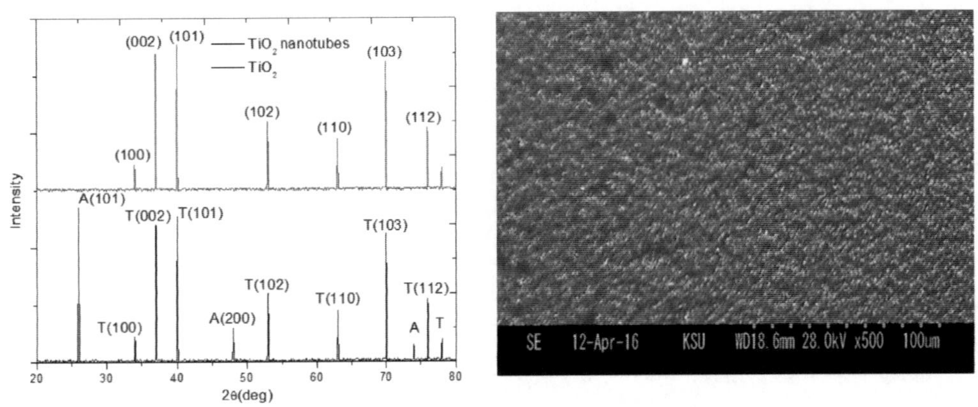

(a) (b)

Figure 3. (a). The XRD pattern of TiO_2 and TiO_2 nanotubes and (b) SEM image of TiO_2 nanotubes

Figure 4 shows the I-V curve of Photo Electrochemical Cell which tasted under the light source by solar simulator at 100mW/cm^2. The parameters which measure in this experimental was open circuit voltage (V_{oc}), short circuit current (I_{sc}) and fill factor (FF) was obtained from I-V curve. The parameters of Photo Electrochemical Cell with TiO_2 nanotubes electrode was $V_{oc} = 0.72$V, $I_{sc} = 2.87$mA, FF = 1.13 and conversion efficiency was 5.78% which higher than the conversion efficiency of TiO_2 electrode. The parameters of Photo Electrochemical Cell with TiO_2 electrode was $V_{oc} = 0.715$V, $I_{sc} = 2.4$mA, FF = 1.05 and conversion efficiency was 4.53%. The both counter electrodes show that the I_{sc} increased significant with change a counter electrode. However, the conversion efficiency of TiO_2 nanotubes was more than that the TiO_2 due to the charge transfer resistance of TiO_2 nanotubes was higher than that the TiO_2.

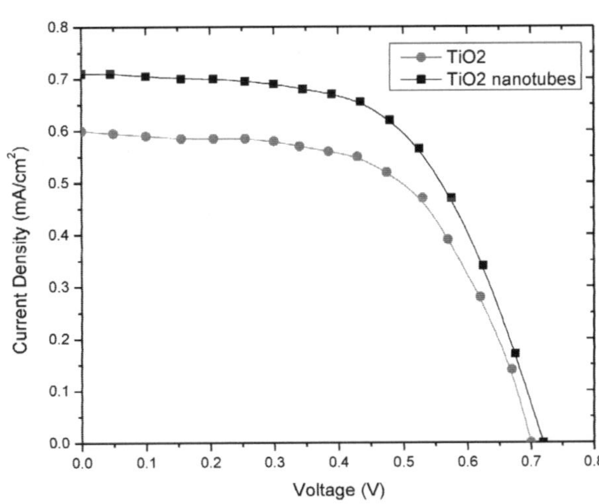

Figure 4. I-V curve of Photo Electrochemical Cell with TiO_2 and TiO_2 nanotubes electrode under testing by solar simulator at 100mW/cm^2

Figure 5 shows electrochemical impedance spectroscopy of TiO_2 nanotubes and TiO_2. The two semicircles can observed from Nyquist plots. The high frequency semicircle shows a charge transfer resistance (R_{CT}) at counter electrode (CE) /electrolyte interface, and the higher frequency represents to the series resistance (R_S)[14].

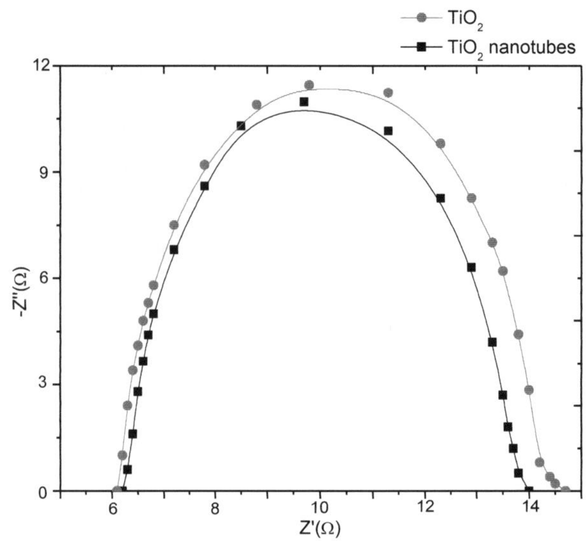

Figure 5. Electrochemical impedance spectroscopy (EIS) Nyquist plots of Photo Electrochemical Cell with TiO_2 nanotubes and TiO_2 electrode

4. Conclusions

The Photo Electrochemical cell with TiO_2 nanotubes electrode from anodization process which was compared with TiO_2 electrode. The energy conversion efficiency of PEC with both electrode samples was investigated with under the solar simulation at AM 1.5 (100 mW/cm^2). The PEC using TiO_2 nanotubes showed a conversion efficiency of 5.78 % which better than the TiO_2 electrode that the conversion efficiency was 4.53% due to the effective electron transfer on the large surface of TiO_2 nanotubes electrode. The TiO_2 nanotubes electrode can be replaced another electrode, and it also can increase efficiency.

Acknowledgement

This research was supported by Basic Science Research Program through the National Research Foundation of Korea (NRF) funded by the Ministry of Education (No. 2016R1D1A3B03935372)

Reference

[1] B. O'Regen and M. Grätzel 1991 A low-cost, Hight-efficiency solar cell on dye-sensitized colloidal TiO_2 films *Nature*. 353 (1991) 737-740

[2] G. Smestad, C. Bignozzi and R. Argazzi 1994 Testing of dye sensitized TiO_2 solar cells I: Experimental photocurrent output and conversion efficiencies *Sol. Energy. Mater. Sol. Cell.* 32(3) (1994) 259-272.

[3] N. Papageorgiou, W.F. Maier and M. Grätzel 1997, *J. Electrochem. Soc.*, 144 (1997), pp. 876–884

[4] K. Tennakonea, P.K.M. Bandaranayakea, P.V.V. Jayaweerab, A. Konnob, G.R.R.A. Kumarab 2002 Dye-sensitized composite semiconductor nanostructures *Physica E: Low-dimensional Systems and Nanostructures.* 63 (2002) 1-7.

[5] N. Meng, K.H.L. Michael, Y.C.L. Dennis and K. Sumathy 2006 An analytical study of the porosity effect on dye-sensitized solar cell performance *Sol. Energy. Mater. Sol. Cell.* 90 (9) (2006) 1331-1344.

[6] Y. Chiba, A. Islam, R. Komiya, N. Koide and L. Han 2006 Conversion efficiency of 10.8% by a dye-sensitized solar cell using a TiO2 electrode with high haze *Appl. Phys. Lett.* 88 (2006) 223505.

[7] S.H. Kim and C.W. Park 2013 Novel application of platinum ink for counter electrode preparation in dye sensitized solar cells *Bull. Korean Chem. Soc.* 34(3) (2013) 831-836.

[8] C.P. Lee, C.A. Lin, T.C. Wei, M.L. Tsai, Y. Meng, C.T. Li, K.C. Ho, C.I. Wu, S.P. Lau and J.H. He 2015 Economical low-light photovoltaics by using the Pt-free dye-sensitized solar cell with graphene dot/PEDOT:PSS counter electrodes *nano energy.* 18 (2015) 109-117.

[9] Y.Y. Li, C.T. Li, M.H. Yeh, K.C. Huang, P.W. Chen and R. Vittal, K.C. Ho 2015 Ordered mesoporous carbon/graphene nano-sheets composites as counter electrodes in dye-sensitized solar cells *Electrochimica Acta.* 179 (2015) 211-219.

[10] S. Lowpa, S. Pimanpang, W. Maiaugree, S. Saekow, P. Uppachai, S. Mitravong and V. Amornkitbamrung 2015 Nanoporous carbon microspheres from carrot juice used as a counter electrode for a dye-sensitized solar cell *Mater. Lett.* 158 (2015) 115–118.

[11] D.J. Kwak, B.H. Moon, D.K. Lee, C.S. Park and Y.M. Sung 2011 Comparison of transparent conductive indium tin oxide, titanium-doped indium oxide, and fluorine-doped tin oxide films for dye-sensitized solar cell application *J. Elec. Eng. Technol.* 6(5) (2011) 684-687.

[12] L. Wei, Y. Yang, R. Fan, Y. Na, P. Wang and Y Dong 2015 Effects of rubrene co-sensitized TiO2 photoanode on the performance of ruthenium dye N719 sensitized solar cells *Thin Solid Films* 592 (2015) 14–23.

[13] J.H. Yang, C.W. Bark, K.H. Kim and H.W. Choi 2014 Characteristics of the Dye-Sensitized Solar Cells Using TiO_2 Nanotubes Treated with $TiCl_4$ *Materials* 2014 7 3522-3532.

[14] Y.M. Sung, H.J. Kim 2007 Sputter deposition and surface treatment of TiO2 films for dye-sensitized solar cells using reactive RF plasma *Thin Solid Film.* 515 (2007) 4996-4999.

A Review on Functionally Gradient Materials (FGMs) and Their Applications

Valmik Bhavar[1,a], Prakash Kattire[1], Sandeep Thakare[1], Sachin patil[1] and Dr. RKP Singh[1]

[1]Kalyani Centre for Technology and Innovation (KCTI), Bharat Forge Limited, Pune, India.

[a]Valmik.bhawar@bharatforge.com

Abstract. Functionally gradient materials (FGM) are innovative materials in which final properties varies gradually with dimensions. It is the recent development in traditional composite materials which retains their strengths and eliminates their weaknesses. It can be formed by varying chemical composition, microstructure or design attributes from one end to other as per requirement. This feature allows FGM to have best material properties in required quantities only where it is needed. Though there are several methods available for manufacturing FGMs, additive based metal deposition (by laser, electron beam, plasma etc.) technologies are reaping particular interest owing to their recent developments. This paper presents evolution, current status and challenges of functionally gradient materials (FGMs). Various manufacturing processes of different types of FGMs are also presented. In addition, applications of FGMs in various fields including aerospace, defence, mining, power and tools manufacturing sectors are discussed in detail.

1. Introduction

Materials are continuously developed from iron, pure metals to composite materials which are in use today. Continuous material development from Bronze Age to present and future scenario is presented in figure 1 [1]. Pure metals have very limited use, since actual application may require contrary property requirement which cannot provide by using single metal. As compared to pure metals, alloys can be stronger and more versatile. Bronze which is alloy of copper and tin was the first alloy that was developed in 4000 BC (Bronze age). Since then, different mixtures of metals and non-metals were tried to combine strengths of multiple materials as per functional requirement.

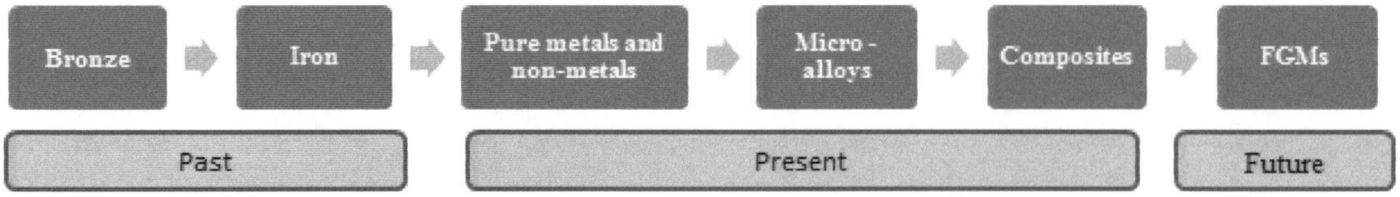

Figure 1. Continuous material development from bronze to FGMs

Composite materials are most advanced form of materials which are made from two or more constituents with significantly different physical and chemical properties from the individual materials. Composite materials allow distinct combinations giving hard, wear resistant surface and soft core as

Content from this work may be used under the terms of the Creative Commons Attribution 3.0 licence. Any further distribution of this work must maintain attribution to the author(s) and the title of the work, journal citation and DOI.
Published under licence by IOP Publishing Ltd

IOP Conf. Series: Materials Science and Engineering **229** (2017) 012021 doi:10.1088/1757-899X/229/1/012021

per functional requirement of application. Heterogeneity, anisotropy, symmetry and hierarchy are the main characteristics of composite materials reaping particular interest for various applications. High strength to stiffness ratio, grater resistance to fatigue, wear and corrosion, high reliability etc. are the advantages of composites over pure or alloyed metals. In spite of all these advantages, composite materials are subjected to sharp transition of properties at the interface which can result in component failure (by delamination) at extreme working conditions.

This drawback of conventional composites eliminated by modified form of composites called functionally gradient materials (FGMs). These materials replaces sharp interface with gradient interface which results in smooth transition of properties from one material to the other. These advanced materials with engineered gradients of composition, structure and specific properties in the preferred direction are superior to homogeneous material composed of similar constituents [1]. The mechanical properties such as Young's modulus of elasticity, Poisson's ratio, shear modulus of elasticity, material density and coefficient of thermal expansion vary smoothly and continuously in preferred directions in FGMs. Bone, teeth, skin and bamboo tree are some examples of naturally occurring functionally gradient materials. The concept functionally gradient material (FGM) was first developed by researchers from japan in 1984. They designed functionally gradient thermal barrier with outside temperature of 2000 K and inside temperature of 1000K across 10 mm thickness. Since then, use of functionally gradient materials has been increased in various fields including aerospace, mining, power and medical. FGMs have numerous advantages that make them suitable in these applications. It includes high fracture toughness, reduction in plane and through the thickness transverse stresses, enhance performance of thermal barrier systems etc. Due to this prominence of FGM, lots of efforts were made to improve FGM manufacturing process and properties of the FGM. There are several manufacturing methods are available for manufacturing FGMs depending on the type of FGM required. These include powder metallurgy, vapour deposition, centrifugal method and solid free form techniques. Out of all these methods, solid free form techniques by using laser, plasma or electron beam as energy source are becoming very popular in the recent years.

This paper presents describes functionally gradient materials (FGMs), their types and methods of manufacturing. Solid free form manufacturing (SFF) techniques of FGMs are also discussed in detail. Finally, few applications of FGMs along with recent research work and challenges of FGMs are presented.

2. Types of Functionally gradient materials (FGMs)

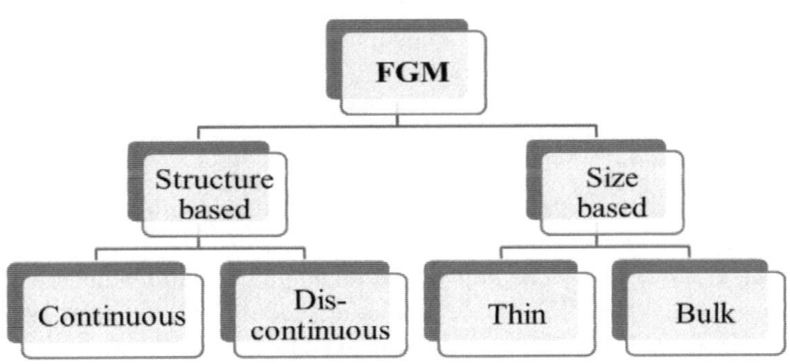

Figure 2. Classification of Functionally gradient materials [2]

Two different criteria's are used to classify functionally graded materials. One is based on structure of material and the other is based on size of functionally graded materials. As shown in figure 2, functionally graded materials can be further divided into two major groups based on structure of materials: Continuously structured and discontinuously structured FGM. In continuous FGM, there is continuous gradient present from one material to the other material. However in case of discontinuous FGM, material gradient is provided in layered fashion. Based on size of materials, FGM's are classified into two major types: Thin FGM and bulk FGM. Thin FGM are having relatively thin sections like surface coating, whereas the bulk FGM's are complete volume of materials. Manufacturing processes like physical Vapor Deposition (PVD), Chemical Vapor Deposition (CVD) and Self propagating High temperature Synthesis (SHS) method are used to manufacture thin FGM. Whereas, bulk FGM's are manufactured by using methods such as powder metallurgy, centrifugal casting and solid freeform/additive manufacturing techniques.

3. Methods of manufacturing FGMs

Several techniques are available to produce functionally graded materials (FGMs). Few of them are described below in detail.

3.1 Vapour deposition technique

Vapor deposition techniques describe a variety of vacuum deposition methods which can be used to produce thin films on the base materials. All these techniques can be used to produce thin FGMs only. Different types of vapour deposition techniques include physical vapour deposition (PVD) and Chemical vapour deposition (CVD). These are energy intensive and produce poisonous gages as their by-products [3]. Other deposition based techniques which can deposit thin functionally gradient coatings are electron beam deposition (EBD), Ion beam deposition (IBD) and Self propagating high temperature synthesis (SHS) [4]. All above mentioned methods are uneconomical to produce bulk type FGMs.

3.2 Powder metallurgy

Powder metallurgy based technique can be used to produce bulk type FGMs with discontinuous (stepwise) structure. The process is carried out by using steps including weighing and mixing of powder according to the pre-designed spatial distribution as per functional requirement, stacking and ramming of the premixed-powders, and finally sintering [5].

3.3 Centrifugal method

Centrifugal method is capable to produce continuously structured bulk type FGMs. It uses force of gravity through spinning of mould to produce functionally graded materials [6]. Difference in material densities and spinning of mould produces FGMs. There are two disadvantages of this method are this method can produce only cylindrical shaped FGMs and there is limit to which type of gradient can be produced.

3.4 Solid free form fabrication/additive manufacturing (AM) techniques

Solid freeform fabrication (SFF)/Additive manufacturing (AM), also known as 3D printing, is a process of joining materials to make objects from 3D model data, usually layer upon layer, as opposed to subtractive manufacturing technology [7]. This tool-less manufacturing method can produce fully dense metallic parts in short time, with high precision. Metal AM processes can be broadly classified into two major groups, - Powder Bed Fusion based technologies (PBF) and Directed Energy Deposition (DED) based technologies. Both of these technologies can be further classified based on the type of energy source used. In PBF based technologies, thermal energy selectively fuses regions of powder bed. Selective laser sintering/melting (SLS/SLM) and electron beam melting (EBM) are main representative processes of PBF based technologies. In DED based technologies focused thermal energy is used to fuse materials (powder or wire form) by melting as they are being deposited. Laser Engineered Net Shaping (LENS)/Direct Metal Deposition (DMD), Electron Beam Free Form Fabrication (EBFFF) and arc based AM are some of the popular DED based technologies.

Most of above mentioned SFF/AM methods are capable to produce functionally gradient materials (FGMs) from thick coatings to complicated FGM bulk parts.

IOP Conf. Series: Materials Science and Engineering **229** (2017) 012021 doi:10.1088/1757-899X/229/1/012021

Advantages offered by AM techniques like higher material utilization, speed of production, design freedom, capability to produce complicated parts and less energy intensiveness are garnering particular interest in manufacturing FGMs for different applications.

Powder bed fusion (PBF) based AM technologies like Selective laser melting (SLM) and Electron beam melting (EBM) are very popular methods for producing complicated parts owing to their high accuracy and surface finish as compared to directed energy deposition (DED) based technologies. However, PBF based technologies are less flexible than DED based technologies as far as functionally gradient material manufacturing are concerned. It is due to fact that material gradient by varying chemical composition of powder is not possible. However these methods can produce bulk FGMs by controlling porosity or by introducing different types of lattice structures in parts to be manufactured.

Directed energy deposition (DED) based AM techniques are most convenient methods to produce FGMs since these methods can produce FGM from thick coatings to bulk parts having continuous or discontinuous gradient. These methods can produce FGMs with better adhesion and mechanical properties than powder bed technologies. Laser metal deposition (LMD) and Electron beam free form deposition (EBFFF)/ Electron beam additive manufacturing (EBAM) are popular methods based on DED based AM systems which can be used to manufacture different kinds of FGMs.

3.4.1. Laser metal deposition (LMD). Laser engineered net shaping (LENS) and direct metal deposition (DMD) are main processes based on DED technology which uses laser beam as power source and raw material in the form of powder. LENS process was originally developed by Sandia national laboratories in 1997 and then licensed to Optomec (USA), whereas DMD process was jointly developed by POM group and University of Michigan [8, 9]. In these process, high power laser beam is used to create a molten pool on base material and then powder material is injected into the molten pool by using nozzles. Delivered powder at laser beam spot is absorbed into the melt pool and creates deposit. As shown in figure 3, the work table can move in x - y direction to obtain desired cross section of sliced model and then subsequent layers can be deposited by incrementing deposition head in z direction to complete the object. Deposition of layers is repeated until the desired three-dimensional component has been additively formed. Metal powder is delivered through nozzles and distributed around the circumference of deposition head either by gravity, or by using inert carrier gas. The entire process is conducted under controlled argon atmosphere where oxygen levels are maintained below 10 ppm.

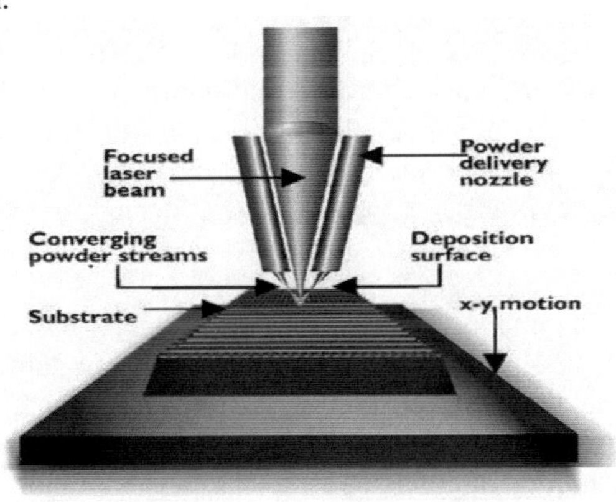

Figure 3. Basic laser engineered net shaping process [10]

Laser based Directed energy deposition (DED) technique of metal AM is the most suitable technology to produce FGMs. All types of FGMs including continuous/discontinuous structured and thin /bulk type can be easily manufactured by using laser metal deposition (LMD). Pre-alloyed powders can be used to produce discontinuous type FGM. Whereas elemental powders can be delivered in precise amounts to the melt zone using separate feeders to generate various alloys and composite materials in continuously graded fashion. With the adoption of this technique, number of

FGMs can be fabricated into complex shapes as the rate of elemental powder deposition can be controlled for each feeder during the fabrication for each layer and the final product can be achieved within hours [11].

3.4.2. Electron beam direct manufacturing. Electron Beam Direct Deposition (EBDM) is another technology based on directed energy deposition (DED) which uses electron beam as power source and raw material in the form of wire. This technology was developed by Sciaky (Chicago, USA) and also known as Electron beam additive manufacturing (EBAM). This process can produce medium to large sized near net shaped components inside vacuum chamber directly from digital model. After manufacturing, component requires finishing operations such as heat treatment and machining. Maximum size of component to be manufactured by EBAM is restricted by vacuums chamber size of the machine. Commercially available welding wires are used as the deposition material. The standard electron beam system is a Sciaky 60 kW / 60 kV welder. The electron beam is electronically focusable and the output power is scalable over a very wide range. This enables a very wide range of deposition rates to be achieved using the same system. Typical deposition rates of EBAM systems are from 3 to 9 Kgs/hrs depending on the material used and part complexity. Additionally, the EBAM system has closed loop control system in which melt pool size is continuously monitored and parameters are adjusted to keep the size constant. This ensures consistent part geometry, uniform microstructure and mechanical properties.

Figure 4. Electron beam additive manufacturing by twin wire deposition [12]

EBAM technology can also produce various types of functionally graded materials (FGMs) by using multiple wire feed nozzles as shown in figure 4 to single EB gun. Two or more wires of different metal alloys can be independently controlled and simultaneously feed to single molten pool to form graded materials. Both coating and bulk type of FGMs can be formed in continuous or discontinuous manner.

3.4.3. FGMs by Arc deposition technologies. Wide ranges of arc based additive manufacturing processes are available where arc (plasma, TIG, MIG) is used as power source and material is used in the form of powder or wire. Plasma transferred arc (PTA) and plasma arc welding (PAW) are free form AM processes which uses plasma arc as power source and raw material in the form of powder and wire respectively. Shaped metal deposition (SMD) is another AM technique which uses tungsten inert gas (TIG) or Metal inert gas (MIG) welding with material in the form of wires for free form fabrications. Since most of such systems are wire feed type, these are also known as Wire assisted additive manufacturing (WAAM) systems. Large number of system configurations can be achieved by integrating conventional welding systems with robots, manipulators or gantries for automation. All of these processes with proper inert gas shielding have strong potential to produce near net shaped medium to large sized parts at much lower cost as compared to laser and electron beam based processes.

IOP Conf. Series: Materials Science and Engineering **229** (2017) 012021 doi:10.1088/1757-899X/229/1/012021

Few welding based AM systems has been developed which can deposit functionally gradient materials. In this case, two filler wires are controlled separately and supplied to the arc (TIG or MIG) for deposition. Several studies have been carried out to demonstrate effectiveness of arc based AM configurations to produce FGMs. Sajan Kapil et. al [13] successfully fabricated Al-Si alloy having gradient in thermal conductivity. It was fabricated by using Hybrid layered manufacturing machine (HLM) which combines 3 axis CNC and gas metal arc welding (GMAW) deposition system. S. Suryakumar et al. [14] demonstrated two different ways to fabricate functionally gradient materials by using weld deposition. FGMs can be produced by varying process parameters or by using double wire feeder which can be guided and controlled separately.

4. Applications

In the recent years, there has been growing interest in the use of functionally gradient materials (FGMs) due to their numerous advantages over composite materials. Owing to graded variation in the composition, the properties of FGMs changes significantly and continuously from one surface to another, thus eliminating interface problems like stress concentrations and poor adhesion [15]. Use of FGMs is increasing in aerospace, defense, nuclear industry, biomedical and electronics sectors.

FGMs are mainly used in those applications where combinations of two extreme properties are required in single component for example hardness and toughness [16]. For example, in case of turbine blade, thermal resistance and anti-oxidation properties are required at high temperature side and mechanical strength and toughness are required at low temperature side. To tackle these requirements turbine blades were used to manufacture by using metal-ceramic composites [17]. However, the property difference between two materials created residual stresses and adhesion issues at interface which may lead to failure. Turbine blade made by using FGM possesses smooth property change from ceramic to metal and diminishes interface problems [18]. Figure 5 shows turbine blade made by using FGM where properties like thermal conductivity and mechanical strength are continuously graded from metal to ceramic region.

The applications like cutting tools and machine parts require heat, wear, mechanical shock and corrosion resistance. Reliability and cost/performance ratio plays major role in these applications [16]. Figure 6 shows composite and FGM cutting tools with metal shank and ceramic tip. Composite tool suffers from sharp transition of properties from metal to ceramic which may result in residual stresses and tool failure. However FGM cutting tool where FGM material used in between metal and ceramic increases thermal strength and tool expected to have long life.

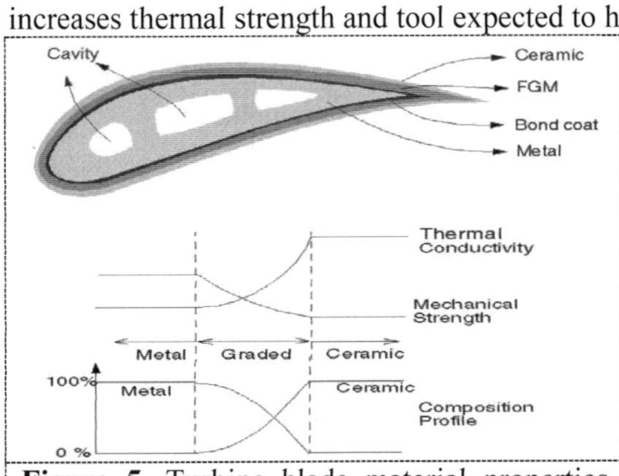

Figure 5. Turbine blade material properties containing FGM [19]

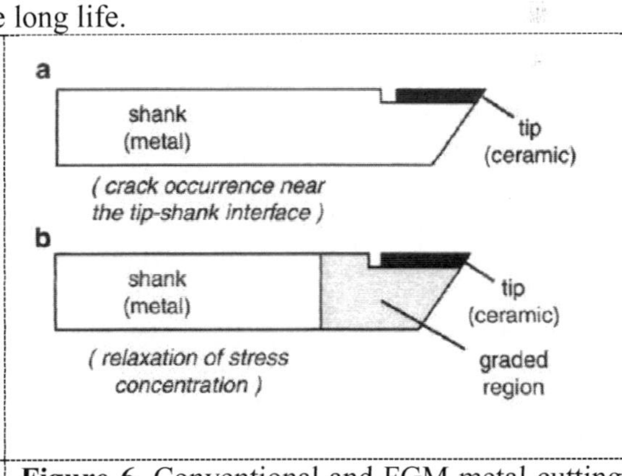

Figure 6. Conventional and FGM metal cutting tool [20]

Metal-ceramic FGM is also used in armors, [21,22] Where hard ceramic front surface blunts the projectile, whereas metallic back surface catches fragments and prevents penetration [23]. Similar FGMs also finds applications as heat resistant valves of internal combustion engines [17]. Functionally graded thermal barrier coatings (FTBC) by using various spraying techniques are popular methods of producing such FGMs [24, 25].

Another emerging area for FGM application is biomedical sector where functionally graded prosthesis joint can increase adhesive strength and reduce pain [26].

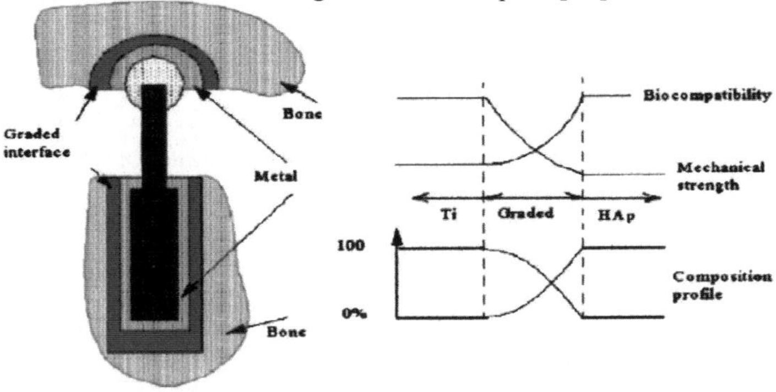

Figure 7. Prosthesis joint by using FGM [27]

Performance of metal prosthesis joint can be enhanced by using functionally gradient material having high biocompatibility at surface. As shown in figure 7, biocompatibility property increases and mechanical strength decreases as we move from metal to bone.

Number of diffusion and deposition based technologies are used to enhance surface properties of components. However, recently it is observed that their performance can be further increased by combining diffusion with deposition processes. Combination of diffusion processes like nitriding, nitro-carburizing etc. with deposition processes like PVD,CVD type coatings of hard material provides functionally gradient effect which improves properties [28]. These treatments are well known as duplex surface treatments where diffusion process (nitriding, boriding etc) are combined with coating processes (like PVD ,CVD or TRD-thermo reactive diffusion). Deposition technologies can produce hard, wear resistant layer on metal surface. However, thickness of this coating is very small and there is sudden change in properties between coating and the substrate material. It can result in premature failure of coating during service conditions by delamination of coating. Thermo chemical diffusion treatment like nitro-carburizing prior to deposition of hard coating can form graded structure from surface to substrate and provides tough and supportive sub-surface for hard coating [29]. It is also observed that such a graded structure can shift failure mechanism from Thus, Duplex surface treatment involving nitro-carburizing and thermo reactive deposition can retain beneficial effects of both treatments and eliminates drawbacks of them by forming graded structure from surface to base.

Along with all these applications, FGMs are also used in piezoelectric actuators [30], functionally graded thermal protection system for hypersonic and supersonic planes [pap-modelling and analysis-206] and functionally graded heated floor systems [31].

5. Recent developments and challenges of FGMs

In case of most of the FGMs, a material property varies in thickness direction [32]. However, modern applications may demand FG materials in which material properties in both thickness and axial directions [33]. Recently, a gradient material in which properties varies in both directions are also developed and extensively studied [34, 35].Such smart materials are known as bidirectional functionally gradient (BDFGMs) materials. Laser metal deposition based AM technique is most suitable to produce such BDFGMs [11] .

Though substantial technology advancement has been made in the field of FGMs, few critical issues still need to be addressed. A proper database of FGMs in terms of parameters and testing is still not available. Conventional testing and measurement method may not suitable to evaluate performance of modern FGMs, so developments of advanced testing methods are required [36]. Most of the processing techniques of FGMs are very costly, so low cost processing technique which can mass produce large sized, complex shape FGMs is still remain as a challenge . The selection of proper material suitable for intended application is the immediate and direct challenge for future technology development in FGM research field.

Acknowledgments
The authors gratefully acknowledge the extended support provided to this work by KCTI (Kalyani Centre for Technology and Innovation) and Bharat Forge limited.

References
[1] Shahistha A, Varghese B, Baby A 2014 A review on functionally graded materials *The International Journal Of Engineering And Science (IJES),* Volume **3**,Issue 6, Pages 90-101, ISSN (e): 2319 – 1813 ISSN (p): 2319 – 1805.
[2] Mahamood R, Akinlabi E, Shukla M and Pityana S 2012 Functionally graded materials: An overview *Proceedings of the World Congress on Engineering 2012* Vol.**3**, WCE, London, U.K.
[3] J. F. Groves, and H. N. G. Wadley 1997 Functionally graded materials synthesis via low vacuum directed vapor deposition Composites Parts B **28**B, (1997), 57-69.
[4] R. Knoppers, J. W. Gunnink, J. Van den Hout, and W. Van Vliet The reality of functionally graded material products *TNO ScienceandIndustry,* The Netherlands, pp 38-43.
[5] Zhu J, Lai Z, Yin Z, Jeon J, and Lee S Fabrication of ZrO2–NiCr functionally graded material by powder metallurgy Materials Chem. Phys. 68 130. Crossref.
[6] Watanabe Y, Inaguma Y, Sato H, and Miura-Fujiwara E 2009 Novel fabrication method for functionally graded materials under centrifugal force: the centrifugal mixed-Powder method *Materials,* Vol. **2**, No. 4, pp.2510-2525.
[7] Bhavar V, Kattire P, Patil V, Khot S, Gujar K, Singh R 2014 A Review on Powder bed fusion technology of metal additive manufacturing *AMSI Conference* 2014, Banglore, India.
[8] Ruan J, Sparks T, Fan Z, Stroble J, Panackal A and Liou F 2006 A Review of Layer Based Manufacturing Processes for *Metals Solid freeform fabrication symposium, January 2006.*
[9] Thymianidis M, Achillas C, Tzetzis D, Iakovou E, Pariente F, Guagliano M. 2012 Modern Additive Manufacturing Technologies: An Upto-Date Synthesis and Impact on Supply Chain Design", *2nd International conference on supply chains,* October -2012.
[10] Hedges M., Calder N 2006 Near Net Shape Rapid Manufacture & Repair by LENS *Cost Effective Manufacture via Net-Shape Processing, pp. 13-1 – 13-14. Meeting Proceedings RTO-MP-AVT-139,* Paper 13. Neuilly-sur-Seine, France: RTO. Available from: http://www.rto.nato.int/abstracts.asp
[11] Popoola P, Farotade G, Fatoba G and Popoola O, Zlatanovic M 2016 Laser Engineering Net Shaping Method in the Area of Development of Functionally Graded Materials (FGMs) for Aero Engine Applications - A Review, from "Fiber Laser", book edited by Mukul Chandra Paul, ISBN 978-953-51-2257-9Chapter 17
[12] (2014) SLM Solution website [online], available at www.sciaky.com, accessed January 2017.
[13] Kapil S, Kulkarni P, Karunakaran K, Joshi P 2014 Development and Characterization of Functionally Graded Materials Using Hybrid Layered Manufacturing *5th International & 26th All India Manufacturing Technology, Design and Research Conference (AIMTDR 2014)*
[14] Suryakumar S, Somashekara M 2013 Manufacture of functionally gradient materials using weld-deposition *Solid Free. Fabr. Symp.* (2013) 1-36.
[15] Bharti I, Gupta N, and Gupta K 2013 Novel Applications of Functionally Graded Nano, Optoelectronic and Thermoelectric Materials *International Journal of Materials, Mechanics and Manufacturing,* Vol. 1, No. 3, August 2013.
[16] Jamian S 2012 Application of Functionally Graded Materials for Severe Plastic Deformation and Smart Materials : Experimental Study and Finite Element Analysis *Thesis submitted to Graduate School of Engineering, Nagoya Institute of Technology, Doctor degree of engineering*
[17] Maciejewski G, Mroz Z 2013 Optimization of functionally gradient materials in valve design under cyclic thermal and mechanical loading *CAMES,* **20**: 99–112, 2013.
[18] Coomar N, Kadoli R 2010 Comparative analysis of steady state heat transfer in a TBC and functionally graded air cooled gas turbine blade", *Sadhana* Vol. **35**, Part 1, February 2010, pp. 1–17. © Indian Academy of Sciences.

[19] Qian X 2003 Design of heterogeneous turbine blade Computer aided design, **35**, 319-329.
[20] Birman V 2014 Functionally graded material and structures *Encyclopedia Therma stresses,*
[21] Liu, L. S., Zhang, Q.-J., and Zhai, P.-C., 2003, "The Optimization Design on Metal/Ceramic, FGM Armor With Neural Net and Conjugate Gradient Method," *Functionally Graded Materials,* VII Proceedings of the Seventh International Symposium on Functionally Graded Materials _FGM2000_, Materials Science Forum Vols. 423–425.
[22] Pan W, Gong J, Zhang L, and Chen L, eds., Trans Tech Publications Ltd., *Uetikon-Zuerich, Switzerland, pp. 791–802.*
[23] Chin ESC 1999 Army focused research team on functionally graded armor composites *Mater Sci. Engg. A259 : 155-161.*
[24] Lee W, Stinton D 1996 Concept of Functionally Graded Materials for Advanced Thermal Barrier Coating *Applications Journal of the American Ceramic Society,* 79,**12** (3003), 12.
[25] Das D, Saini A, Pathak M, Kumar S 2012 Application of functionally graded materials as thermal insulator in high temperature engineering components *IJREAS* Volume **2**, Issue 2
[26] Quin, X., and Dutta, D. 2004 Feature-Based Design for Heterogeneous Objects *Comput.-Aided Des., 36, pp. 1263–1278.*
[27] Birman V, Byrd L Modeling and Analysis of Functionally Graded Materials and Structures".
[28] Zlatanović M 2003 Combined plasma surface treatments for wear and corrosion protection *8th International Tribology Conference, 8. - 10. Oktober 2003., Belgrade, Serbia 25.*
[29] Epler L 2012 New Duplex Surface Treatment Dramatically Improves Die Life, Part Quality & Cost Saving *Die Cast Engineer Magazine, September 2012.*
[30] Li F, Takagi K, Ono M, Pan W, Watanabe R, Almajid A, and Taya M 2003 Fabrication and Evaluation of Porous Piezoelectric Ceramics and Porosity Graded Piezoelectric Actuators *J. Am. Ceram. Soc.,* 86, pp. 1094– 1098.
[31] Takeuch K, Kawazoe M, and Kanayama K. 2003 Design of Functionally Graded Wood-Based Board for Floor Heating System With Higher Energy Efficiency Functionally Graded Materials *VII Proceedings of the Seventh International Symposium on Functionally Graded Materials (FGM2000),* Materials Science Forum Vols. 423–425
[32] Hao D, Wei C 2016 Dynamic characteristics analysis of bi-directional functionally graded Timoshenko beams *Composite structures (Sciencedirect),* **141**, (2016), 253-263.
[33] Simsek M 2015 Bi-directional functionally graded materials (BDFGMs) for free and forced vibration of Timoshenko beams with various boundary conditions *Composite structures (Sciencedirect)*, **133**, (2015), 968-978.
[34] Qian LF, Batra RC 2005 Design of bidirectional functionally graded plate for optimal natural frequencies. *J Sound Vib* 2005;**280**:415–24.
[35] Enab T Performance improvement of total knee replacement joint through bidirectional functionally graded material *International Journal of Mechanical & Mechatronics Engineering IJMME-IJENS* Vol:**14** No:02.
[36] Udupa G, Rao S, Gangadharan K 2014 Functionally graded composite materials: An overview", *International conference on advances in manufacturing and material engineering, AMME 2014.*

IOP Conf. Series: Materials Science and Engineering **229** (2017) 012022 doi:10.1088/1757-899X/229/1/012022

Ion Implantation of Silver Nanoparticles on Electrodeposited Polycarbazole Via Plasma Sputter Type Negative Ion Sources

M C Marquez[1,*], V Mascarinas[2] and H Ramos[2]

[1]School of Chemical Engineering and Chemistry, Mapua University, Muralla St., Manila, Philippines, 1002
[2]Plasma Physics Laboratory, National Institute of Physics, University of the Phlippines-Diliman Diliman, Quezon City, Philippines, 1101

*menandro.marquez@gmail.com

Abstract. The discovery of conducting polymer has brought tremendous advancement in developing various polymeric materials, Carbazole pendants of poly(ethyl methacrylate) was cross-linked via an electrochemical route. The obtained film of poly(carbazole ethyl methacrylate) electrodeposited on the surface of indium tin oxide (ITO) was modified by implanting silver nanoparticles on its surface. This was done using Plasma Sputter-type Negative Ion Sources (PSTNIS) The modified and unmodified films of polycarbazole were characterized to assess its properties. Implanting silver nanoparticles on the surface of the cross-linked polymer abruptly changed its surface roughness, absorbance in the visible region and its current-voltage characteristic. A more pronounced diode-like characteristic was observed with a turn-on voltage of ~0.4V. Investigation and tailoring the properties of electropolymerized carbazole attached to poly (ethyl methacrylate) backbone with implanted silver nanoparticles could lead to important materials with impact in optoelectronic devices.

1. Introduction

The increasing demand for advanced technologies resulted researchers to develop new techniques in synthesizing or modifying nanostructured materials. Conducting polymers are one of the most interesting materials with semiconducting properties. Since the discovery of this group of materials by Heeger and coworkers, numerous researches have been devoted to conductive polymers [1, 3]. Among the numerous conducting polymers prepared to date, polycarbazole is by far one of the most extensively studied. As a result of its good intrinsic properties, polycarbazole could be a promising material for several applications including, electrochemical biosensors, conductive textiles and fabrics and optoelectronic devices. One useful and viable technique in depositing conducting polymer on a suitable substrate is electrodeposition. This is an advantageous process since the polymer is formed and deposited on the surface of the working electrode in just one process. The chemical and physical properties of the film can then be controlled by adjusting synthesis parameters such as current density, monomer concentration, electrolyte type, electrolyte concentration, pH and number of cycles [2, 3]. The intrinsic properties of electrodeposited polycarbazole are highly dependent on electropolymerization conditions. With the advent of surface modification techniques, plasma modification through ion implantation emerged as the most promising one since it allows reaching a high metal filing factor beyond the limit of metal solubility. In addition this technique allows easier

Content from this work may be used under the terms of the Creative Commons Attribution 3.0 licence. Any further distribution of this work must maintain attribution to the author(s) and the title of the work, journal citation and DOI.
Published under licence by IOP Publishing Ltd

control on the dosage of ions that one would want to implant [4, 5, 6]. In this study, Ag nanoparticles were implanted using negative ion sources of plasma and the effect of this implantation on the optical, morphological and electrical properties of the conducting polymer film were observed.

2. Methodology

2.1 Materials

An indium tin oxide (ITO) coated glass slide purchased from Ossila Ltd. With sheet resistivity of 30Ω-m was used as the substrate. 2-(9*H*-carbazol-9-yl)ethanol, methacrylic acid, 4-(*N,N*-dimethylamino)pyridine, *N,N'*-dicyclohexylcarbodiimide, dichloromethane and chloroform were all purchased at Sigma-Aldrich. All reagents were used as received without further purification.

2.2 Electropolymerization of Polycarbazole

The substrate was prepared and cleaned by following the conventional procedure of cleaning ITO coated glass slide. Briefly, the ITO coated glass slide was washed in a detergent solution (Hellmanex, Ossila Ltd) then ultrasonicated in acetone, ethanol, propanol and deionized water for five minutes in particular order. The ITO was dried at 110C and blown with dry nitrogen prior to its use. Polycarbazole was electrodeposited using the following parameters: 1mg/mL precursor material concentration, 0.1M of TBAPF6 as supporting electrolyte in chloroform, ITO as working electrode, platinum rod as counter electrode and Ag/AgCl as reference electrode, cycled for 15 times at 25mV/s. After the electrodeposition, the film was washed with chloroform and dried at 60C for 5-10 minutes and vacuum dried for another 30 minutes.

2.3 Ion Implantation of Silver Nanoparticles

The PSTNIS was first evacuated using a rotary pump and turbomolecular pump to reach a base pressure of $4x10^{-5}$ torr. The chiller is then switched on and set to 140°C before applying current to the tungsten filament. Then the argon gas was allowed to flow inside the chamber with a flow rate of 5sccm. Next, the valve was throttled to control the pressure inside the chamber after introducing gas into it. The operating pressure was kept around 1mtorr. Steady state plasma was finally produced by applying a current of 9A to the tungsten filament. A discharge voltage of 50V was also applied to ignite the plasma. The target potential used for deposition were -300V, -400V, and -500V.

3. Results and Discussion

3.1 Electrodeposition

The cyclic voltammograms obtained with poly(carbaole ethyl methacrylate) on the surface of the ITO coated glass slide electrode is depicted in Figure 1A. The voltammetric curves are indicative of oxidation proceeding on every surface under consideration in two stages. The anodic peak observed at positive potentials is particularly intense in the anodic branch of the first spinning cycle and can be attributed to the initial monomer oxidation. The succeeding cycles of cycling the potential led to the formation of dimers and eventually to oligomers and/or polymers. The dimer of carbazole is proposed to be formed by the loss of an electron (e–) and of a proton for every carbazole molecule.

3.2 Characterization

3.2.1 FTIR and XRD. The FTIR spectrum of the monomer, poly(carbazole ethyl methacrylate) (PCzEMA) and electrodeposited poly(carbazole ethyl methacrylate) (ePCzEMA) is shown in Figure 1C . While band centered at 832cm-1 in ePCzEMA only, it can also be observed in the spectra of 9H carbazole and PCzEMA the presence of peaks ranging from 730-750cm-1 indicating vibrational modes of 1: 2 disubstituted benzene rings of carbazole species. However, upon electrodeposition this peak suddenly disappeared and the presence of a new band centered at 833cm-1 is observed. This peak is mainly attributed to the crosslinking of the carbazole units due to the polymerization of benzene rings along the polymer backbone. The band centered at 1708 in PCzEMA and ePCzEMA is attributed to C=O, a characteristic peak of ethyl methacrylate.

Shown in Figure 1B is the x-ray diffractogram of the ePCzEMA with Ag. It can be seen from this diffractorgram the presence of the broad peak from 2 values of 15-35 which is mainly attributed to the amorphous nature of ePCzEMA. On the other hand sharp peaks for the (111) of Ag was observed signifying the presence of Ag on the polymer. Peaks for the ITO substrate were also evident in the diffractogram, a result that is inevitable since the deposited ePCzEMA forms only a thin layer on top of the ITO.

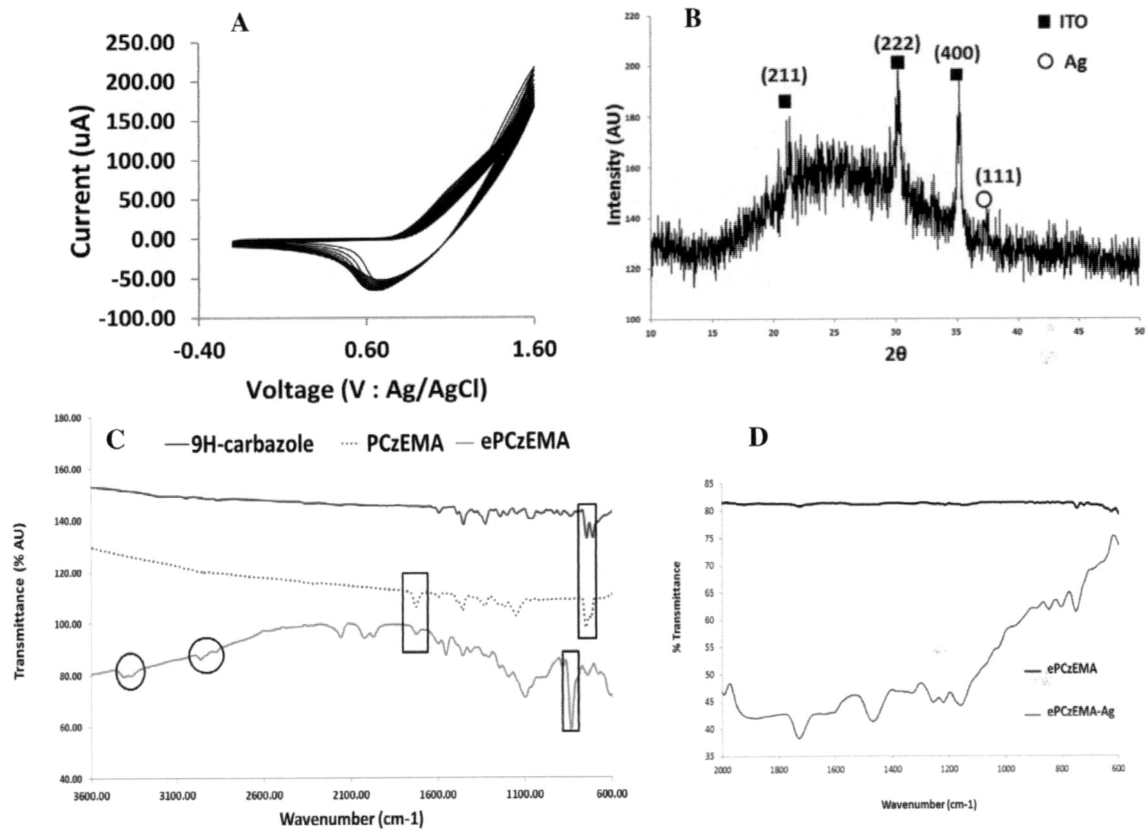

Figure 1 A) Cyclic voltammogram of ePCzEMA deposition, B) XRD diffractogram and C) FTIR of monomer, prepolymer and ePCzEMA, and D) FTIR of ePCzEMA-Ag on ITO

3.2.2 AFM and SEM. The surface morphology and topography of the electrodeposited film were observed using SEM and AFM. Figure 2 shows the AFM images for the substrate, Ag nanoparticles deposited on the substrate, pristine ePCzEMA and Ag implanted ePCzEMA. From this image the surface roughness were obtained and found that the ITO substrate has an rms value of ~2.5nm, the Ag implanted ITO with rms value of 9.5nm, pristine ePCzEMA ~ 6nm and Ag-ePCzEMA ~ 7.6nm. These results corroborate with the SEM images depicted in Figure 2E and 2F showing the surface morphology of pristine ePCzeMA and Ag-ePCzEMA. From these images it is evident that the implantation of Ag led to a rougher texture with areas that appear to be bright and protruded relative to the surface.

To observe the effect of the target potential to the topography and roughness of the film, 300V, 400V and 500V were applied. Figure 3 shows the AFM images with varying target potential. Increasing the target potential led to a rougher surface since it was believed that the ions have more energy that could mechanically impact the surface of ePCzEMA upon landing.

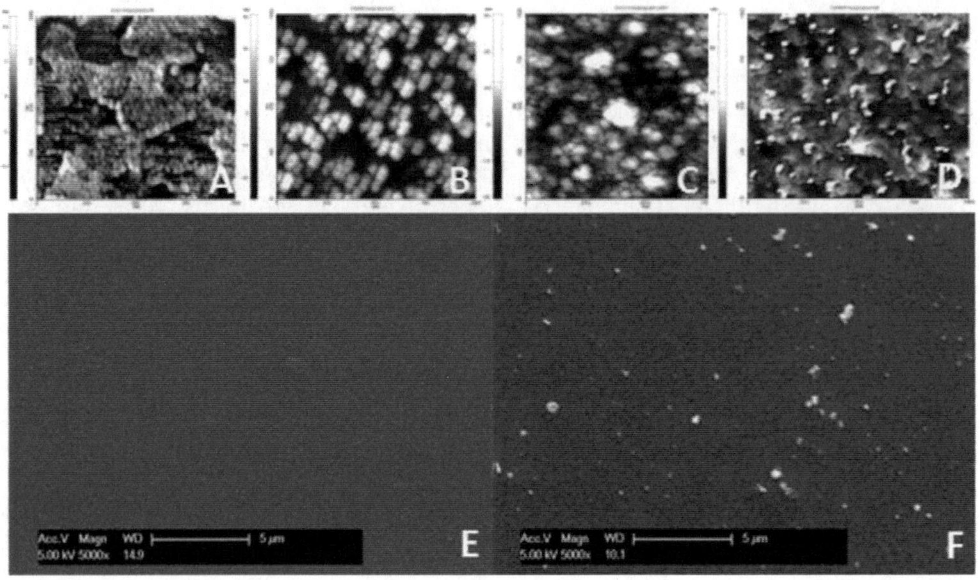

Figure 2.Atomic Force micrographs of a) pristine ITO coated glass slide, b) Ag on ITO c) ePCzEMA on ITO, d) Ag-ePCzEMA on ITO e) SEM of ePCzEMA on ITO, and f) SEM of Ag-ePCzEMA on ITO

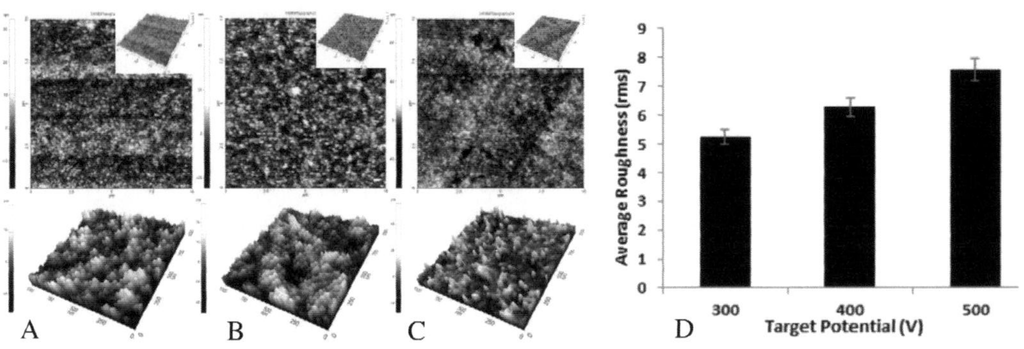

Figure 3. Atomic Force Micrographs of ePCzEMA with implanted Ag at different target potentials A) 300V B) 400V C) 500V and D) root mean squared value for each film

3.2.3 UV-Vis and I-V Curve. As can be seen from Fig. 4A, the UV-visible spectrum of ePCzEMa shows strong and sharp absorption peaks at 325 which may be derived from the π–π* transition of the short conjugated chain. Furthermore optical band gap was calculated based on the absorbance spectra of the polymers. The band gap of the polymer was estimated using the relation.

$$\alpha h\nu = B(h\nu - E_g)^n \quad (1)$$

where α is the absorption coefficient, hv is the energy of absorbed light, n = 1/2 for direct allowed transition and B is proportionality constant. Energy gap (Eg) was obtained by plotting $(\alpha h\nu)^2$ vs. hv and extrapolating the linear portion of $(\alpha h\nu)^2$ vs. hvto zero. Employing the method stated above yields an estimated 3.7 eV optical bandgap for ePCzEMA. It can be noted from this result that the signal for the UV-Vis spectrum of the pristine ePCzEMA was increased upon the implantation of Ag on it. The presence of Ag nanoparticles on the polymer matrix could provide more excitation of carriers due to plasmon resonance effect as observed in systems with small metallic nanoparticles. The I-V curve of the device with ePCzEMA-Ag sandwiched to ITO and thermally evaporated Al had been obtained and

IOP Conf. Series: Materials Science and Engineering **229** (2017) 012022 doi:10.1088/1757-899X/229/1/012022

is presented in Figure 4B. Applied potential from -0.2V – 1.0 V was used to observed the electrical properties of pristine ePCzEMA and ePCzEMA-Ag. Both I-V curves exhibit a nonlinear trend and exhibit more of a diode like characteristic. The shape of this I-V curve is attributed to the Schottky contact formed between the ITO and the polymer, together with an ohmic contact that could formed between the Al and some silver nanoparticles. The rectifying behavior is more prominent to that of the sample with Ag primarily because of the enhanced contacts between the ITO and Ag as well as the improved charge excitation offered by the Ag nanoparticles on the ePCzEMA matrix.

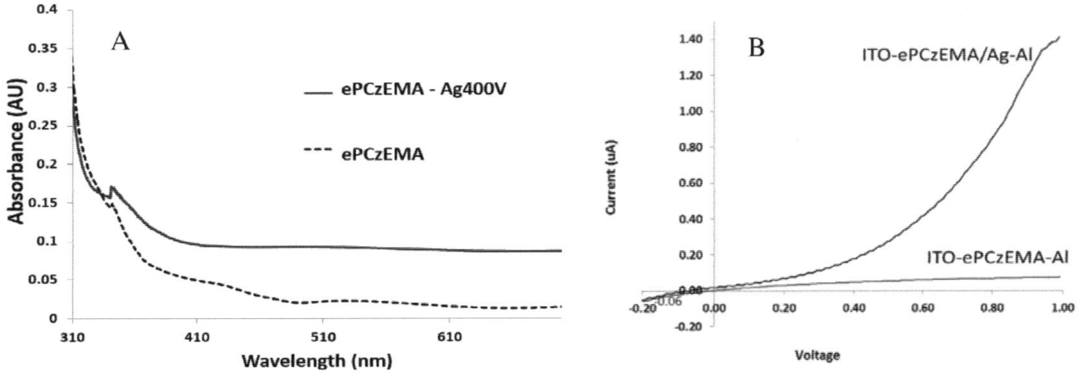

Figure 4. A) UV-Vis spectra and B) I-V curve of ePCzEMA, with and without implanted Ag

4. Conclusion

Silver nanoparticles have been deposited on the surface PCzEMA via plasma sputter type negative ion sources. Characterization results based on FTIR, XRD and UV-Vis showed that structures and optical property of the carbazole pendants have been intensified but preserved. The increased in target potential for plasma deposition altered the morphology and topography of the electropolymerized carbazole Lower voltage of 300V produced a smoother surface compared to 500V target potential. The optical and electrical characteristics Ag modified ePCzEMA were enhanced and this could wave to the development of functional materials necessary for applications in optoelectronics field.

Acknowledgement

The authors would like to thank the Philippine Council for Industry, Energy and Emerging Technology Research and Development - Department of Science and Technology (PCIEERD-DOST) of the Philippine Government for the financial support they have extended for this study.

References

[1] Zhu X., Xu L., Wang M., Wang Z., Liu R., Zhao J. (2011) Electrosynthesis and Characterization of a Soluble Blue Fluorescent Copolymer Based on Pyrene and Carbazole, Internation Journal of Electrochemical Science 6 pp 1730-1746

[2] Crespilho F. N., Zucolotto V., Oliveira, O. N., & Nart, F. C. (2006). Electrochemistry of Layer-by-Layer Films: A Review. Int. J. Electrochem. Sci., 1, pp 194-214.

[3] Diamant Y., Chen J., Han H., Kamenev B., Tsybeskov L., Grebel H. (2005) "Electrochemical polymerization, optical and electrical characterizations of polycarbazole on single wall carbon nanotubes, Synthetic Metals 151, pp 202–207

[4] Taoud H.I., Bernede J.C., Bonnet A., Morsli M., Godoy A. (1997) Comparison of polycarbazole obtained by oxidation of carbazole either in solution or in thin film form, Thin Solid Films 304 pp 48-55

[5] Stepanov A.L.. (2009) Fabrication of Metal Nanoparticles In Polymers By Ion Implantation Nanostructured Materials for Advanced Technological Applications, pp 153-162.

[6] Stepanov A.L.. (2010) Synthesis of Silver Nanoparticles in Dielectric matrix by Ion Implantation : A Review, Advanced Materials Science 26, pp1-29

IOP Conf. Series: Materials Science and Engineering **229** (2017) 012023 doi:10.1088/1757-899X/229/1/012023

Isocyanate-modified Nano-SiO_2 and Corresponding Process Optimization

X F Ye[1], X Y Huang[1], G Yu[1], J N Wang[1], D C Chen[1,*], H W Hu[1] and M L Chang[1]

[1]School of Materials Science and Energy Engineering Foshan University, Foshan, China

*chendc@fosu.edu.cn

Abstract. Nano-SiO_2 is hydrophilic, and the surface of nano-SiO_2 should be modified to improve its dispersion in a polymer matrix. Here, triallyl isocyanurate was selected to modify the surface of the nano-SiO_2 powders. Because the reaction between isocyanate groups (-NCO) and hydroxyl groups (-OH) on surface of the nano-SiO_2 does not need any pre-treatment, one single and simple step can facilitate the graft. By optimizing the specific modification process via orthogonal experiments, optimal processing conditions that yield good graft modification effects are obtained. This enhances the lipophilicity of the nano-SiO_2. The Fourier-transform infrared spectrometer (FTIR) results indicate that the isocyanate groups (-NCO) in triallyl isocyanurate and the hydroxyl groups (-OH) on surface of the nano-SiO_2 undergo a chemical reaction. Further, the precipitation data show that modified nano-SiO_2 is well dispersed in the organic phase, and its hydrophobic property is strong. This fulfills the goal of modifying nano-SiO_2 powders.

1. Introduction

SiO_2 is a non-metal functional material that is odorless, non-toxic and pollution-free. Due to its special particle sizes, nano-scale SiO_2 has features such as small particle sizes and high specific surface area. It also has surface effects, small-size effects, quantum-tunneling effects, macroscopic quantum tunneling effects and special photoelectric effects. As a result, nano-SiO_2 is often used as an effective insulating material, a catalyst carrier, a gas filter, or a high-grade paint filler. It is widely applied in rubbers, plastics, coatings, oil paint, or cosmetics [1–6].

Figure 1 shows that on the surface of SiO_2, there are unsaturated residual bonds and hydroxyl groups (-OH) in different bonding states. While adjacent hydroxyl groups (-OH) are connected via hydrogen bonds, hydrogen atoms in isolated hydroxyl groups show strong electropositivity, and they tend to attract electronegative atoms. This can facilitate a dehydration condensation reaction with any molecule containing a hydrogen group. The presence of surface hydroxyl groups facilitates chemical adsorption including hydrogen bonds with water. Meanwhile, the compatibility between the nanoparticle and the polymer matrix is poor because the nanoparticle atoms are in a highly active state, have a relatively high specific surface energy, and easily agglomerate. Consequently, nanoparticles that are dispersive and polymer-compatible with a uniform particle size distribution is an active area of research—especially in processing the surface hydroxyl groups (-OH) [7–8].

Content from this work may be used under the terms of the Creative Commons Attribution 3.0 licence. Any further distribution of this work must maintain attribution to the author(s) and the title of the work, journal citation and DOI.
Published under licence by IOP Publishing Ltd

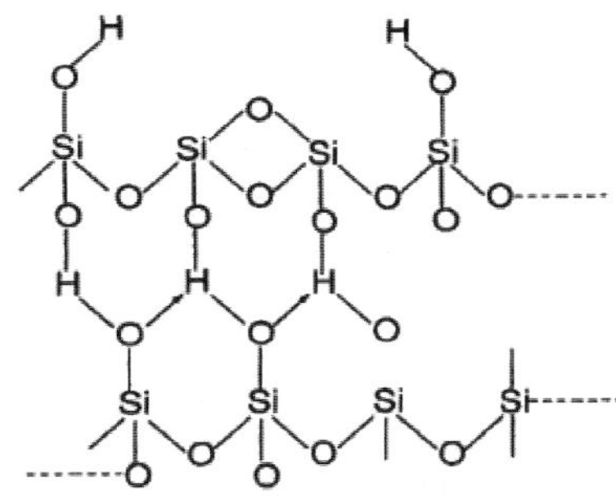

Figure 1. A schematic view of the SiO2 surface.

There are currently many nano-SiO_2 surface modification processes, but each process has its own disadvantages [9–12]. For example, while the modification method utilizing silane coupling is easy to use, the related large-scale production is discouraged due to the expensive price of the silane-coupling agent. Modification using polymer graft polymerization offers poor grafting efficiency due to the existence of homo-polymerization, which influences the mechanical performance of the material. Moreover, esterification can hydrolyze the ester groups resulting in relatively poor thermal stability.

This research utilizes the chemical reaction between the isocyanate groups (-NCO) in triallyl isocyanurate and the hydroxyl groups (-OH) on the surface of the nano-SiO_2 to modify the surface of the nano-SiO_2. The modified surface of the nano-SiO_2 has acrylate groups, which isolate nano-SiO_2 particles from each other to prevent agglomeration and produce stably modified nano-SiO_2. Meanwhile, the modified system may further undergo a polymerization reaction to fabricate nano-SiO_2 composites.

Though studies on the inorganic nanomaterials using isocyanate for modification have been reported previously [13], these studies often use two-step reactions. Most commonly, toluene diisocynate (TDI) is used as the modifier and the isocyanate groups (-NCO) undergo a grafting reaction with the hydroxyl groups (-OH) in the nano-SiO_2. Another -NCO group is then utilized for connection to the various desired groups. Therefore, this method is relatively complicated, and the grafting efficiency is affected by the process parameters. Here, triallyl isocyanurate was used to modify nano-SiO_2 resulting in outstanding advantages including no need for pre-treatment. That is, the acrylate groups with chain-extending polymerization ability can be connected to nano-SiO_2 in a single step. As a result, the processing process is simplified with superior grafting modification effects.

2. Experiments

2.1. Major Experimental Reagents and Instrument

2.1.1. Major reagents in the experiments. The major reagents used in this research are listed in Table 1.

Table 1. Experimental reagents.

Name	Size	Manufacturer
Nano-SiO_2	Particle size 15±5 nm	Shenzhen Jingcai Chemical Industry Co., Ltd.
Toluene	Analytically pure	Xilong Chemical Co., Ltd.
Acetone	Analytically pure	Guangzhou Haizhu Chemical Reagent Co. Ltd.
Absolute ethyl alcohol	Analytically pure	Tianjin Yongda Chemical Reagent Co. Ltd.
Triallyl isocyanurate	Analytically pure	Sinopharm Chemical Reagent Co. Ltd.
Dibutyltin dilaurate	Analytically pure	Tianjin Fuchen Chemical Reagent Factory
High purity nitrogen	Purity ≥ 99.99%	Foshan MS Gas Co. Ltd.

2.1.2. Major instrument in the experiments. The major instrumentation used in this research is shown in Table 2.

Table 2. Experimental instrument.

Name	Style	Manufacturer
Digital thermostatic water bath	HH-2	Jintan Fuhua Instrument Co. Ltd
Power electric mixer	JJ-1	Jintan Honghua Instrument Factory
Water-cycling vacuum pump	SHZ-D(III)	Gongyi Yuhua Instrument Factory
Electric centrifuge	TDL-50	Jintan Medical Instrument Factory
Numerical control ultrasonic cleaning machine	KQ-200DB	Kunshan Ultrasonic Instrument Co. Ltd
Electronic analytical balance	GR-200	A&D (Guangzhou) Company Limited
Electric blast drying oven	101-1AS	Shanghai Xunneng Electric Device Co. Ltd.

2.2. Specific Experimental Method

The number of hydroxyl groups (-OH) on surface of the nano-SiO_2 is reduced by using the isocyanate groups (-NCO) in triallyl isocyanurate to react with the hydroxyl groups (-OH) on surface of the nano-SiO_2. Meanwhile, it is possible for the nano-SiO_2 to undergo the chain-extension reaction by introducing the acrylate groups into the surface of the nano-SiO_2. The specific reaction principles and procedure are explained as follows.

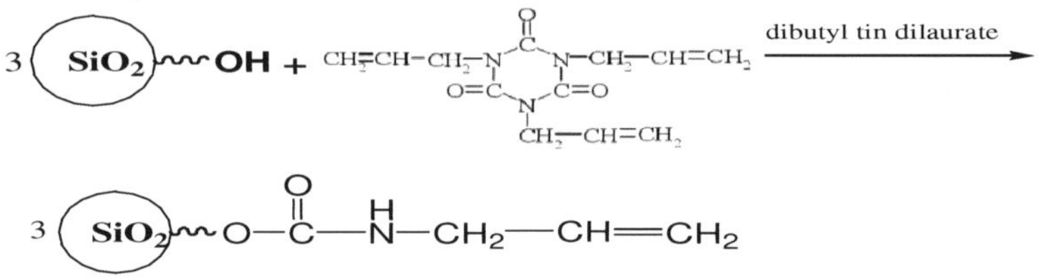

(1)

(1) The nano-SiO_2 was placed on a watch glass, and then dried in 100–130°C oven for 2-6 hours.

(2) The dried nano-SiO_2 was added to a flask with toluene. This was dispersed with shaking and ultrasonication for 20–50 min, which made the nano-SiO_2 be evenly dispersed.

(3) The triallyl isocyanurate and dibutyltin dilaurate were then added for reaction at 50–90°C under N_2 aeration and stirring for 0.5–5 hours, after which, the reaction solution was cooled to room temperature.

(4) The reaction system was centrifuged, and the precipitate was washed with acetone, suction-filtered, and heated at 80°C for 8 h to obtained the well-dispersed nano-SiO_2 product.

2.3. Tests and Characterization Methods

2.3.1. IR characterization. The reaction conditions between isocyanate groups (-NCO) in triallyl isocyanurate and hydroxyl groups (-OH) on surface of the nano-SiO_2 were determined via the KBr method and FTIR (manufacturer: Nicolet; stype: FTIR-850; test range: 4000 cm−1–900 cm−1, precision: 0.01 cm−1). This technique characterized the structure of the nano-SiO_2 before and after modification.

2.3.2. Calculation method of percentage grafting. The grafting percentage is used as the parameter to compare the modification effects of triallyl isocyanurate on nano-SiO_2. Specifically, the weight of nano-SiO_2 before and after modification is measured (W_1 is the weight of the SiO_2 modified by triallyl isocyanurate, and W_2 is the weight of nano-SiO_2 before modification). The percentage grafting of the reaction is calculated: $(W_1-W_2)/W_2\times 100\%$.

2.3.3. Precipitation experiment

2.3.3.1. Precipitation in water. Nano-SiO_2 before and after modification were weighed and placed in different tubes. After deionized water was added to each tube, the tubes were vibrated and ultrasonic-dispersed for 30 min to observe the precipitation of nano-SiO_2 in water. If the modified nano-SiO_2 precipitates in water, then it is very hydrophilic, and the amount of residual -OH groups on its surface is relatively large, indicating that the modification effect is not ideal. In contrast, if there is no significant precipitation in water, then the lipophilicity of the modified nano-SiO_2 is good and the amount of residual -OH groups on the surface is small indicating a good modification effect.

2.3.3.2. Precipitation in ethyl alcohol. Nano-SiO_2 before and after modification were weighed and placed in different tubes. Ethyl alcohol was then added into each tube, and the tubes were vibrated and ultrasonic-dispersed for 30 min to observe the precipitation condition of nano-SiO_2 before and after modification in the ethyl alcohol. If the modified nano-SiO_2 is evenly dispersed in the ethyl alcohol, then its lipophilicity is good, and the amount of residual -OH groups is small indicating good modification. In contrast, if the modified nano-SiO_2 precipitates in the ethyl alcohol, then its hydrophilicity is good and the amount of residual -OH groups is relatively large indicating poor modification.

3. Results and discussion

3.1. Analysis of IR Spectra

Figure 2 shows an IR spectrum of the nano-SiO_2 before modification, and Figure 3 shows the IR spectrum of nano-SiO_2 modified by triallyl isocyanurate. Figure 2 shows that the absorption peak near 1118 cm−1 corresponds to the asymmetric stretching mode of the Si-O-Si bond; the absorption peak at 3462 cm−1 is attributed to the stretching vibration of the hydroxyl groups (-OH) on the surface of the nano-SiO_2 as well as the hydroxyl groups (-OH) in the physically adsorbed water. The absorption peak at 1636 cm−1 is generated due to the bending vibration of the hydroxyl groups (-OH) in the physically adsorbed water on surface of the nano-SiO_2.

Figure 3 shows the IR spectrum of the nano-SiO_2 modified by triallyl isocyanurate. The stretching vibration of the hydroxyl groups (-OH) still exists at 3462 cm−1, but the antisymmetric stretching vibration of the C-H bond occurs near 2955 cm−1 indicating the existence of organic materials on the SiO_2 surface. This also shows that triallyl isocyanurate is already grafted to the SiO_2 particles. Further, the adsorption peak at 1636 cm−1 shows a decreasing trend compared to Figure 2, which indicates that fewer hydroxyl groups (-OH) on surface of the nano-SiO_2 were modified by triallyl isocyanurate.

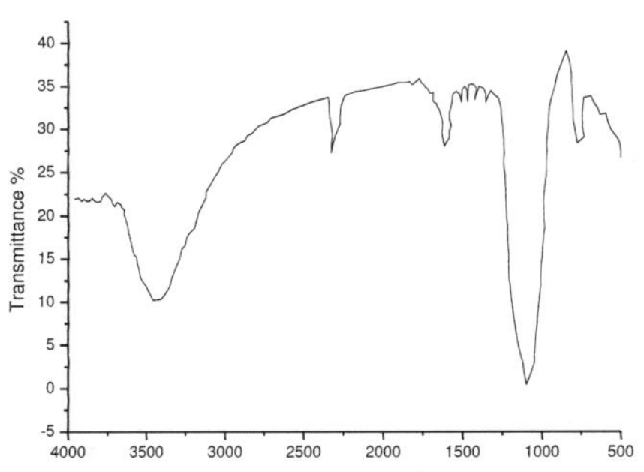

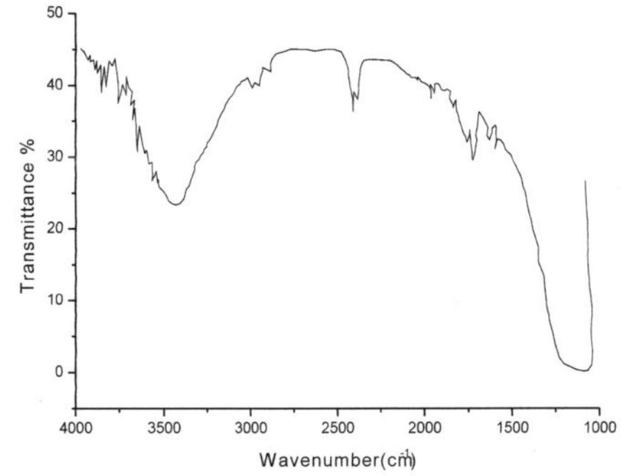

Figure 2. IR spectrum of nano-SiO_2 before modification. **Figure 3.** IR spectrum of nano-SiO_2 after modification.

3.2. Level and Factors of Orthogonal Experiments

We used time (h), temperature (°C) and the weight ratio of triallyl isocyanurate to the nano-SiO_2 to identify the optimal processing conditions. We compared the percentage grafting and evaluated the modification effect of triallyl isocyanurate on the nano-SiO_2. We also used a three-factor, four-level orthogonal experiment. The factors and levels can be found in Table 3, and the orthogonal experiment design is illustrated in Table 4.

Table 3. Factor level table.

Run	Time/h	Temperature/°C	Weight ratio of triallyl isocyanurate to nano-SiO_2
1	1.5	50	1:0.6
2	2	60	1:0.8
3	3	70	01:1
4	4	80	1:1.2

Table 4. Orthogonal experiments.

No.	Time/h	Temperature/°C	Weight ratio of triallyl isocyanurate to nano-SiO_2	Percentage grafting/%
Exp. 1	1.5	50	1:0.6	5.1
Exp. 2	1.5	60	1:0.8	7.5
Exp. 3	1.5	70	1:1	15.1
Exp. 4	1.5	80	1:1.2	16.3
Exp. 5	2	50	1:0.8	6.1
Exp. 6	2	60	1:0.6	8.2
Exp. 7	2	70	1:1.2	16.0
Exp. 8	2	80	1:1	18.6
Exp. 9	3	50	1:1	6.3
Exp. 10	3	60	1:1.2	9.4
Exp. 11	3	70	1:0.6	16.9
Exp. 12	3	80	1:0.8	19.7
Exp. 13	4	50	1:1.2	5.9
Exp. 14	4	60	1:1	10.7
Exp. 15	4	70	1:0.8	13.3
Exp. 16	4	80	1:0.6	15.4
Avg. 1	11.000	5.850	11.400	
Avg. 2	12.225	8.950	11.650	
Avg. 3	13.075	15.325	12.675	
Avg. 4	11.325	17.500	11.900	
Range	2.075	11.650	1.275	

Figures 4, 5, and 6 illustrate the effect of reaction time (h), temperature (°C), and the weight ratio of triallyl isocyanurate to the nano-SiO_2 on the percentage grafting of the nano-SiO_2, respectively. Typically, a greater range has a larger influence on the level variation of each factor. If the range is small, then the influence of the level variation of the specific factor on the experimental result is also small. The range analysis results show that the order of the influence of the three factors on the nano-SiO_2 modified by isocyanate groups (-NCO) is: temperature > time > the weight ratio of triallyl isocyanurate to the nano-SiO_2. The best conditions are 3 h, 80°C, and a 1:1 weight ratio of triallyl isocyanurate to the nano-SiO_2 (Table 4).

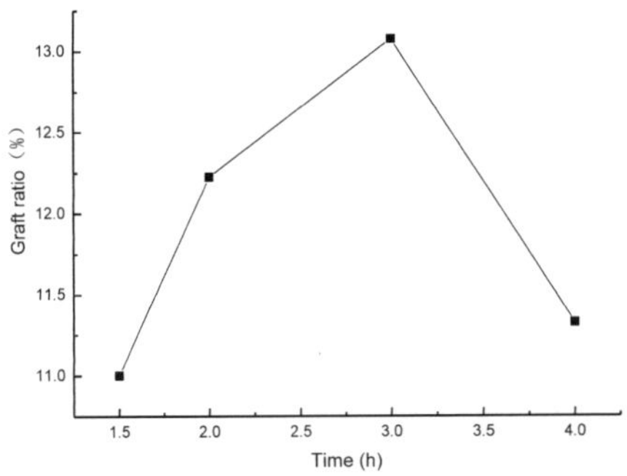

Figure 4. Influence of time on percent grafting.

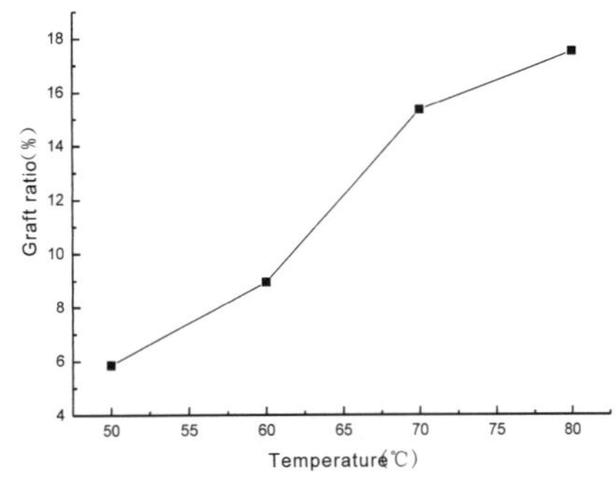

Figure 5. Influence of temperature on percent grafting.

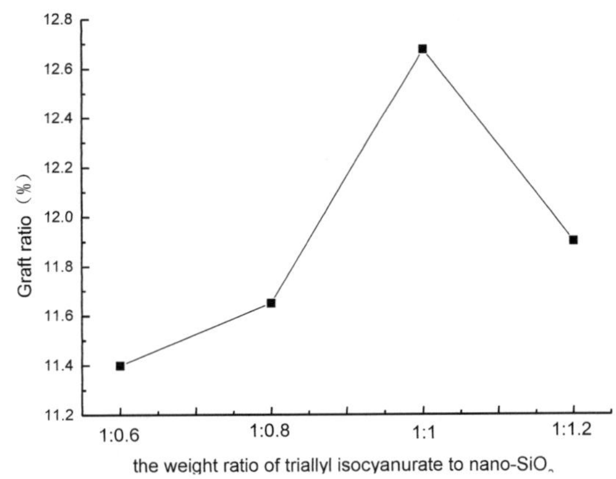

Figure 6. The influence of the weight ratio of triallyl isocyanurate to nano-SiO_2 on the percentage grafting.

3.3. Precipitation Data

Precipitation experiments were performed with nano-SiO_2 modified by triallyl isocyanurate under the optimal conditions as well as control nano-SiO_2 before modification as described in section 2.3.2. Then, 0.1 g nano-SiO_2 before and after modification were weighed and placed in dry and clean tubes a and b, respectively (nano-SiO_2 before modification was placed in tube a, and the nano-SiO_2 after modification was placed in tube b). Then, 10 mL de-ionized water was added to tubes a and b, respectively, and dispersed with sonication for 30 min to observe the precipitation of nano-SiO_2 before and after modification in the water. Similarly, 0.1 g nano-SiO_2 before and after modification were weighted and placed in dry and clean tubes c and d, respectively (the nano-SiO_2 before modification was placed in tube c, and the nano-SiO_2 after modification was placed in tube d). Then, 10 mL ethyl alcohol was injected into tube c and tube d, respectively, and dispersed with 30 min of sonication to observe the precipitation of nano-SiO_2 before and after modification in the ethyl alcohol.

3.3.1. Precipitation experiments in water. Figure 7 shows that unmodified SiO_2 can be dispersed evenly in the water (tube a) to form a suspension solution. No precipitates were found on the bottom of the tube indicating that a large amount of hydrophilic groups (-OH) exist on the surface of unmodified SiO_2, and the unmodified SiO_2 shows strong hydrophilicity. In contrast, the nano-SiO_2 modified by triallyl isocyanurate totally floats on the water surface (tube b). This is due to the capillary effect. A thin layer of water film is attached to the wall of the tube above the water level in the tube. The modified nano-SiO2 is hydrophobic because the modified nano-SiO_2 is attached to the water film. This indicates that the amount of hydrophilic groups (-OH) on surface of modified SiO_2 is greatly reduced.

3.3.2. Precipitation experiments in ethyl alcohol. Figure 7 shows that the unmodified SiO_2 in ethyl alcohol precipitates quickly (tube c), and the upper layer of the solution in tube c is clear indicating that the unmodified nano-SiO_2 has low lipophilicity. The number of hydrophilic groups (-OH) on the surface of the unmodified nano-SiO_2 is relatively large. Meanwhile, the nano-SiO_2 modified by triallyl isocyanurate may be dispersed evenly in ethyl alcohol (tube d) to form a suspension solution indicating that the modified nano-SiO_2 shows reduced hydrophilicity, and the number of hydrophilic groups (-OH) on its surface decreases.

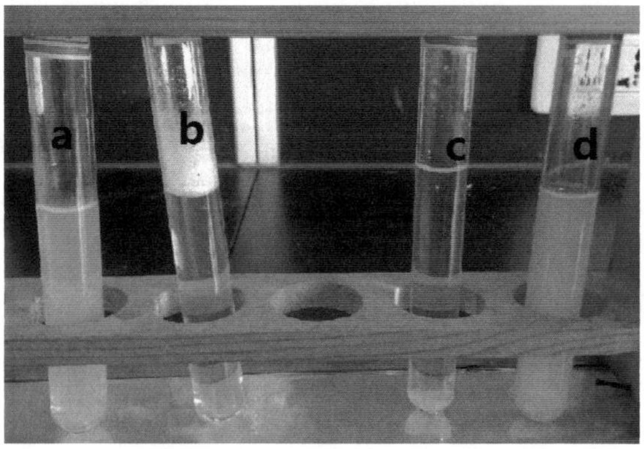

Figure 7. Precipitation data

Figure 8 shows the precipitation conditions of SiO_2-containing ethyl alcohol obtained from tube c and tube d after being placed in 10 mL measuring cylinders, respectively. For every 30 min, the precipitation volumes of solution c and solution d were recorded. In particular, curve A shows the precipitation conditions of nano-SiO_2 modified by triallyl isocyanurate, and curve B shows the precipitation condition of unmodified nano-SiO_2. The figure shows that the modified powder has a significantly smaller precipitation rate than unmodified powder due to the reduction in its hydrophilicity. After 2.5 hours, the precipitation of the modified powder basically stops, and the precipitation volume of the modified powder is greater than that of the unmodified powder.

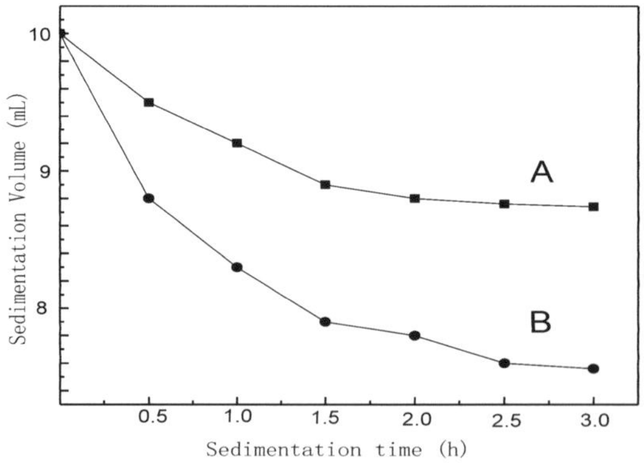

Figure 8. Variation of the precipitation volume of nano- SiO_2 in ethyl alcohol before and after modification.

Clearly, many hydroxyl groups (-OH) exists on surface of the nano- SiO_2 before modification, and the unmodified nano- SiO_2 shows strong hydrophilicity and tends to aggregate in the organic phase. Moreover, unmodified nano-SiO_2 has large nano-SiO_2 particles that precipitate relatively fast. The small particles precipitate fast at the beginning and these induce large particle precipitation. After modification, the number of hydroxyl groups (-OH) on the surface decreases, and the modified nano-SiO_2 is lipophilic. They then disperse uniformly. After precipitation, the modified nano- SiO_2

precipitates very slowly, and it eventually reaches a dynamic equilibrium. The precipitation volume is constant throughout.

4. Conclusions and issues future work

(1) The IR data indicate that under these experimental conditions, isocyanate groups (-NCO) in triallyl isocyanurate and hydroxyl groups (-OH) on the surface of the nano-SiO_2 undergo a chemical reaction. The number of hydroxyl groups (-OH) on the surface of the nano- SiO_2 is reduced.

(2) The optimal conditions for triallyl isocyanurate modification of the nano- SiO_2 is 3 h at 80°C with a 1:1 weight ratio of triallyl isocyanurate to nano-SiO_2.

(3) The precipitation experiments show that the modified nano- SiO_2has an extremely strong hydrophobicity, and it can be dissolved in organic solvent indicating that the number of the hydroxyl groups (-OH) on surface of the nano- SiO_2 is greatly reduced.

Acknowledgments

This work was financially supported by Foundation for Distinguished Young Talents in Higher Education of Guangdong, China (2015KQNCX178). The key platform funded projects of Guangdong provincial department of education(gg041002) and Foshan city science and technology innovation special fund (2014AG10009).

References

[1] Alireza K, Mohammad H M and Payam H 2016 *J. Constr. Build. Mater* **113** 188-201
[2] Wang Y Q, Gao X N, Chen P, Huang Z W, Xu T, Fang Y T and Zhang Z G 2016 *J. Appl. Therm. Eng* **96** 699-707
[3] Hosseini P, Mohamad M I, Nekooie M A, Taherkhani R and Booshehrian A 2011 J. Basic Appl. Sci **12** 2768–2782
[4] Bahadori H and Hosseini P 2012 *J. Civ. Eng. Manage* **18** 416-425
[5] Hosseinpourpia R, Varshoee A, Soltani M, Hosseini P and Tabari H 2012 *J. Constr.Build. Mater* **31** 105-111
[6] Pourjavadi A, Fakoorpoor S M, Hosseini P and Khaloo A 2013 *J. Cem. Concr. Compos* **37** 196-204
[7] Shan W and Liao M Y 2006 *J. Chin. Polym. Bull* **3** 1-9
[8] Wang X Y, Yao S W and Zhang W G 2005 *J. Electroplat. & Finishing* **24** 42-44
[9] Hou P K, Cheng X, Qian J S, Zhang R and Cao W 2015 *J. Cement. Concrete. Comp.***55** 26-33
[10] Tai Y L, Qian J S and Zhang Y C 2008 *J. Chem. Eng. J.* **144** 354-361
[11] Wang L J, Lu A H and Xiao Z Y 2009 *J. Appl. Surf. Sci.* **17** 7542-7546
[12] Zhang M L, Ding L G, Jing X Y and Hou X Q 2003 *J. Chen. Eng.* **10** 12-14
[13] Yang M, Gao J, Wang Y X and Zhao Z L *J. Chem. Res. Appl.* **21** 477–480

IOP Conf. Series: Materials Science and Engineering **229** (2017) 012024 doi:10.1088/1757-899X/229/1/012024

Fabrication and Characterization of Carbonized Rice Husk/Barium Titanate Nanocomposites

G J H Melvin[1, *], Z Wang[2], Q-Q Ni[3], N J Siambun[1] and M M Rahman[1]

[1] Faculty of Engineering, Universiti Malaysia Sabah, Jalan UMS, 88400 Kota Kinabalu, Sabah, Malaysia.
[2] Institute of Carbon Science and Technology, Shinshu University, 4-17-1 Wakasato, Nagano 380-8553, Japan.
[3] Department of Functional Machinery and Mechanics, Shinshu University, 3-15-1 Tokida, Ueda 386-8576, Japan.

*melvin.gan@ums.edu.my

Abstract. Carbon materials were prepared by carbonizing rice husk (RHs) at 2500°C. Few- and multi-layer graphene were obtained from this carbonization process. Barium titanate (BTO) nanoparticles were fabricated by using sol-gel method. Then, the BTO nanoparticles were grafted onto the surface of carbonized rice husk (CRH) to fabricate CRH/BTO nanocomposites. The nanocomposites were characterized using scanning transmission electron microscopy (STEM), transmission electron microscopy (TEM), X-ray diffraction (XRD), Raman measurement, and X-ray photoelectron spectroscopy (XPS). Based on the broadening of (1 1 0) peak from XRD result, the average crystalline size of BTO nanoparticles were calculated to be 16.5 nm. Coexistence of cubic and tetragonal phase of BTO nanoparticles is expected, based on the XRD and Raman results. From XPS result, carbon, barium, titanium, and oxygen peaks were also observed. The combination of CRH with BTO can integrate the properties of these two components to form nanocomposites for broad applications.

1. Introduction

Since the discovery of fullerene in 1985 [1]; the first synthesis in 1976 [2] and the structural characterization in 1991 [3] of carbon nanotubes (CNTs); and the extraction of graphene from bulk graphite in 2004 [4], the interest in carbon materials has rapidly increased. Furthermore, carbon materials possess outstanding properties that allow them to be exploited for various applications and research studies. Not limited to chemical process or reaction to obtain carbon materials, they also can be derived from agricultural waste materials, such as rice husks (RHs), through carbonizing them using heat [5-7]. Moreover, unique structure and morphology of carbon materials derived from waste materials also can be obtained at high temperature [8, 9]. The utilization of waste materials to obtain valuable carbon materials can promotes greener and low cost synthesis process.

Barium titanate (BTO) is one of the widely used materials for electric ceramics and has been applied for wide range of applications [10-12]. Various methods can be carried out to fabricate BTO particles. For instance, Bi et al [13] and Huang et al. [14] used solvent-thermal method to fabricate BTO/CNT compound. Meanwhile, Wang et al. [15] fabricated BTO and carbon black compound by using sol-gel method. Sol-gel method is a broadly used and common fabrication approach, which offer

Content from this work may be used under the terms of the Creative Commons Attribution 3.0 licence. Any further distribution of this work must maintain attribution to the author(s) and the title of the work, journal citation and DOI.
Published under licence by IOP Publishing Ltd

a lot of advantages such as high purity, better homogeneity, lower temperature of preparation, and low manufacturing cost [16, 17].

In this study, the RHs were carbonized at high temperature of 2500°C (CRH). Then, the BTO nanoparticles (sol-gel method) were grafted onto the surface of CRH to fabricate CRH/BTO nanocomposites. Their structure, morphology, elemental composition and crystallization were evaluated and discussed. This type of material can be exploited for various applications, such as electromagnetic wave absorber [10-12], dielectric materials or microelectronic applications [18], energy storage [19], and so on.

2. Experimental

The carbonization of RHs was carried out in two steps. First, the raw RHs were pre-carbonized at 800°C, under the presence of argon gas, for 1 h using furnace. Then, the pre-carbonized RHs were loaded into graphite box, further heated at 2500°C, 0.5 h using graphite resistance furnace operating in a high purity argon gas atmosphere at 760 Torr. The samples in both steps were cooled down naturally and then grounded into powder. RHs carbonized at 2500°C will be denoted as CRH, hereafter.

BTO nanoparticles were prepared by using sol-gel method based on Refs. [10-12], with some modifications, and then CRH were mixed together to fabricate CRH/BTO nanocomposites. Barium acetate $Ba(CH_3COO)_2$, titanium tetraisopropoxide $Ti[(CH_3)_2CHO]_4$, acetic acid, ethanol, and pure water were used as the chemical reagents. $Ba(CH_3COO)_2$ was dissolved in mixed solution of 20 ml ethanol and 3 ml acetic acid (solution 1), stirred in water bath at 60°C for 0.5 h. Simultaneously, $Ti[(CH_3)_2CHO]_4$ was mixed in 10 ml ethanol (solution 2). Solution 1 and 2 were combined and 1 ml pure water was added to perform hydrolysis. Then these two solutions were stirred at 60°C for 1.0 h to form a sol, which contained Ba/Ti molar ratio 1:1. Next 0.2 g CRH were added to the final solution, sonicated for 0.5 h, and stirred furthermore for about 0.5 h, 90°C until gel was formed. The black color gel was then dried at 100°C for 12 h to make an xerogel. Finally, the product was calcined at 800°C for 2 h under the presence of argon gas. Samples for BTO nanoparticles were prepared using similar process excluding the addition of CRH.

The samples were observed using scanning transmission electron microscopy (STEM; Hitachi HD-2300A) and transmission electron microscopy (TEM; JEM-2100F, JEOL) operated at 200 kV. X-ray diffraction (XRD; Rigaku SmartLab) was conducted to characterize the crystal structure. Raman spectroscopy measurement was conducted on a Raman spectrometer (Renishaw inVia Raman spectroscope) with 532 nm laser excitation. X-ray photoelectron spectroscopy (XPS; PHI Quantera II) was performed to characterize the elemental composition.

3. Results and discussions

STEM image is shown in figure 1(a). BTO nanoparticles can be observed to be grafted onto the surface of CRH. Noticeably, the BTO nanoparticles partially bundled up to form a bigger agglomerate. The BTO nanoparticles were dispersed throughout the CRH surface. Magnified TEM images were depicted in figure 1(b) and (c), for clearer observation. The BTO nanoparticles are in spherical shape and also grafted onto the surface of CRH. Figure 1(c) further reveals that the BTO nanoparticle form a single crystal with crystal spacing of ~0.28 nm corresponds to (1 1 0) crystal plane [10-12]. RHs carbonized at 2500°C produced clean surface, clear edges, and relatively high crystallinity of few- and multi-layer graphene. Furthermore, from figure 1(a)~(c), mixture of flattened graphene, corrugated graphene, and partially carbon cage can be observed. Moreover, no amorphous carbon was observed on the graphene surface. This can be attributed to realignment or restructure of vapor carbon at high temperature [9].

XRD patterns of the CRH/BTO are shown in figure 2(a). All peaks of the CRH/BTO can be assigned to the cubic perovskite structure (JCPDS no. 31-0174) [10-12]. Strong intensity of (110) peak can be observed from CRH/BTO at $2\theta = 31.6°$. The typical (002) peak of carbon materials, which in our case CRH, can be detected at $2\theta = 26.2°$. Using the Scherrer's equation as shown below

IOP Conf. Series: Materials Science and Engineering **229** (2017) 012024 doi:10.1088/1757-899X/229/1/012024

$$D = \frac{0.9\lambda}{\beta \cos\theta} \tag{1}$$

where, D is the particle size, λ is the wavelength of the Cu Kα radiation (0.15406 nm), θ is the diffraction angle, and β is the full width at half maximum of the peak, average crystalline size of BTO nanoparticles were calculated to be 16.5 nm from the broadening of BTO/CNT (110) peak. From the TEM images, the diameter of BTO nanoparticles are about 10~15 nm, and this shows high similarity with the calculation using equation 1.

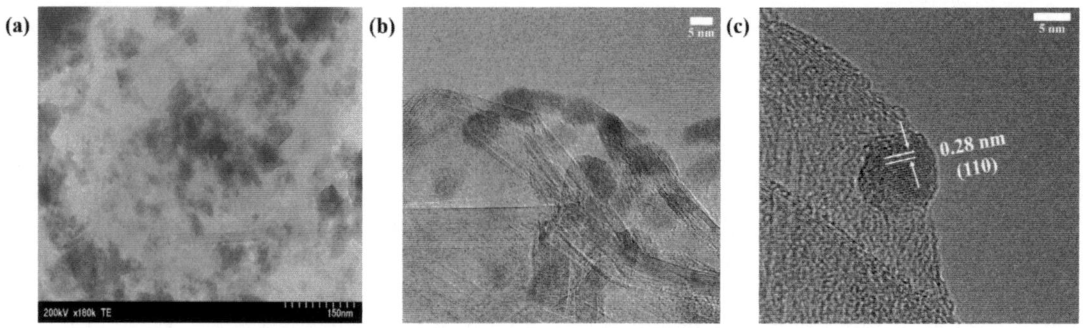

Figure 1. (a) STEM, (b) and (c) TEM images of CRH/BTO nanocomposites.

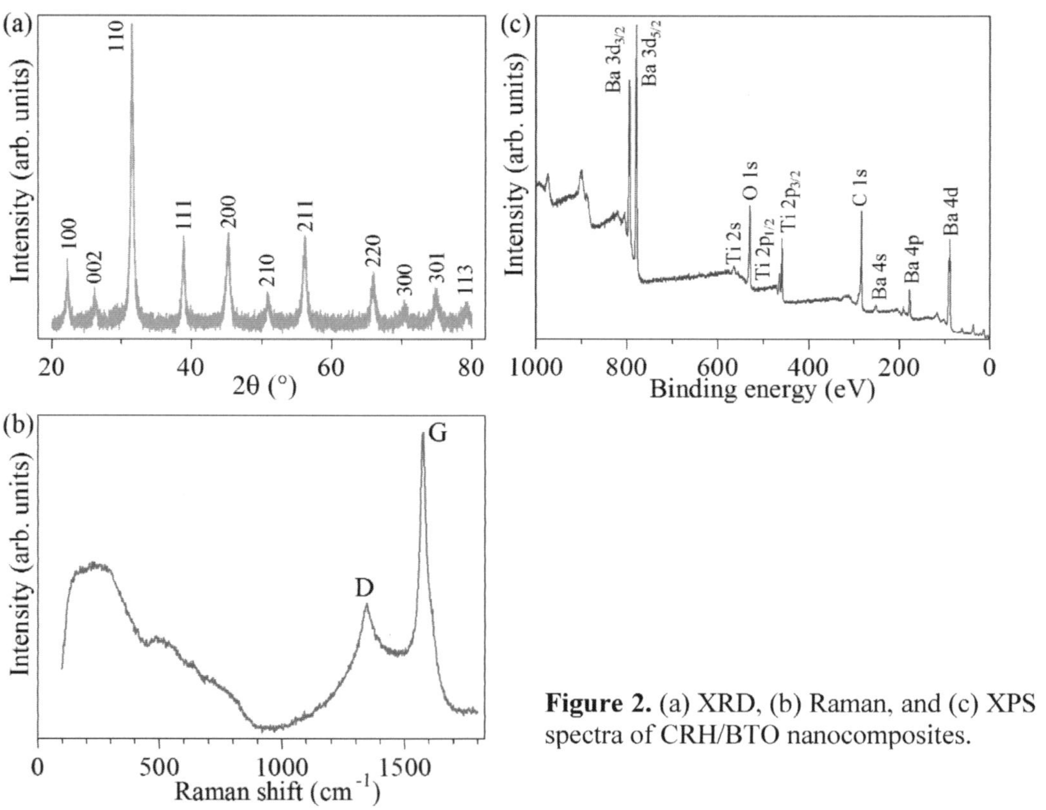

Figure 2. (a) XRD, (b) Raman, and (c) XPS spectra of CRH/BTO nanocomposites.

Raman spectroscopy is a nondestructive and an effective tool to probe structure at the atomic scale on the basis of vibrational symmetry. Generally, cubic BTO has no Raman active modes, but such modes are expected for the noncentrosymmetric tetragonal structure [12, 20]. The Raman spectra of CRH/BTO is depicted in figure 2(b). The Raman peaks at low wavenumber (100~1000 cm^{-1}) can be assigned to more than one phonon mode, which are suggestive of a tetragonal phase [12, 20]. Although XRD patterns reveal a cubic perovskite structure (figure 2(a)), a certain degree of tetragonal phase mixed with cubic phase on the atomic scale is expected [12, 20]. From the XRD and Raman results, coexistence of cubic and tetragonal phase is thinkable. The typical characteristic bands of carbon materials were also observed from CRH/BTO nanocomposites, D band (defect) at ~1345 cm^{-1} and G band (graphite band) at ~1575 cm^{-1}. The relative intensity of the D and G bands (I_D/I_G) obtained from the Raman spectra can be associated to the amount of structural defects. Based on Ref. [12], the I_D/I_G for pristine CNT (d = 20-30 nm) and BTO/CNT is 1.02 and 1.09, respectively. The I_D/I_G for CRH/BTO is 0.6, and relatively lower than those mentioned before. This might be attributed to the high crystallinity of CRH.

XPS evaluation was conducted to determine the elemental composition of CRH/BTO nanocomposites, and the results are illustrated in figure 2(c). The XPS wide survey spectra of the CRH/BTO confirmed the presence of carbon (C 1s) at 284 eV, oxygen (O 1s) at 530 eV, barium (Ba 3d, Ba 4d, Ba 4p, Ba 4s), and titanium (Ti 2p, Ti 2s). The presence of Ba 3d, Ti 2p, and O 1s peaks further proves the success formation of BTO nanoparticles. The Ba $3d_{5/2}$ and Ti $2p_{3/2}$ appeared at 779.2 eV and 458.4 eV, respectively, and are in good agreement with those reported [10, 12, 20]. The atomic percentage ratios for the CRH/BTO are 27.6% for C 1s, 11.7% for Ba 3d, 10.6% for Ti 2p, and 50.1% for O 1s. The atomic percentage ratio of Ba 3d and Ti 2p is about 1:1, similar to the molar ratio of Ba/Ti in the fabrication process. Notably, no other elemental compositions are detected, proving the high purity of CRH/BTO nanocomposites.

4. Conclusions

BTO nanoparticles were fabricated by using sol-gel method and then immobilized onto the surface of RHs carbonized at 2500°C, which consist of few- and multi-layer graphene. Their structure, morphology, elemental composition and crystallization evaluations were carried out through STEM and TEM observation, XRD and Raman measurement, and XPS evaluation. The BTO nanoparticles exist in spherical shape and successfully grafted onto the surface of CRH. The average crystalline size of BTO nanoparticles were calculated to be 16.5 nm, and is in good agreement with the TEM observation. Mixture of cubic and tetragonal phase is expected from the BTO nanoparticles. CRH/BTO nanocomposites can be exploited by utilizing both properties from separate material for various applications.

Acknowledgements

This work was supported by grant (SLB0120-SG-2016) from Universiti Malaysia Sabah (UMS), Malaysia. The authors also would like to acknowledge members of Institute of Carbon Science and Technology, Japan for their technical supports.

References

[1] Kroto H W, Heath J R, O'Brien S C, Curl R F and Smalley R E 1985 *Nature* **318** 162-3
[2] Oberlin A, Endo M, Koyama T 1976 *J. Cryst. Growth* **32** 335-49
[3] Iijima S 1991 *Nature* **354** 56-8
[4] Novoselov K S, Geim A K, Morozov S V, Jiang D, Zhang Y, Dubonos S V, Grigorieva I V and Firsov A A 2004 *Science* **306** 666-9
[5] Guo Y, Yu K, Wang Z and Xu H 2000 *Carbon* **41** 1645-87
[6] Wang L, Wang X, Zou B, Ma X, Qu Y, Rong C, Li Y, Su Y and Wang Z 2011 *Bioresource Technol.* **102** 8220-4
[7] Kumagai S, Sato M and Tashima D 2013 *Electrochim. Acta* **114** 617-26
[8] Saito Y and Arima T 2007 *Carbon* **45** 248-55
[9] Wang Z, Ogata H, Morimoto S, Fujishige M, Takeuchi K, Hashimoto Y and Endo M 2015 *Carbon* **90** 154-9
[10] Melvin G J H, Ni Q-Q and Wang Z 2016 *Phys. Status Solidi A* DOI: 10.1002/pssa.201600541

[11] Ni Q-Q, Melvin G J H and Natsuki T 2015 *Ceram. Int.* **41** 9885-92
[12] Melvin G J H, Ni Q-Q and Natsuki T 2014 *J. Alloy. Comp.* **615** 84-90
[13] Bi C, Zhu M, Zhang Q, Li Y and Wang H 2011 *J. Nanosci. Nanotechnol.* **11** 1030-6
[14] Huang X, Chen Z, Tong L, Feng M, Pu Z and Liu X 2013 *Mater. Lett.* **111** 24-7
[15] Wang G, Chen X, Duan Y and Liu S 2008 *J. Alloy. Compd.* **454** 340-6
[16] Kavian R and Saidi A 2009 *J. Alloy. Comp.* **468** 528-32
[17] Li W, Xu Z, Chu R, Fu P and Hao J 2009 *J. Alloy. Comp.* **482** 137-40
[18] Wan Y-J, Zhu P-L, Yu S-H, Yang W-H, Sun R, Wong C-P and Liao W-H 2017 *Compos. Sci. Technol.* **141** 48-55
[19] Wang G, Huang X and Jiang P 2017 ACS Appl. Mater. Interfaces DOI: 10.1021/acsami.6b14454
[20] Brutchey R L and Morse D E 2006 *Angew. Chem.* **118** 6714-6

IOP Conf. Series: Materials Science and Engineering **229** (2017) 012025 doi:10.1088/1757-899X/229/1/012025

The Prediction Model of Dam Uplift Pressure Based on Random Forest

Xing Li[1,2], Huaizhi Su[1,2] and Jiang Hu[3]

[1] College of Water Conservancy and Hydropower Engineering, Hohai University, Nanjing 210098, China
[2] State Key Laboratory of Hydrology-Water Resources and Hydraulic Engineering, Hohai University, Nanjing 210098, China
[3] State Key Laboratory of Hydrology-Water Resources and Hydraulic Engineering, Nanjing Hydraulic Research Institute, Nanjing 210029, China

370876835@qq.com

Abstract. The prediction of the dam uplift pressure is of great significance in the dam safety monitoring. Based on the comprehensive consideration of various factors, 18 parameters are selected as the main factors affecting the prediction of uplift pressure, use the actual monitoring data of uplift pressure as the evaluation factors for the prediction model, based on the random forest algorithm and support vector machine to build the dam uplift pressure prediction model to predict the uplift pressure of the dam, and the predict performance of the two models were compared and analyzed. At the same time, based on the established random forest prediction model, the significance of each factor is analyzed, and the importance of each factor of the prediction model is calculated by the importance function. Results showed that: (1) RF prediction model can quickly and accurately predict the uplift pressure value according to the influence factors, the average prediction accuracy is above 96%, compared with the support vector machine (SVM) model, random forest model has better robustness, better prediction precision and faster convergence speed, and the random forest model is more robust to missing data and unbalanced data. (2) The effect of water level on uplift pressure is the largest, and the influence of rainfall on the uplift pressure is the smallest compared with other factors.

1. Introduction

As the high dam reservoir and high dam constructed in the complex foundation is increasing, in order to ensure the safe operation of the dam and give full play to the role of water conservancy project, people pay more and more attention to the dam safety monitoring[1]. The monitoring data of deformation, seepage, stress, strain and temperature of the dam and dam foundation can be obtained through the dam safety monitoring. Analysis and comprehensive evaluation of these monitoring data can timely understand the working state of the dam, find hidden dangers and take appropriate engineering measures to avoid the occurrence or expansion of the accident. Analysis of monitoring data reveals the internal relationship between the actual operating state of the dam and the various factors. It provides scientific basis and practical information for the safety of flood control and maintenance of dam and is an important basis of evaluation the dam workability state [2] .

Uplift pressure on the dam foundation refers to the uplift force produced by osmotic pressure and the tail water on the dam foundation which produced by the upstream and downstream head difference.

Content from this work may be used under the terms of the Creative Commons Attribution 3.0 licence. Any further distribution of this work must maintain attribution to the author(s) and the title of the work, journal citation and DOI.
Published under licence by IOP Publishing Ltd

Uplift pressure on dam body and dam foundation can cause damage. It not only has the characteristics of it changes with time and space distribution, and it will also cause the changes of other forces, such as reduce the shear strength of the sliding surface, dissolution the cementing material of dam, erosion and dissolution the rock filling[3] . For concrete dam, the size of uplift pressure of dam foundation can judge the effect of grouting and drainage system. If the design value is exceeded, is a big danger for the dam safety. Dam foundation uplift pressure is an important load in the bottom of the dam. It has obvious influence on the stability, stress and deformation of the dam. It is very important to observe and analyse the uplift pressure of dam foundation due to the effect of upward uplift pressure can decrease the effective weight of the dam, decrease the sliding resistance and threaten the safe operation of the dam.

At present, the uplift pressure prediction models include statistical model, deterministic model, hybrid model, grey model, neural network model and so on. Wu Zhongru analyzed the effect of each component and the established a statistical model of concrete dam foundation uplift pressure[4]; On the basis of three-dimensional seepage finite element calculation and combined with safety monitoring data, Zhang Qianfei et al have put forward the dam seepage coefficient inversion method [5]; Based on the analysis of the influencing factors of uplift pressure, Jiang Yu and Wang Zuqiang established the monitoring and forecasting model of uplift pressure by stepwise regression[6]; Gu Chongshi et al. Studied the application of gray relational grade and fuzzy clustering analysis principle in the prediction of pressure in Xin'anjiang dam based on the pressure observation data[7]; WANG Wei et al. Improved particle swarm optimization for dam foundation pressure prediction by introducing incentive factors and penalty factors[8]; Zhou Jian et al. Using artificial neural network to quantitatively determine the influence ratio of the influence of the pressure on the dam [9]. Rankovic´ V et al. Developed the support vector regression identification model for prediction of dam structural behavior [10]; Mata J. used the artificial neural network and multiple linear regression models to interpret the dam behavior [11]; Demirkaya S. used ANFIS to analyze the deformation of arch dam [12] .

In the previous model, grey prediction model in the system under the condition of stationary time series prediction accuracy is higher, under the system of the volatile situation prediction accuracy is low, and only for small samples to forecast, forecast time is short. The neural network model[13] has the advantages of nonlinear, self-adaptive and robust computing and information processing capabilities. Modelling and control of complex systems can be solved with uncertainty, serious nonlinear and time-varying lag, but this model has shortcomings of slow convergence speed and training time is too long. It is difficult to meet the system requirements in real-time, and in learning with limited samples to ensure accuracy. When learning a large number of samples, the model will fall into "dimension disaster" and the generalization is not high. Statistical forecasting model is a traditional forecasting method, which has a good adaptability, but the choice of the model factor has great randomness. The regression analysis [14] needs more original data, the amount of calculation is large, it is difficult to extrapolate and the prediction accuracy need to be improved. In view of the advantages and disadvantages of these models, this paper introduces the Random Forest (RF) algorithm to predict the dam uplift pressure. The basic idea of RF is multiple weak classifier together to form a strong classifier [15], which play a complementary role between weak classifier, this can reduce the impact of a single classifier error and improve classification accuracy and stability. As a natural nonlinear modelling tool, RF [16] has a very good effect on the prediction of multi variables, thus being applied to many fields. A large number of theoretical and practical examples show that the RF has strong data mining ability and very high prediction accuracy[17][18][19], and has very good tolerance to outliers and noise and not prone to fitting, even is known as one of the best algorithm.

2. The principle of random forest algorithm

Random forest algorithm is a kind of integrated learning algorithm, which is put forward by master Leo Breiman of statistical learning[16]. It is a combination of a number of random trees, each of

IOP Conf. Series: Materials Science and Engineering **229** (2017) 012025 doi:10.1088/1757-899X/229/1/012025

which is independent of each other, and the introduction of randomness in the selection of training samples and the growth of trees in order to reduce the variance of tree structure classifier.

Random forest is a classifier composed of several classification decision trees, and the independent and identically distributed random vectors are used to determine the growth process in each classifier. The final result of the model is determined by the majority vote of all the trees[16].

As shown in Figure 1 and 2, the random forest is a multi-decision tree classifier, a subset of data subsets is randomly sampled from the original training set as the training subspace in each classifier's construction. Then construction on this random subspace, in the process of construction, each of the feature selection is based on the characteristics of random subspace, and feature selection is based on the information gain metric, finally a decision tree model is formed on each bag. Then the final training model can be regarded as a collection of these trees, and the classification result is decided by the classification results of the majority tree model. The characteristics of random forests are mainly concentrated in three aspects: first, the training sample subset is randomly selected, that is the base classifier training set is formed by many times sampling with replacement. Second, the feature subset is selected randomly, that is the feature set is based on random sampling before considering the importance of feature. Third, let all the tree models grow freely, without pruning. Because of the randomness processing of these aspects, the random forest is improved greatly compared with the decision tree. If the sample set has a slightly larger amount of data in the original decision tree, it is easy to form a luxuriant foliage situation, the generated model will be very complex, that will inevitably lead to overfitting of the data. The idea of random forest is to train a decision tree has the decision ability in a certain aspect. The decision tree has no problem of over complexity and over fitting. Generally speaking, it is a weak decision tree, but many aspects of the weak classifier set can form a powerful classifier.

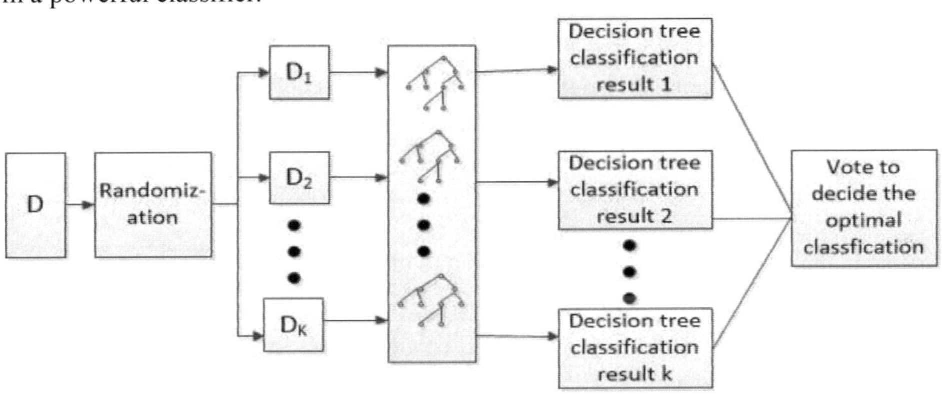

Figure 1. The generation steps of RF

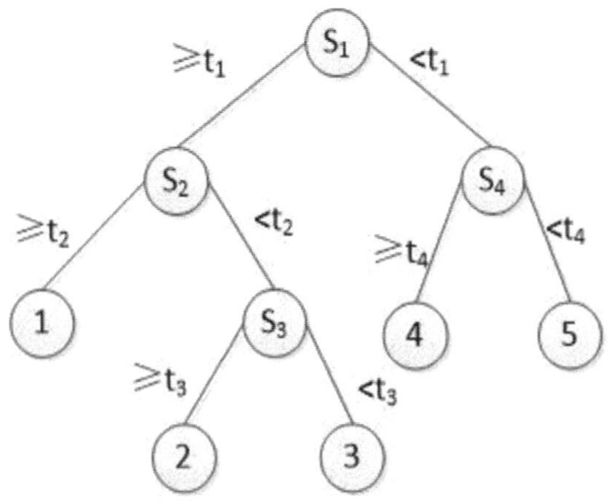

Figure 2. Single risk classification tree

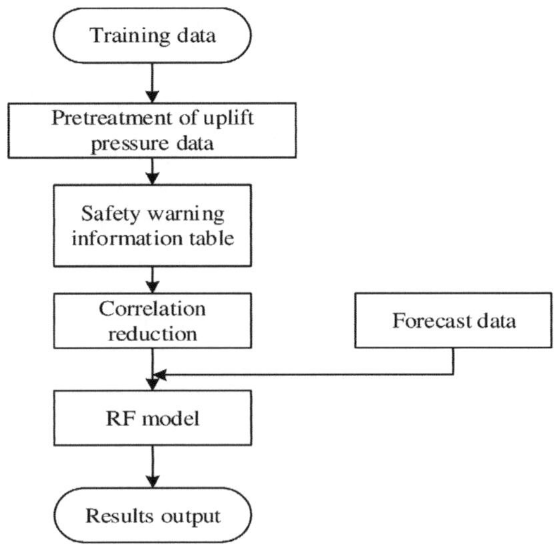

Figure 3. Flow chart of uplift pressure prediction based on random forest

3. Random forest model

Random forest is a classifier composed of a series of tree Classifiers $h(x,\theta_k),k=1,\dots$, here θ_k is an independent and identically distributed random vector, and each tree vote an equal vote to get the most popular class for the input variable x , and k is the number of trees, x is the input sample vector, θ_k the random vector is actually the parameter vector of the k tree, it was determined independently and identically through learning on the basis of independent and identically distributed bootstrap sets. When at the prediction stage the k tree determined by θ_k is used to predict the input vector x . The prediction value y of an input vector x is determined by voting on the output of all the trees $h(x,\theta_k),k=1,\dots$

Random forests can be directly used for regression analysis. The main research of statistical regression analysis is the dependency relationship of a variable to one or multiple independent variables. It is helpful to understand the typical values of dependent variables with independent variables when the other independent variables are fixed. In general, regression analysis is used to estimate the conditional expectation of the dependent variable in the case of a given independent variable. The goal of the estimator is a function called the regression function to describe the independent variables of the probability distribution. In other words, it is assumed that there exists a certain correlation between random variable y and x as a dependent variable. The $F(y|x)$ express the distribution function of y when the x get a certain value, if the regulation of $F(y|x)$ change with the value of x is mastered, then we can fully grasp the relationship between the random variables y and x . It is often not possible to obtain a complete distribution function, as a result we can investigate the mathematical expectation of $F(y|x)$ as an approximation. The mathematical expectation of random variable y (that is, the mathematical expectation of $F(y|x)$)depends on the value of x, which is a function of x, the function is denoted as $u_{y|x}$. It is the regression function about x and y .

The random forest used for regression analysis is obtained by training the tree with random vector θ as parameters. The difference between the random forest for classification and the random forest for regression is that the target of the tree predictor $h(x|\theta_k)$ is a numerical value rather than a class label, the output of forest is numerical value.

For the random forest used for regression analysis, it is assumed that the training set is independent extracted from random variable Y and X. Like all predictor its mean square random error can be expressed as $E_{X,Y}(Y-h(X))^2$.

The random forest predictor is derived from the average value of these trees $h(x,\theta_k)$ about k. So when the number of trees in the forest increases indefinitely, almost everywhere we get

$$E_{X,Y}(Y-av_k h(X,\theta_k))^2 \to E_{X,Y}(Y-E_\theta h(X,\theta))^2 \tag{1}$$

Therefore, the random forest regression function is: $Y=E_\theta h(X,\theta)$, in practice, can be used in the case of sufficiently large k approximation formula: $Y=av_k h(X,\theta_k)$ instead.

At this point, the generalization error of random forest is represented by PE*. The average generalization error of a decision tree is:

$$PE(tree)=E_\theta E_{X,Y}(Y-h(X,\theta))^2 \tag{2}$$

Assuming that for all the θ we can get $EY=E_X h(X,\theta)$, there is

$$PE(forest)\le \overline{\rho}PE(tree) \tag{3}$$

Here $\overline{\rho}$ is the weighted correlation coefficient between the residue of $Y-h(x,\theta)$ and $Y-h(x,\theta')$, θ and θ' are independent of each other. Prove:

$$PE(forest)=E_{X,Y}\left[E(Y-h(X,\theta))\right]^2=E_\theta E_{\theta'}E_{X,Y}(Y-h(X,\theta))(Y-h(X,\theta')) \tag{4}$$

On the right side of the above formula is a convergent series, it can be written in the following form: $E_\theta E_{\theta'}(\rho(\theta,\theta')sd(\theta)sd(\theta'))$

Here $sd(\theta)=\left(E_{X,Y}(Y-h(X,\theta))^2\right)^{\frac{1}{2}}$, the weighted correlation coefficient can be defined as:

$$\overline{\rho}=E_\theta E_{\theta'}(\rho(\theta,\theta')sd(\theta)sd(\theta'))(E_\theta sd(\theta))^{-2} \tag{5}$$

Then we can get: $PE(forest)=\overline{\rho}(E_\theta sd(\theta))^2\le\overline{\rho}PE(tree)$

It shows that the generalization error of random forest is less than $\overline{\rho}$ times than that of the membership tree, by introducing random variables x and y through bagging integration can reduce $\overline{\rho}$ to obtain a better generalization error. The modeling process of the uplift pressure prediction model based on random forest algorithm is shown in Figure 3.

4. Case study

A dam is a concrete gravity arch dam, the maximum dam height is 76.3m. The dam is divided into 28 dam block (from left to right numbered #3~#30). In order to monitor the dam foundation uplift pressure distribution and water blocking effect of dam foundation curtain, 55 uplift pressure holes are fixed up in the dam foundation. In order to verify the feasibility of the proposed algorithm in this paper, the uplift pressure of the dam in the #8 dam block is taken as an example, the proposed method and support vector machine model are respectively used to analyze the uplift pressure. A total of 230 sets of measured data of the effect and the environment were collected from January 2003 to May 2005. The first 200 sets of data are used as the modeling data of the prediction model, and the latter 30 sets of data are used as the prediction interval. Only part of the data was listed in the table.1 due to the limited space of this paper.

Table1. Initial information table of uplift pressure prediction

Date	Upstream water level(m)	Downstream water level(m)	Temperature(°C)	Rainfall(mm)	Uplift pressure(m)
2003/1/6	111.23	58.35	0.6	0	112
2003/1/13	111.01	58.42	8.0	0	112
2003/1/20	110.84	58.49	5.7	0	112
2003/1/27	110.79	58.49	2.1	0	112
2003/2/3	110.78	58.49	4.7	0	112
2003/2/10	110.95	58.49	7.4	10	112
2003/2/17	110.89	60.42	7.0	0	112
2003/2/24	110.03	58.47	8.5	0	110.98
...	...	...	...	...	...

The following 18 parameters [20][21] are selected as the input layer of the RF regression prediction model in this paper: (1) water pressure component H_1: Upstream water level of monitoring day; (2) water pressure component H_2: the average upstream water level of 1~4 days before the monitoring day; (3) water pressure component H_3: the average upstream water level of 5~10 days before the monitoring day; (4) water pressure component H_4: the average upstream water level of 11~20 days before the monitoring day; (5) water pressure component H_5: the average upstream water level of 21~30 days before the monitoring day; (6)water pressure component H_6: Downstream water level of monitoring day; (7) temperature component T_1 : $sin\left(\frac{2\pi it}{365}\right)-sin\left(\frac{2\pi it_0}{365}\right)$;(8) temperature component $T_2 : cos\left(\frac{2\pi it}{365}\right)-cos\left(\frac{2\pi it_0}{365}\right)$; (9) temperature component T_3: $sin\left(\frac{4\pi it}{365}\right)-sin\left(\frac{4\pi it_0}{365}\right)$; (10) temperature component $T_4 : cos\left(\frac{4\pi it}{365}\right)-cos\left(\frac{4\pi it_0}{365}\right)$; (11) rainfall component P_1 : rainfall of monitoring day; (12) rainfall component P_2 : rainfall 1 days before the monitoring day; (13) rainfall component P_3 : rainfall 2 days before the monitoring day; (14) rainfall component P_4 : the average rainfall of 3~4 days before the monitoring day; (15) rainfall component P_5 : the average rainfall of 5~15 days before the monitoring day; (16) rainfall component P_6 : the average rainfall of 16~30 days before the monitoring day; (17) time effect S_1 : $\theta-\theta_0$; (18) time effect S_2 : $ln\,\theta - ln\,\theta_0$.

Model parameter selection: in this paper, the model of dam uplift pressure based on random forest is constructed by calling the random forest package in R. The model mainly needs to set two parameters: (1) n_{tree} is the number of the tree, which represents the number of single tree in the model. The greater value of the n_{tree}, the smaller the fitting effect of the prediction. In general, the n_{tree} is not less than 100. (2) m_{try} is the split nodes of the decision tree of the variable which indicates the number of the candidate variables are selected randomly in each split. The greater the value of m_{try}, the smaller the difference between the sub prediction models. This parameter is important. In general, the default value of the parameter is the variable number for secondary root in the classification model and a third of the number of variables in the regression model. In order to achieve better prediction results, in this paper, we select the optimal values of m_{try} and n_{tree} through successive calculation in the process of model building. The obtained results through calculation shown in Fig.4 and Fig.5.

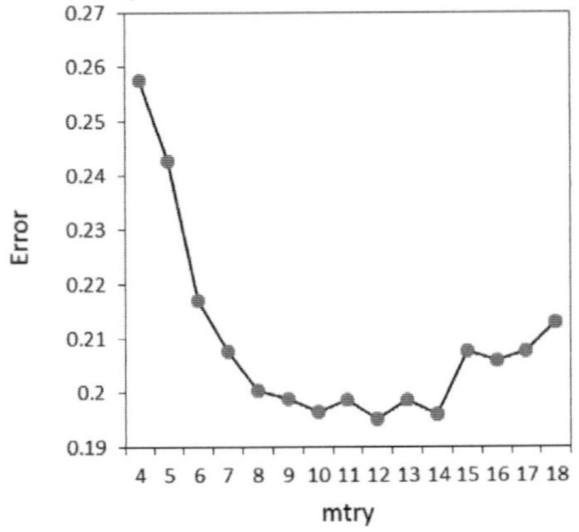

Figure 4. Relationship between m_{try} value and mean square deviation

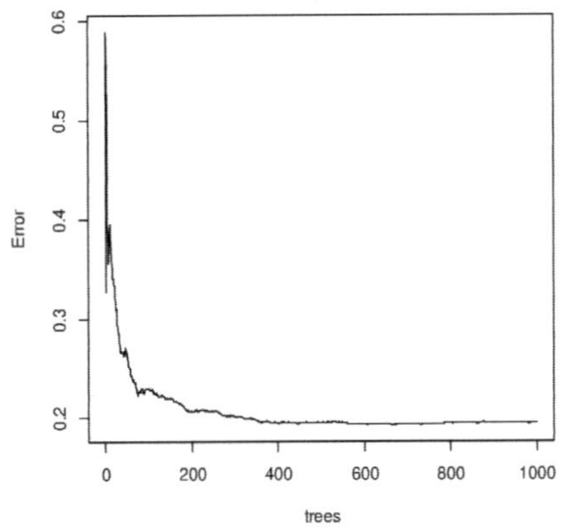

Figure 5. Relationship between n_{tree} value and OOB error

From the figure, it is known the error is small when m_{try}=12, $n_{tree}=400$, so the parameters of prediction model take the above value to calculate.

In order to evaluate the accuracy of RF in dam safety monitoring of uplift pressure prediction, the model of support vector machine (SVM) is selected as a comparison. SVM has a very strong nonlinear processing ability[2][10][22] , is also a kind of efficient and reliable artificial intelligence algorithm, the comparison of the two models are more meaningful. Through the calculation of R, the process graph of learning and prediction effect of random forest model is shown in figure 6, and the residuals graph of the RF model is shown in figure 7; the process graph of learning and prediction effect of SVM model is shown in figure 8, and the residuals graph of the SVM model is shown in figure 9. The degree of importance of the independent variables of the RF prediction model calculated by the built-in function is shown in figure 10.

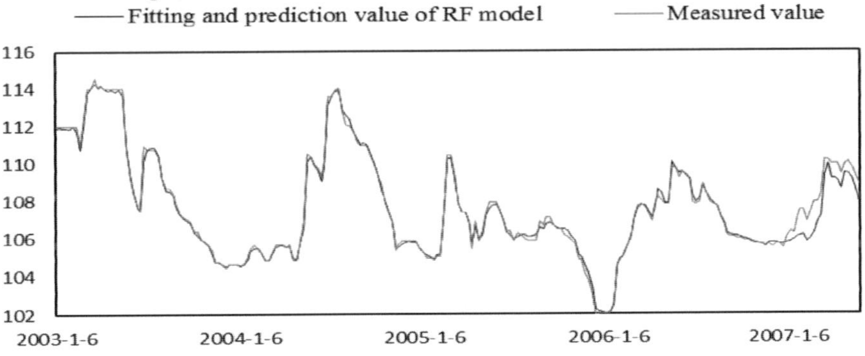

Figure 6. Fitting prediction process line of RF model

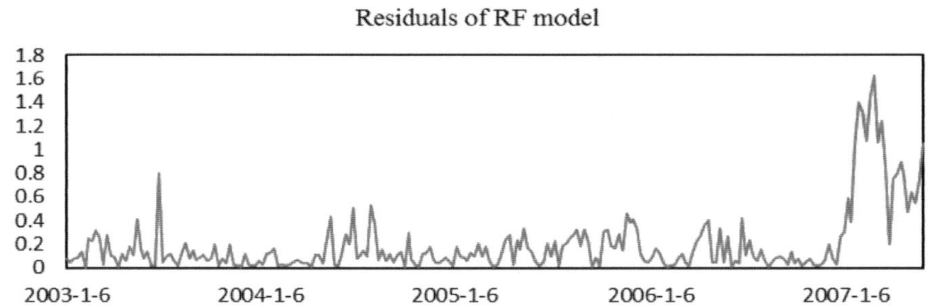

Figure 7. Fitting prediction residuals of RF model

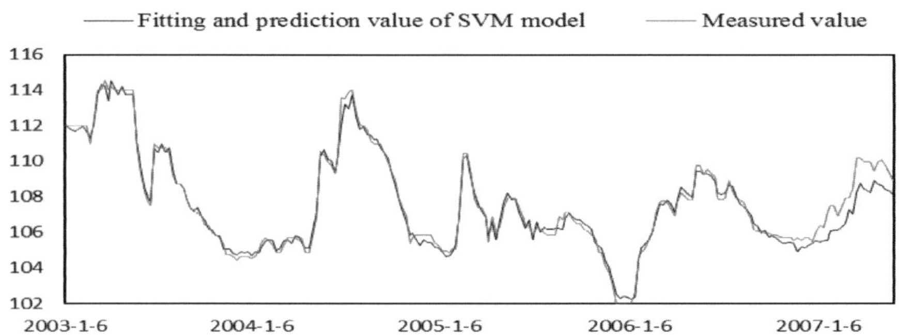

Figure 8. Fitting prediction process line of SVM model

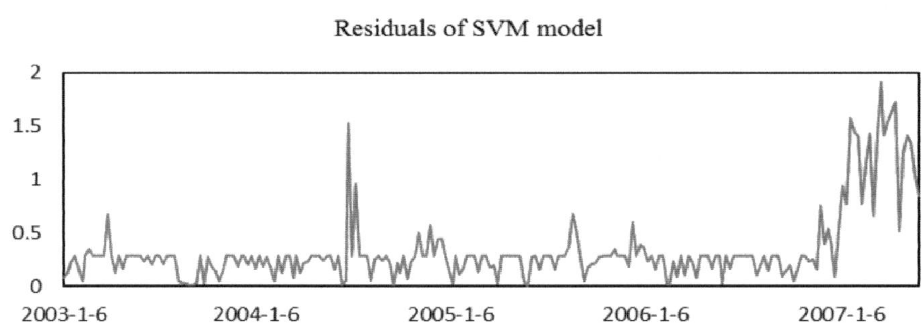

Figure 9. Fitting prediction residuals of SVM model

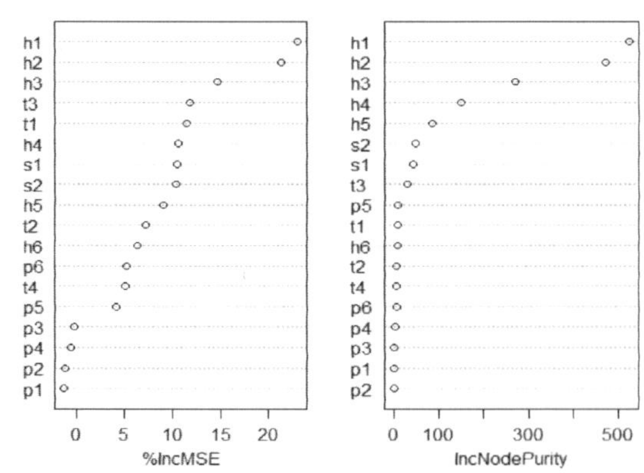

Figure 10. Independent variable importance degree of RF model

In this paper, we compare the fitting and prediction accuracy of the two models by using the mean square error σ of the residuals(s). The expression of σ [23] is shown in equation (6), the results of comparison is shown in table 2.

$$\sigma=\left(\frac{1}{n}\sum_{i=1}^{n}\left(y_i-\overline{y_i}\right)^2\right)^{\frac{1}{2}} \tag{6}$$

n is the number of samples, y_i is the monitoring data, $\overline{y_i}$ is the calculation value of the model.

Table2. Comparison of model residual mean square error

Time interval	Mean square error of model	
	RF model	SVM model
Mean square error of fitting time	0.175	0.292
Mean square error of prediction time	0.817	1.109

it can be drawn from the figures and table above:

(1) It can be drawn from the importance degree figure of variables: The upstream water level has the greatest influence on uplift pressure, the upstream water level rises, the uplift pressure increases, the upstream water level drops, the uplift pressure decreases. The effect of rainfall on uplift pressure is the minimum, this is also consistent with the actual situation. However, due to the lack of the measured data of the non-load factors, in this case only considered the influence of load factors. If the non-load factors are taken into account, the prediction results will be closer to the actual situation.

(2) The residual of the SVM model is larger than that of RF model from the fitting and prediction effect of the two models. As can be seen from table 2, the mean square deviation of RF model is less than the corresponding value of SVM model, which indicates that the RF model has a better fitting accuracy. Moreover, the prediction mean square error of the RF model is smaller than that of the SVM model, which shows the RF model has a better prediction effect.

Footnotes should be avoided whenever possible. If required they should be used only for brief notes that do not fit conveniently into the text.

5. Conclusion

In this paper, the random forest algorithm is applied to the uplift pressure prediction model of dam safety monitoring, construction of uplift pressure prediction model by RF algorithm, to explore the new way to predict the uplift pressure, in order to provide reference for dam safety management, normal operation of the dam and disaster early warning. It can be drawn from the practical examples and analysis:

(1) The uplift pressure prediction model based on random forest, the training error is taken as the constraint condition of the optimization problem, and the minimum confidence range is used as the optimization objective, the theoretical basis is clear, and the generalization ability is obviously superior to other machine learning methods such as support vector machine. It can capture the global nonlinear dynamic characteristics of the uplift pressure in the process of dam work, and obtain excellent results in fitting and forecasting applications.

(2) Random forests method also can be used to pre-treatment of dam safety monitoring data, it can extract the implicit rules of between independent variables and dependent variable depends on the model of the random forest, and extract the independent variables that have great effect on the dependent variables in the case of large data sets, remove the independent variables that have little effect to simplify the input variables and improve the computational efficiency.

(3) Due to the lack of measured data of non-load factors, only the influence of load factors is considered in the process of building an uplift pressure prediction model. If the measured data of non-

load factors can be obtained in the future, in the process of modelling, the non-load factors can be taken into account, the model can get a higher prediction precision, and the warning effect of the uplift pressure prediction model will be greatly improved.

Acknowledgements
This research has been partially supported by National Natural Science Foundation of China (SN: 51579083, 51409167, 41323001).

References
[1] Wu Zhong-ru, W.A.H.S., Dam health diagnosis and evaluation. *Smart Materials and Structures*, 2005. **14(3)**: p. S130.
[2] Su, H., Z. Chen and Z. Wen, Performance improvement method of support vector machine-based model monitoring dam safety. *Structural Control and Health Monitoring*, 2016. **23(2)**: p. 252-266.
[3] Zhu, K., et al., The Analysis of the Concrete Gravity Dam' s Foundation Uplift Pressure under the Function of Typhoon. *Mathematical Problems in Engineering*, 2016. **16**: p. 1-9.
[4] Wu Zhong-ru. Safety Monitoring Theory & Its Application of Hydraulic Structures. *Higher Education Press*, 2003.
[5] Zhang Qian-fei, Wang Jian, Wu Zhong-ru. Divisional back analysis of dam percolation parameter based on artificial neural network model. *Water Resources and Power*. 2001, **19(4)**: 4-7.
[6] Jiang Yu, Wang Zu-qiang, Li Yu-qi. Data analysis method of uplift pressure monitoring for concrete gravity dam. *Pearl River*, 2011, **32(2)**: 47-50.
[7] Jia Cai-hong,Gu Chong-shi. Cluster analysis-based new method for prediction of uplift pressure of dams. *Journal of Hohai University (Natural Sciences)*, 2002, **30(3)**: 110-112.
[8] Wang Wei, Shen Zhen-zhong. Prediction method of dam foundation pressure based on improved particle swarm optimization algorithm. *Yellow River*, 2009 (7): 121-123.
[9] Zhou Jian, Song Han-zhou. Application of improved neural network model into analysis of uplift pressure under dam-foundation. *Hydrogeology and Engineering Geology*, 2005, **32(3)**: 38-41.
[10] Ranković, V., et al., Development of support vector regression identification model for prediction of dam structural behaviour. *Structural Safety*, 2014. **48**: p. 33-39.
[11] Mata J. Interpretation of concrete dam behaviour with artificial neural network and multiple linear regression models. *Eng Struct* 2011;**3 (3)**:903–10.
[12] Demirkaya S. Deformation analysis of an arch dam using ANFIS. In: Proceedings of the second international workshop on application of artificial intelligence and innovations in engineering geodesy, *Braunschweig*, Germany; 2010. p. 21–31.
[13] Kao, C. and C. Loh, Monitoring of long-term static deformation data of Fei-Tsui arch dam using artificial neural network-based approaches. *Structural Control and Health Monitoring*, 2013. **20(3)**: p. 282-303.
[14] De Sortis, A. and P. Paoliani, Statistical analysis and structural identification in concrete dam monitoring. *Engineering Structures*, 2007. **29(1)**: p. 110-120.
[15] Breiman, L., Manual on setting up, using, and understanding random forests v3. 1. Statistics Department University of California Berkeley, CA, USA, 2002. 1.
[16] Breiman, L., Random Forests. *Machine Learning*, 2001. **45(1)**: p. 5-32.
[17] DONG, L., X. LI and K. PENG, Prediction of rockburst classification using Random Forest. *Transactions of Nonferrous Metals Society of China*, 2013. **23(2)**: p. 472-477.
[18] Chen, X. and H. Ishwaran, Random forests for genomic data analysis. *Genomics*, 2012. **99(6)**: p. 323-329.
[19] Mihai Mihailescu, D., Computer aided diagnosis method for steatosis rating in ultrasound images using random forests. *Medical Ultrasonography*, 2013. **15(3)**: p. 184-190.
[20] Su, H.Z., J. Hu and Z.R. Wu, A study of safety evaluation and early-warning method for dam global behavior. *Structural Health Monitoring*, 2012. **11(3)**: p. 269-279.
[21] Su, H., et al., Time-varying identification model for dam behavior considering structural reinforcement. *Structural Safety*, 2015. **57**: p. 1-7.
[22] Kuang, F., W. Xu and S. Zhang, A novel hybrid KPCA and SVM with GA model for intrusion detection. *Applied Soft Computing*, 2014. **18**: p. 178-184.
[23] Su, H., Z. Wen and Z. Wu, Study on an Intelligent Inference Engine in Early-Warning System of Dam Health. *Water Resources Management*, 2011. **25(6)**: p. 1545-1563.

IOP Conf. Series: Materials Science and Engineering **229** (2017) 012026 doi:10.1088/1757-899X/229/1/012026

A Simulation Modeling Approach Method Focused on the Refrigerated Warehouses Using Design of Experiment

G S Cho

Department of Port Logistics System, Tongmyong University, 428, Sinseon-ro, Nam-gu, Busan 48520, Republic of Korea.

gscho@tu.ac.kr.

Abstract. For performance optimization of Refrigerated Warehouses, design parameters are selected based on the physical parameters such as number of equipment and aisles, speeds of forklift for ease of modification. This paper provides a comprehensive framework approach for the system design of Refrigerated Warehouses. We propose a modeling approach which aims at the simulation optimization so as to meet required design specifications using the Design of Experiment (DOE) and analyze a simulation model using integrated aspect-oriented modeling approach (i-AOMA). As a result, this suggested method can evaluate the performance of a variety of Refrigerated Warehouses operations.

1. Introduction

There has been growing interest in improving the operating efficiency of warehouses, which is one of the key facilities used in the logistics sector. Such interest reflects the intention of the sector in increasing its competitive edge, reducing costs, and improving services as a way to bolster the competitiveness of logistics companies [1]. Warehouses are core facilities for undertaking integrated logistics operations comprising the storage, loading, and unloading of goods, and operations of information systems. They are used to increase added value through expanding the function and role of storage facilities and to ensure flexibility in supply. In particular, refrigerated warehouses (i.e., insulated warehouses with refrigerating systems) have become more important owing to the changes in and expansion of the nation's food consumption practices. Although cold-storage warehouses have been investigated from various perspectives, studies on the systems needed for the conversion of these warehouses into refrigerated distribution centers are still limited. Based on the comparison of the costs mobilized for building a non-refrigerated warehouse against the costs for a refrigerated warehouse, and the comparison of procurement and operating costs for a non-refrigerated vehicle and a refrigerated vehicle, the study analyzed the factors contributing toward the high-cost structure of refrigerated logistics systems [2]. Previous studies have addressed ways to bring about advancements in the cold chain systems and accordingly presented simulation models [3-4]. However, there should be more suggestions on how to build a simulation model that takes into consideration the characteristics of refrigerated warehouses. Moreover, there should be more proposals on how to increase the efficiency of the operating system needed for converting the warehouses into refrigerated distribution centers [5]. This study aimed to analyze the operating system so the cold-storage warehouses can be converted into refrigerated distribution centers. The purpose of the study was to propose a standard platform for the analysis of refrigerated distribution centers methodology, based on applied test planning methods, and the operation of such distribution centers.

Content from this work may be used under the terms of the Creative Commons Attribution 3.0 licence. Any further distribution of this work must maintain attribution to the author(s) and the title of the work, journal citation and DOI.
Published under licence by IOP Publishing Ltd

2. Current status of refrigerated warehousing in Korea

An investigation into the present status of refrigerated warehouses in South Korea reveals the existence of 804 such facilities that are being operated by 120 businesses across the country, of which about 15% warehouses are located in Busan that handle more than 50% of the nation's refrigerated capacity [1]. The figure indicates that the city has large-scale cold storage facilities and they play an important role. This is because, in general, the closer the facilities are to import hubs of marine products, the greater is the efficiency they can offer to those implementing the distribution processes. Therefore, the greatest number of refrigerated warehouses are located and currently under operation in Busan. However, the city's cold storage facilities are also being operated with an emphasis on storage function, as previously mentioned. Of the 120 cold-storage operators in South Korea, 32 companies (or 27%) have completed at least 31 years of operation, thus having advanced in age. Thus, it is necessary for the local refrigerated distribution processes to revitalize the operations of cold storage logistics completely to ensure improvement. Until now, only a very small number of studies have been conducted on the conversion of cold storage warehouses into refrigerated distribution centers. The results show that the South Gyeongsang Province ranks the highest with its 178 cold-storage companies, followed by Busan, Seoul and Gyeonggi Province, and South Jeolla Province with 120, 102, and 93 cold storages, respectively; additionally, 14 new refrigerated warehousing companies have entered the market in South Jeolla Province alone in the past year, followed by 4 in the Incheon area, 3 in the Daegu and South Gyeongsang Province, respectively, and 1 in the South Chungcheong Province. Busan is the only city showing a net reduction of one.

3. Operation of Cold Chain Systems

As illustrated in figure 1, a cold chain system refers to a temperature-controlled distribution system designed to ensure the quality and safety of products throughout the process of producing, storing, transporting, marketing, and consuming temperature-susceptible items, such as agricultural, livestock, marine products, foods, flowers and plants, and medicines. The cargo refrigeration system aims at achieving extended shelf life for retaining the freshness and value of products by preserving them at an appropriate temperature(s). A cold chain system is divided into the main part (including precooling, packaging, transporting, storage, and low-temperature facilities near consumers) and supplementary part (other functions) [2].

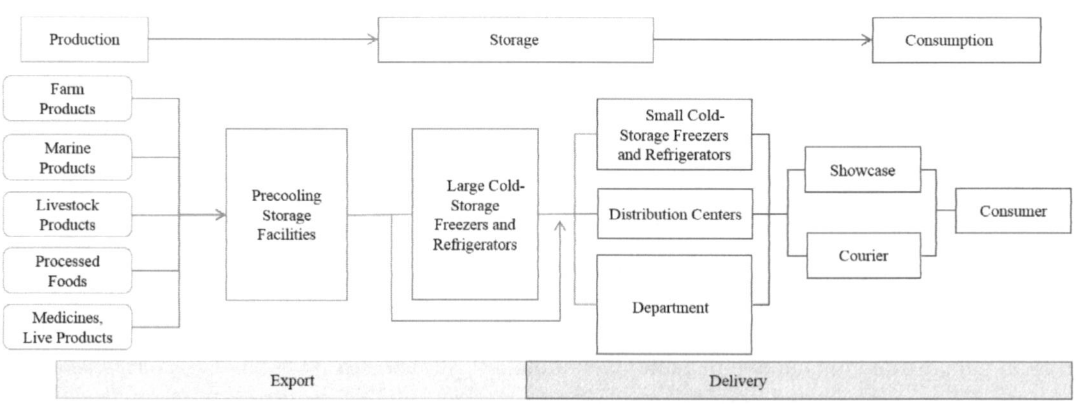

Figure 1. Definition of cold chain system.

This study proposed a platform for an operating system for refrigerated distribution centers that are integrated with a cold chain system.

4. Analysis of refrigerated distribution center design

This study first implemented a crosscutting concern analysis of the refrigerated distribution center's operating system design by transmitting each piece of information for each class and executing the feedback. When it comes to building a simulation model, an object-oriented model is implemented vertically, thereby facilitating an efficient representation of the simulated design and model construction [6]. The downside of object-oriented simulation designs and models is that the exclusive consideration of vertical elements might lead to a failure in modularizing the crosscutting concern existing between the modules because the sole emphasis is being placed on vertical elements. Based on the previous theories, this study (a) built a simulation model that modularizes common areas by utilizing the i-AOMA and (b) achieved redundancy elimination, modification of the established model, and an increased efficiency during reuse in figure 2 [7].

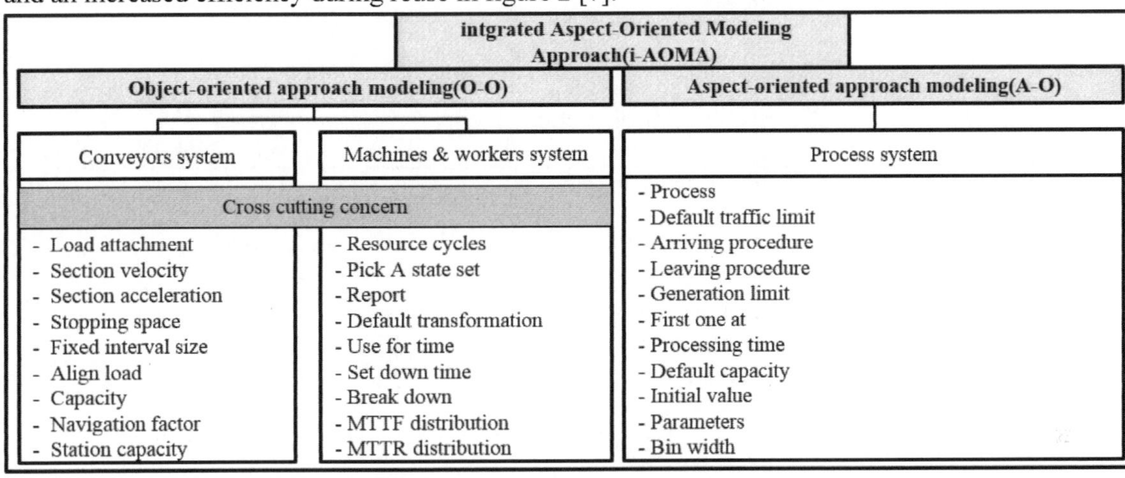

Figure 2. Concept of the i-AOAM.

The task of refrigerated warehouse simulation modeling was divided into two parts: modeling the operating equipment system and warehouse operating system. The division was followed by the implementation of inheritance-class creation and inheritance within the operating equipment system, which was materialized by the object-orientation technique. Furthermore, many output variables are available for the simulation model while designing refrigerated distribution centers based on design parameters obtained through the i-AOAM. However, the most important among them is the operating efficiency of the refrigerated distribution centers. Hence, the distribution centers' storage rate was selected as the output parameters for determining the optimum way to operate. Concerning the determination of the values of input parameters for obtaining the best output parameters through simulation, a significant amount of time and cost is required as the number of input parameter combinations increases, which leads to a significant increase the number of areas to be explored. Thus, an efficient test plan was implemented as a solution. This study chose to apply the Taguchi methodology to reduce the number of simulation tests to be performed and to estimate the optimal input parameters. The most prominent features of the methodology include the test planning based on the use of orthogonal arrays and determination of optimal condition based on the analysis of SN ratio [8-9]. For designing a refrigerated distribution center, this study utilized the 'A Center' located in the Busan area as the focal point of investigation. The automated cold-storage warehouse of the A Center is integrated with an automated storage and retrieval system (AS/RS) and a cold chain system.

Table 1 shows seven input/output parameters established for the refrigerated distribution center. Table 2 summarizes the test values that were used for simulations of the distribution center.

Table 1. Input/output parameters for the refrigerated distribution centers.

Input parameters	Range of parameter values
Speed at which forklifts drive within the warehouse premise	20 km/h–40 km/h
Speed at which forklifts drive within the	20 km/h–40 km/h

loading-dock premise	
Unit transport capacity of forklifts (no. of pieces)	1–3
The no. of forklifts operating within the premise of loading dock	1–3
No. of transport units	1–3
Cold-storage capacity by region (no. of pieces to be refrigerated)	10–20

Table 2. Simulation level-values of refrigerated distribution centers.

Control Factor	Details of Control Factor	Level (Level Values)
A	Refrigerating capacity	Level 6 (10, 12, 14, 16, 18, 20)
B	Number of forklifts within the premise of the warehouse	Level 3 (1, 2, 3)
C	Speed at which the forklifts run within the premise of warehouse	Level 3 (20, 30, 40)
D	Speed at which the forklifts run within the premise of the loading dock	Level 3 (1, 2, 3)
E	Speed at which the forklifts run within the premise of the loading dock	Level 3 (20, 30, 40)
F	Number of transportation systems/equipment	Level 3 (1, 2, 3)

The refrigerated distribution center's storage ratio has a larger-the-better property, with the signal-to-noise (SN) ratio computed in Equation (1) as follows:

$$SN_i = -10\log\left\{\frac{1}{n}\sum_{j=1}^{n}\frac{1}{y_{ij}^2}\right\} \tag{1}$$

The orthogonal array in line with this study is $L_{12}(4^1 \times 3^6)$. Using this array will require only 12 tests to be conducted. In the $L_{12}(4^1 \times 3^6)$array, A (Level-4 factor) was placed in Row 1, while the remaining Level-3 factors (B through G) were placed in Rows 2 through 7, respectively. Under the 12 actual test conditions, the storage rate is computed to be the ratio between (a) the actual storage volume in the refrigerator throughout the simulation process, to (b) the refrigeration capacity of the cold-storage cargo, as shown in Equation (2) below:

Storage rate = Actual storage volume/refrigeration capacity (2)

The model built to calculate the storage rate was subjected to a simulation that lasted 365 days, with each day comprising 8 hours. After 2 weeks (or 80 hours) of simulation, the system was found to have stabilized. The main effect diagram showed that the total refrigeration capacity of the warehouse increased as the rack-specific storage capacity increased from 20 to 40, causing the within-warehouse storage rate to decrease. The results also showed that an increase in the number of forklifts from 3 to 6 within the premise of warehouse led to a decrease in the storage rate due to the interference between the forklifts. Contrarily, an increase in the number of forklifts operating inside the loading dock was found to have led to an increase in storage rate. Concerning the number of transportation systems, three systems enjoyed greater storage rate than two systems; however, four systems showed an opposite trend. Furthermore, the analysis showed that no significant effects were exerted on the storage rate by either the speed of forklifts operating inside the warehouse or the ones inside the loading dock. Analysis of Variance (ANOVA) was performed to determine the statistical significance of the control factors. Tables 3 and 4 summarize the results of ANOVA and the optimal conditions for control factors, respectively.

Table 3. ANOVA for SN ratio.

Source	DF	SS	MS	F	P
B	4	5.66264	2.31194	156.36**	0
C	3	0.34235	0.09993	14.85*	0.009
D	2	0.29837	0.08182	11.21*	0.031
G	3	0.15235	0.0573	5.28*	0.1
Error	3	0.05592	0.0083		
Total	15	6.51163			

Table 4. Optimal conditions for control factors.

Control Factor	Level Value
Storage capacity (B)	30
Speed at which the transportation equipment moves within the premise of warehouse (C)	45
The number of RGVs inside the automated refrigerated warehouse (D)	4
The number of transportation equipment (G)	3

Based on the results, the design control factors for the refrigerated distribution center were applied to a cold-storage distribution center located in the Busan area. The center, a 7-storied building (1-storey in basement and 6-storey above the ground), has a storage capacity of approximately 31,000 pallets, gross floor area of 24,380 m2, and operates facilities such as freight elevators, passenger elevators, and emergency power generators. The center has a nominal capacity of 90,000 tons. It is equipped with eight vertical conveyors that move the cargo between the loading dock located on the first floor and the automated freezers and refrigerators in the cold storage areas in the upper floors. Such system design allows the handling of 2,000 pallets, per day. Compared to the data from other refrigeration systems, it was found that this center's storage capacity decreased by 50%, but its processing time increased by 50%.

5. Conclusion

This study aimed to propose an efficient way of designing refrigerated warehouses, considering the diverse operating and storage methods of automated cold-storage warehouses and based on the characteristics of the cargo to be refrigerated. The Taguchi methodology was utilized for designing such a system to construct and analyze the refrigerated warehouse simulation model. Specifically, the study aimed to propose a way to revitalize the operations of refrigerated warehouses in the Busan area. As the first step toward achieving this goal, the operating process to be applied within the warehouse was examined and the results were used for designing a simulation model. This application was followed by the performance of simulation analysis, based on the results obtained by using the Taguchi methodology, and the optimal plans for operating the warehouses were established. The results also revealed that the areas storing mostly multiple accounts of an identical product had a greater average process volume compared to that of other areas with dissimilar qualities. These areas also had an extremely high transportation equipment operation rate inside the warehouse. Based on the aforementioned, additional investments for new warehouses and new warehouse construction were not found to be essential. However, effective mobilization of transportation equipment inside the warehouse was found crucial. The investigative approach considered by this study can offer advantages, such as reduced time in not only designing and operating refrigerated warehouses but also undertaking the redesign. In terms of simulation implementing techniques, this study carried out an examination of interfaces between the modules by using object-oriented techniques based on the re-utilization of the integration-oriented modules developed here. The examination allowed the study to design and materialize an operating model that took into consideration the operating characteristics of refrigerators dealing with a wide range of cold-storage cargo. The results were obtained from the establishment of additional systems, facility expansion, and the resulting changes in input data. The results were also obtained through the easy establishment of a simulation model when the system was subjected to reconfiguration. Future studies could use the response surface methodology (RSM) to estimate the optimal combination(s) of various implementation conditions that are needed for building and operating a simulation model. The combination(s) estimated might be utilized when investigating a way to operate the facilities actively, given an increase in the level of practical challenges.

Acknowledgements

This work was supported by the National Research Foundation of Korea(NRF) grant funded by the Korea government(MSIP) (No. 2015R1D1A1A01061115).

References

[1] Cho G. S and Kim H. G 2013 *Int. J. Industrial Engineering: Theory, Applications and Practice* **19** 24
[2] Cho G. S, Lee S. H, Lee D. W and Kim H. G 2011 *J. Korea Manage. Engineers Society* **16** 115
[3] Han G. S 2015 *Korea Logistics Review* **25** 65
[4] Ryou O. H, Kang Y. S and Lee Y. H 2015 *J. Korean Institute of Industrial Engineers* **41** 173
[5] Thomas B and Jarot S. S 2017 *IOP Conf. Series: Materials Science and Engineering* **190** 012012
[6] Kuznetsova N S, Yudin A S and Voitenko N V 2017 *J. of Physics: Conf. Series* **830** 012046
[7] Cho G. S and Kim M. S 2015 *ICIC Express Letters* **9** 1361
[8] Brightman H 1978 *J. Decision Science* **69** 166
[9] Myers R 2009 *Response Surface Methodology: Process & Product Optimization using designed experiments* New York Wiley

IOP Conf. Series: Materials Science and Engineering **229** (2017) 012027 doi:10.1088/1757-899X/229/1/012027

Manufacturing a Porous Structure According to the Process Parameters of Functional 3D Porous Polymer Printing Technology Based on a Chemical Blowing Agent

C J Yoo[1], B S Shin[2,3,*], B S Kang[1], D H Yun[1], D B You[3] and S M Hong[4]

[1]Interdisciplinary Department for Advanced Innovative Manufacturing Engineering, Pusan National University, 63-2, Jangjeon-dong, Geumjeong-gu, Pusan, South Korea
[2]Department of Optics and Mechatronics Engineering, Pusan National University, 63-2, Jangjeon-dong, Geumjeong-gu, Pusan, South Korea
[3]Department of cogno-mechatronics Engineering, Pusan National University, 63-2, Jangjeon-dong, Geumjeong-gu, Pusan, South Korea
[4]Convergence Research Center of 3D Laser-Aided Innovative Manufacturing Technology, Pusan National University, 63-2, Jangjeon-dong, Geumjeong-gu, Pusan, South Korea

bosung@pusan.ac.kr

Abstract. In this paper, we propose a new porous polymer printing technology based on CBA(chemical blowing agent), and describe the optimization process according to the process parameters. By mixing polypropylene (PP) and CBA, a hybrid CBA filament was manufactured; the diameter of the filament ranged between 1.60 mm and 1.75 mm. A porous polymer structure was manufactured based on the traditional fused deposition modelling (FDM) method. The process parameters of the three-dimensional (3D) porous polymer printing (PPP) process included nozzle temperature, printing speed, and CBA density. Porosity increase with an increase in nozzle temperature and CBA density. On the contrary, porosity increase with a decrease in the printing speed. For porous structures, it has excellent mechanical properties. We manufactured a simple shape in 3D using 3D PPP technology. In the future, we will study the excellent mechanical properties of 3D PPP technology and apply them to various safety fields.

1. Introduction

Recently, interest in 3D printing technology is rapidly increasing in the field of future manufacturing technologies. Three-dimensional printing technology is being applied to various fields, such as the medical field and tissue engineering, because of various benefits [1-2]. We have been conducting research on a new porous polymer printing (PPP) technology applications based on the 3D printing technology [3-4]. The porous polymer has excellent mechanical properties compared to conventional polymers [5-6]. These characteristics of the porous polymers are utilized in various fields such as functional materials, inner buffer, etc. [7-8].

In this study, we produced chemical blowing agent (CBA) hybrid filament by combining polymers and CBA, and we suggested a 3D PPP technology based on CBA for a process that is uncomplicated,

Content from this work may be used under the terms of the Creative Commons Attribution 3.0 licence. Any further distribution of this work must maintain attribution to the author(s) and the title of the work, journal citation and DOI.
Published under licence by IOP Publishing Ltd

and reduces time and expense. In addition to this technology, we manufactured a simple 3D structure and experimentally confirmed the variation in porosity according to the process parameters.

2. Background

2.1. CBA-based 3D PPP process

Three-dimensional PPP refers to the layering process which simplifying processes, and forming porosity. In other words, it refers to the technology of stacking and manufacturing products with one process in a single piece of equipment by the thermos-chemical foaming process of thermoplastic polymer material. This process has the advantage of being relatively simple, is less time-consuming, and inexpensive. Figure 1 shows the overall process of the CBA-based 3D PPP. The CBA hybrid filament is mixed with a polymer and a CBA is manufactured; next, the structure is stacked as the pores were formed by the thermal decomposition reaction of the CBA.

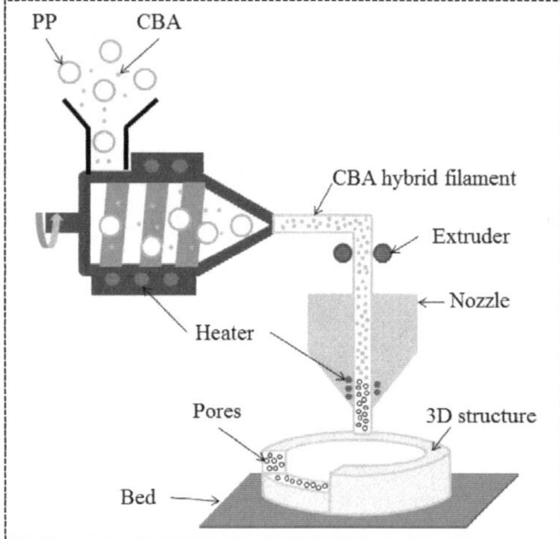

Figure 1. The overall process of CBA-based 3D PPP (porous polymer printing).

2.2. Blowing Agent

Blowing agents are largely classified into physical blowing agents (PBAs) and CBA. PBAs supply gas by applying a physical force to the molten polymer and changing the physical state. Generally, it is used for gases such as nitrogen, dichlorotetrafluoroethane, and n-pentane [9]. On the other hand, CBAs generally supply the gas by changing the chemical condition through a thermal decomposition reaction. The CBA is easy to introduce into polymers and has the advantage of facilitating the process in general equipment [10]. The endothermic CBA absorbs heat continuously during disassembly, but the exothermic CBA releases heat during disassembly [11]. Figure 2 shows the chemical reaction equation of CBA. The CBA used in the experiment is a yellow powder with a formula $H_4N_4C_2O_2$, and the porous structure is formed by releasing the gases N_2 and CO_2 during disassembly.

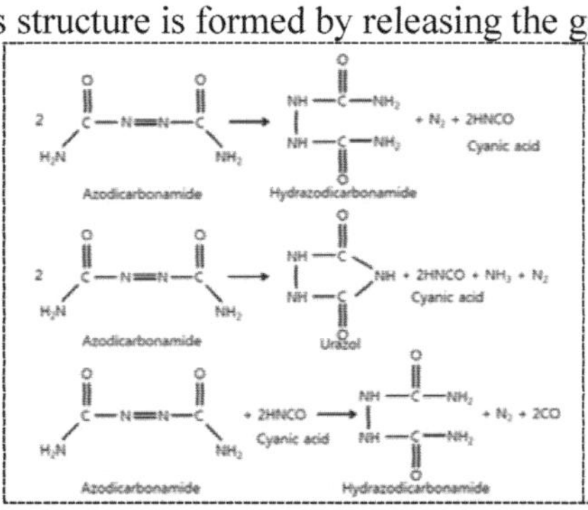

Figure 2. Thermal deposition reaction of ADC [12-13].

3. Experiment

3.1. Polypropylene

Compared to other polymers, PP is relatively inexpensive and has excellent mechanical properties at room temperature [14]. The polymer used in this experiment is Homo PP H930D produced by SK Global Chemical. The characteristics are indicated in Table 1.

Table 1. The characteristics of polypropylene.

	Melt index [g 10 min^{-1}]	Vicat temperature [°C]	Heat distiortion temperature [°C]
Polypropylene	3.5	158	112

3.2. Chemical Blowing Agent

The CBA used in experiment is ADC($H_4N_4C_2O_2$) of Cellcom-AC series produced by Kumyang. The properties of this agent are indicated in Table 2.

Table 2. The properties of chemical blowing agent.

	Particle diameter [µm]	Thermal decomposition temperature [°C]	Amount of gas [ml g^{-1}]
CBA	3–20	200–205	280–300

3.3. Experiment equipment

The following are descriptions of the equipment used in this experiment. We used the extrusion equipment called "Filabot Wee", which can manufacture the CBA hybrid filament for 3D PPP at a regular thickness by combining polymers and CBA. Also, we used the 3D printing equipment from the FDM scheme which was called "Willybot MS"; it is possible to melt the CBA hybrid filament at the nozzle temperature and to produce a 3D product based on the input data.

3.4. Manufacturing of the CBA hybrid filament

Figure 3 shows the CBA hybrid filament manufacturing process. For the density of CBA to be 5 wt.%, we mixed the prepared ingredients of PP pellet and CBA powder. Manufacturing the CBA hybrid filament made of polymers and CBA by extruder is shown in Figure 3 (a). The thickness of the CBA hybrid filament is manufactured regularly as 1.60–1.75mm; therefore, the pre-existing FDM equipment can be used, as shown in Figure 3 (b) and (c). At this point, the temperature setting is 160 °C, because the CBA should not cause a thermal decomposition reaction and the PP only needs to melt for manufacturing the CBA hybrid filament.

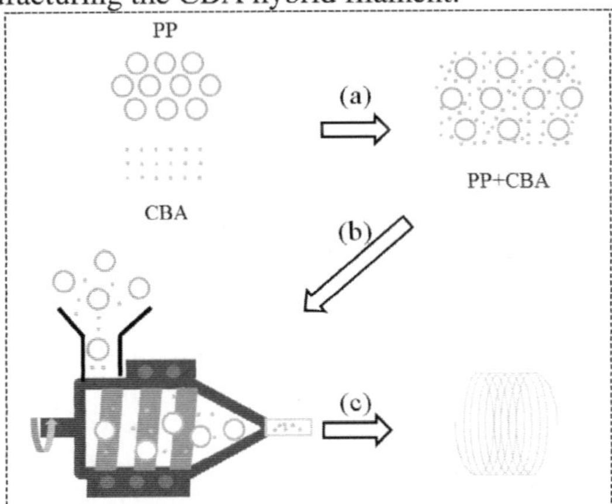

Figure 3. CBA hybrid filament manufacturing process.

3.5. Formation of porous structure

3D printing is performed using the fabricated CBA hybrid filament, and using the heater temperature of the nozzle, the thermal decomposition reaction of CBA is used to form a porous structure. At this point, the heater temperature of the nozzle must be at a temperature greater than 200–205 °C sufficient enough for the thermal decomposition reaction of the CBA.

3.6. Conditions of experiment

We conducted experiments on variations of the pore according to the printing speed at the nozzle temperature near the foaming temperature of the CBA. Likewise, we conducted experiments on variations of the pore depending on the temperature change of the nozzle at a constant printing speed. The conditions of the experiment are shown in Table 3 and Table 4.

Table 3. An experimental condition about printing speed.

CBA density [wt%]	Nozzle temperature [°C]	Printing speed [mm min^{-1}]
5	220	150
5	220	300
5	220	450
5	220	600
5	220	750
5	220	900
5	220	1050

Table 4. An experimental condition about nozzle temperature.

CBA density [wt%]	Printing speed [mm min^{-1}]	Nozzle temperature [°C]
5	600	190
5	600	200
5	600	210
5	600	220
5	600	230
5	600	240
5	600	250

4. Results and Discussions

The process parameters affecting the formation of the porous structures in the 3D PPP include CBA density, nozzle temperature, and printing speed. The porosity graph showed in a ratio of the weight difference between the weight of the foamed polymer and the unfoamed polymer.

4.1. Printing speed

Figure 4 shows the result of changes in printing speed in the same conditions as the CBA density and nozzle temperature. Figure 4 (a)–(c) is white, indicating that thermal decomposition reaction of CBA was completely conducted. Figure 4 (d)–(g) is yellow, indicating that as the printing speed increases, the CBA is totally extruded before the thermal decomposition reaction is triggered by the temperature of the nozzle. As the printing rate decrease, we can see layers forming more porosity. In addition, the graph in Figure 5 shows changes in the porosity.

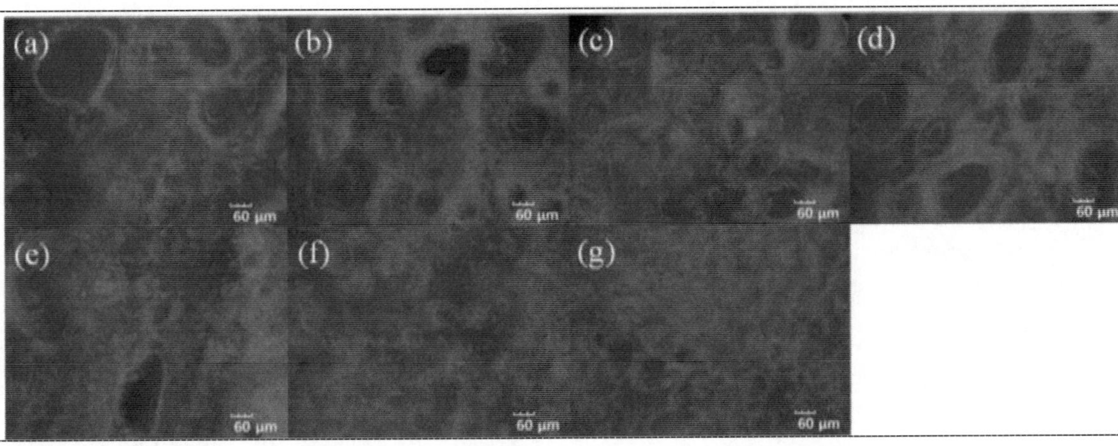

Figure 4. An inner cross section observed with an optical microscope at a printing speed of (a) 150 mm/min, (b) 300 mm/min, (c) 450 mm/min, (d) 600 mm/min, (e) 750 mm/min, (f) 900 mm/min, and (g) 1050 mm/min.

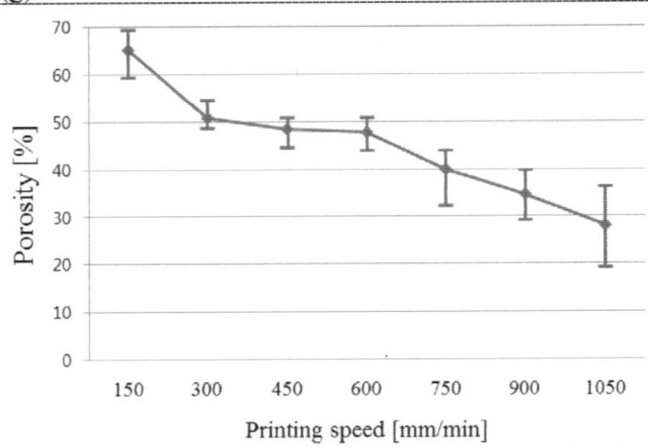

Figure 5. The porosity graph of the printing speed.

4.2. Nozzle temperature

Figure 6 shows the result of changes in nozzle temperature under the same condition as CBA density and printing speed. The temperature at which the blowing agent causes thermal decomposition reaction is 200–205 °C. Figure 6 (a) shows that the temperature is below the temperature that causes the thermal decomposition reaction, and therefore, does not form a porous structure. Figure 6 (b)–(e) is yellow, indicating that the temperature is low enough to produce sufficient thermal conduction for the CBA hybrid filament. Figure 6 (f) and (g) is white, which is the color of the PP, indicating that sufficient heat conduction has occurred in the CBA hybrid filament. As opposed to the printing rate, the higher the nozzle temperature, the more porosity is created and laminated. Similarly, the graph in Figure 7 shows the variation of porosity.

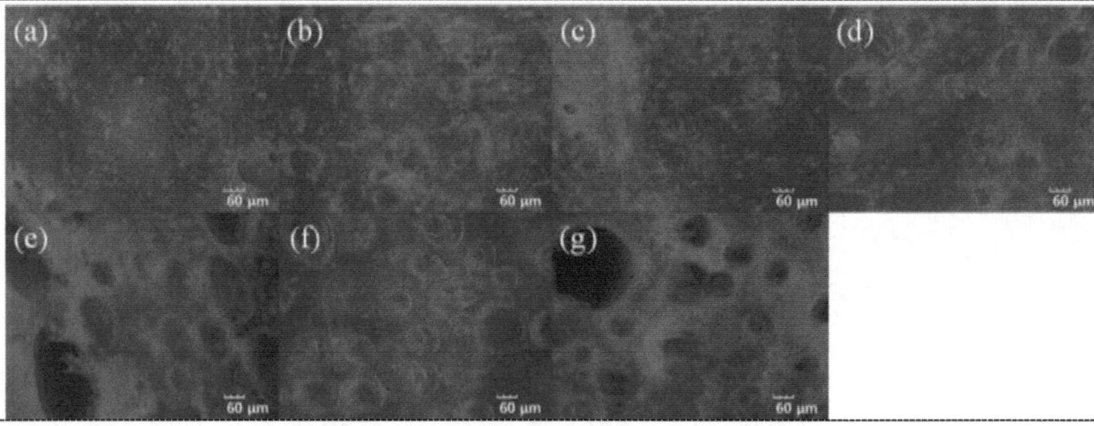

Figure 6. An inner cross section observed with an optical microscope at a nozzle temperature of (a) 190 °C, (b) 200 °C, (c) 210 °C, (d) 220 °C, (e) 230 °C, (f) 240 °C, and (g) 250 °C.

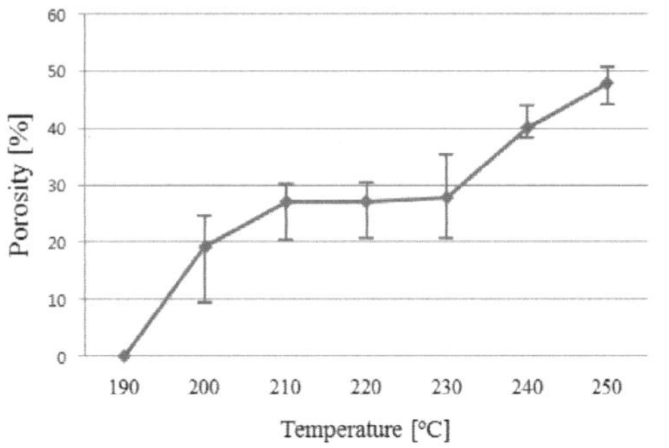

Figure 7. The porosity graph of the nozzle temperature.

4.3. Simple 3D structure
Figure 8 is a simple 3D shape manufactured using the 3D PPP technology. In the experiment, we set the temperature of the nozzle at 250 °C and the printing speed at 600 mm/min. In addition, the internal cross section of the simple 3D shape is shown in Figure 9. Most of the pores were created in varying sizes, and we were able to ascertain that the pores were formed within the pores.

Figure 8. The simple 3D shape manufactured by 3D PPP.

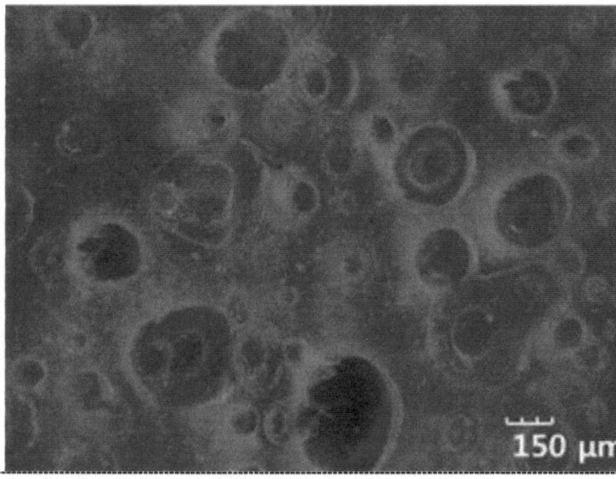

Figure 9. An internal cross section of the structure formed by 3D PPP.

5. Conclusions

In this study, based on the results of according to the printing speed and the nozzle temperature in the process parameter of the CBA-based 3D PPP, and a basic study was conducted on the porous structure of PPP. Overall, the size of the pores ranged from 10 to 300 μm. At the slight temperature difference about 10 to 20 degrees Celsius, the size of the pores was largely unrelated to the process parameter. But, if the temperature difference is very large, the size of pores will be affected by something such as the shrinkage of the polymer, cooling time and etc. At the same nozzle temperature, the slower the printing speed, the greater the porosity. At the same printing speed, the higher the nozzle temperature, the greater the porosity. In this process, compared to unformed PP, formed PP is lightweight because it forms a porous structure. The porous structure is suitable for mechanical shock mitigation because deformation will occur. In the future, we intend to use 3D PPP to conduct research on mechanical properties and, their application to various fields.

Acknowledgments
This research was supported by the Ministry of Trade, Industry and Energy of the Republic of Korea (10047652).

References

[1] Ventola CL 2014 *P&T* **39** 704
[2] Pfister A, Landers R, Laib A, Hubner U, Schmelzeisen R and Mulhaupt R 2004 *Journal of Polymer Science Part A* **42** 624
[3] Yoo C J, Kim H S, Park J H, Yun D H, Shin J K and Shin B S 2016 *Journal of Korean Society for Precision Engineering* **33** 333
[4] Yoo C J, Shin B S, Kang B S, Gwak C Y, Park C, Ma Y W and Hong S M 2017 *Procedia Engineering* **184** 10
[5] Jung D W, Jeong J H, Park C B and Shin B S 2013 *International Journal of Precision Engineering and Manufacturing* **15** 1127
[6] Park C, Shin B S, Kang M S, Ma Y W, Oh J Y and Hong S M 2015 *International Journal of Precision Engineering and Manufacturing* **16** 1385
[7] Liu P S and Liang K M 2001 *Journal of Materials Science* **21** 5059
[8] Thomas A 2010 *Angewandte Chemie International Edition* **49** 8328
[9] Han C D, Kim Y W and Malhotra K D 1976 *Journal of Applied Polymer Science* **20** 1583
[10] Robledo-Ortiz JR, Zepeda C, Gomez C, Rodrigue D and Gonalez-Nunez R 2008 *Polymer Testing* **27** 730
[11] Heck RL 1998 *Journal of Vinyl & Additive Technology* **4** 113
[12] Reyes-Labarta J A and Marcilla A 2008 *Journal of Applied Polymer Science* **107** 339
[13] Reyes-Labarta J A and Marcilla A 2008 *Journal of Applied Polymer Science* **110** 3217
[14] Chun S J and Lee S Y 2014 *Journal of the Korean Wood Science and Technology* **42** 682

IOP Conf. Series: Materials Science and Engineering **229** (2017) 012028 doi:10.1088/1757-899X/229/1/012028

Experimental Characterization of Electron Beam Welded SAE 5137H Thick Steel Plate

Prakash Kattire[1,*], Valmik Bhawar[1], Sandeep Thakare[1], Sachin Patil[1], Santosh Mane and Dr. Rajkumar Singh[1]

[1]Bharat Forge Ltd, Pune, Maharashtra, India

*prakash.kattire@bharatforge.com

Abstract. Electron beam welding is known for its narrow weld zone with high depth to width ratio, less heat affected zone, less distortion and contamination. Electron beam welding is fusion welding process, where high velocity electrons impinge on material joint to be welded and kinetic energy of this electron is transformed into heat upon impact to fuse the material. In the present work electron beam welding of 60 mm thick SAE 5137H steel is studied. Mechanical and metallurgical properties of electron beam welded joint of SAE 5137H were evaluated. Mechanical properties are analysed by tensile, impact and hardness test. Metallurgical properties are investigated through optical and scanning electron microscope. The hardness traverse across weld zone shows HV 370-380, about 18% increase in the tensile strength and very low toughness of weld joint compared to parent metal. Microstructural observation shows equiaxed dendrite in the fusion zone and partial grain refinement was found in the HAZ.

1. Introduction

In conventional welding high depth application of welding requires many passes and high preheating temperature generating more heat affected zone (HAZ) which affects joint mechanical strength. In case of electron beam welding permits deep penetration in single pass, generating narrow heat affected zone affecting less on the mechanical property [1]. Electron beam welding is fusion welding process for joining metals which uses highly focused beam of electrons as heat source. Very high velocity of electrons and high power density causes almost instantaneous local melting and vaporization of work-piece material. The electron beam is thus able to establish a ‘keyhole’ delivering heat, deep into material being welded. This produces narrow, parallel, fusion welding zone allowing butt welding in single pass for material thickness ranging from 0.025 mm to 300 mm [2-3]. Electron beam welding is initially employed in nuclear industry but rapidly spread into the aerospace, automobile and electronic industries. Nowadays it is employed over wide variety of application.

EBW is used for joining numerous metallic material including steel, aluminum, copper, titanium, nickel and magnesium and refractory metals. The process produces high integrity welds, with minimal thermal distortion and freedom from component oxidation.

SAE 5137H steel is used for auto component manufacturing. To improve manufacturability of intricate auto components they can be split and welded by electron beam welding. In automobile industry electron beam welding is used for welding of speed gear, torque convertor, shaft assemblies, differential gears, fuel injectors, planet carriers, turbocharger impellers etc.

Content from this work may be used under the terms of the Creative Commons Attribution 3.0 licence. Any further distribution of this work must maintain attribution to the author(s) and the title of the work, journal citation and DOI.
Published under licence by IOP Publishing Ltd

Benjamin Joseph et al. (2012) [3] conducted study on 316L (N) austenitic stainless steel by varying welding parameter and mechanical and metallurgical properties were analysed. Mechanical properties of the weld metal are better than parent material. The variations in the mechanical properties were related with variation in the cooling rate of the weld metal. Walter D. [4] investigated electron beam welding structural steel heavy sections of high pressure valves and disc of turbine shafts. He addressed problem of porosity in electron beam welding of heavy sections by oscillatory deflection of focused electron beam spot and low welding speed. Takashi Inoue et al. (1993) [5] studied electron beam weldability of heavy steel plate. The study resulted into development of steel which can be electron beam welded to give good toughness. In the present study electron beam welding of SAE 5137H steel was done and mechanical and metallurgical testing were carried out to understand variation in the properties of the weld.

2. Experimentation

2.1. Experimental setup

The welded joint in butt configuration, was produced by an electron beam welding machine, Model 4500 of PTR Strahltechnik GmbH, located at Langenselbold, Germany (Figure1). The machine has 150 kV generators and can produce very tall weld geometry. The work chamber pressure is maintained at 8 x 10^{-4} mbar and beam generator maintained at 1 x 10^{-5} mbar pressure. Different energy levels were obtained by varying accelerating voltage and beam current. Some trials were done to fix welding speed at which full penetration can be obtained in single pass.

2.2. Material Details

The SAE 5137H steel is used for the experimentation. The elemental composition of SAE 5137H is determined by spectroscopy and is shown in table1. Two blocks of 100 x 100 mm with thickness of 60 mm is used for experimentation. The edges and surfaces of the blocks are machined such that zero gap is maintained between the surfaces while welding. Butt joint is used for welding. Blocks edges and surface were rinsed and polished with acetone before welding.

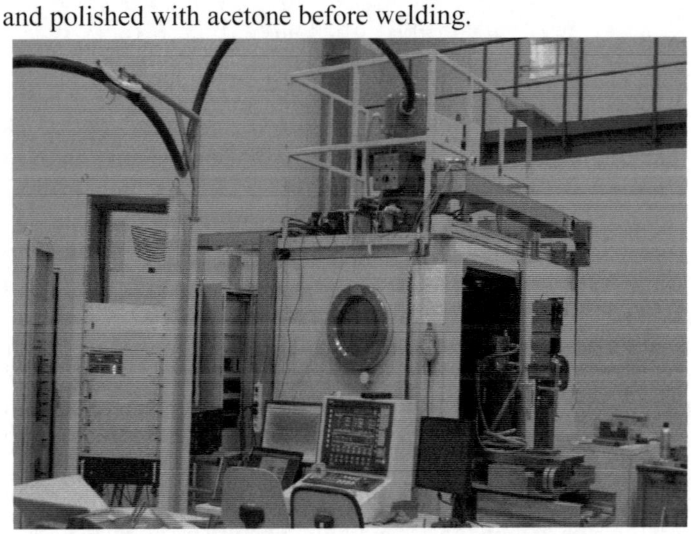

Figure1. Experimental setup at PTR Germany.

Figure2. Electron beam welded sample.

2.3. Experimental details

2.3.1. Weld Block preprocessing. EB welding is carried out in vacuum and as result the process allows parts to be joined without any adverse oxidizing effects. It follows therefore, that the blocks should be free of any contaminants, such as cutting lubricant, anti-rusting oil etc. when they are assembled for welding. Hence, the pre-weld preparation stage is all important to the success of the operation. Following procedure is followed

- Parts are thoroughly degreased before assembly and joint faces are abraded and acetone washed
- To avoid deflection of electron beam by stray magnetism blocks are demagnetized

2.3.2. Preheating. Preheating involves heating the material to be welded either locally or in it's entirely, to a specified temperature, before starting the welding operation. The main reason for preheating is to reduce the thermal gradient while welding and thus sudden quenching resulting in cold cracking [6]. The most critical factor in determining preheating temperature is carbon equivalent. As per structural welding code AWS D1.1 the empirical formula used to determine carbon equivalent is given below in equation 1.

$$\%\,CE = \%C + \%\frac{(Mn+Si)}{6} + \%\frac{(Cr+Mo+V)}{5} + \%\frac{(Ni+Cu)}{15} \quad (1)$$

SAE 5137H materials weldability is poor because of higher content of alloying element like carbon and manganese. With equation (1) carbon equivalent (CE) for SAE 5137H is 0.9%. The approximate recommended preheating temperature for CE above 0.6% is 205 to 375 °C. So keeping this in mind preheating is done in the electron beam welding machine vacuum chamber with the electron beam to 300- 350°C with the predetermined optimized parameters as given in table 2. Usually steel that requires preheating must be kept at this temperature between weld passes, but characteristics of electron beam welding is that it can weld 60mm weld depth in single depth so preheating between passes is not required. Using same setup full penetration welding is done using the optimized parameters mentioned in table 2. Figure 2 shows the welded specimen. After welding the block was slowly cooled in furnace (post heating 250° C) to room temperature to reduce the risk of cracks. After welding sample was stress relived for 570°C for 3hrs in separate furnace to reduce the residual stresses generated in welding.

2.4. Characterization.

Welded block was visually inspected and tested with die penetrant for surface defects. The block was also radiographed for internal soundness. Tensile samples were prepared with the weld oriented transverse to the gauge length as shown in the figure. 3 (ASTM E8). Tensile test samples were tested on MTS machine. For microstructural examination samples were prepared using standard metallographic method. 2% nital is used for etching the sample and microstructure of base material (BM), heat affected zone (HAZ) and fusion zone (FZ) were examined by optical microscope and Carl Zeiss make EVO18 scanning electron microscope (SEM). The Vickers micro hardness test was taken along the cross section of the welding across the weld with the load of 200g and time of 10s using

Clemex make micro hardness tester. Impact toughness studies using Charpy V- notch was done with the V- notch at the center of the weld.

Table1. Chemical composition of SAE 5137H.

Element (% Wt)	C	Mn	Si	S	P	Cr	Mo	Ni
SAE5137	0.38	1.19	0.28	0.025	0.015	1.23	0.042	0.11

Table2. Optimized electron beam welding parameter.

Welding Plan	Accelerating voltage (kV)	Beam current (mA)	Welding speed (mm/s)	Focus current (mA)	Welding distance (mm)
Preheating	150	18	40	995	200
Welding	150	65	10	685	200

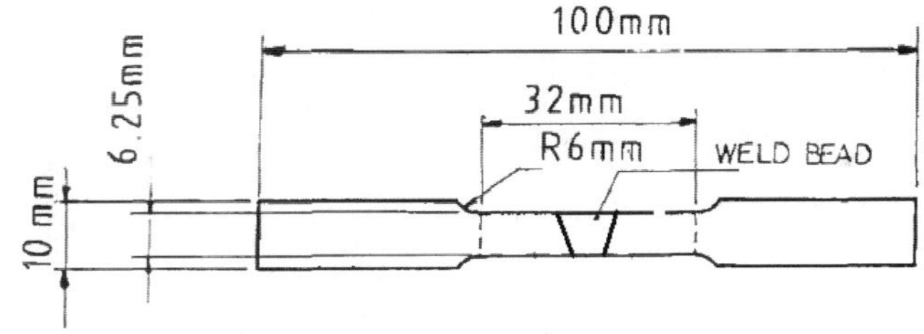

Figure3. Tensile sample used for testing EB welds.

3. Result and discussion

3.1 Micro hardness

The Transverse micro-hardness of cross section of weld zone was measured and plotted in figure 4. Figure 4 shows hardness variation in the fusion zone, HAZ and parent material. The hardness value of fusion zone is higher than HAZ and parent metal because of presence of fine dendtric microstructure

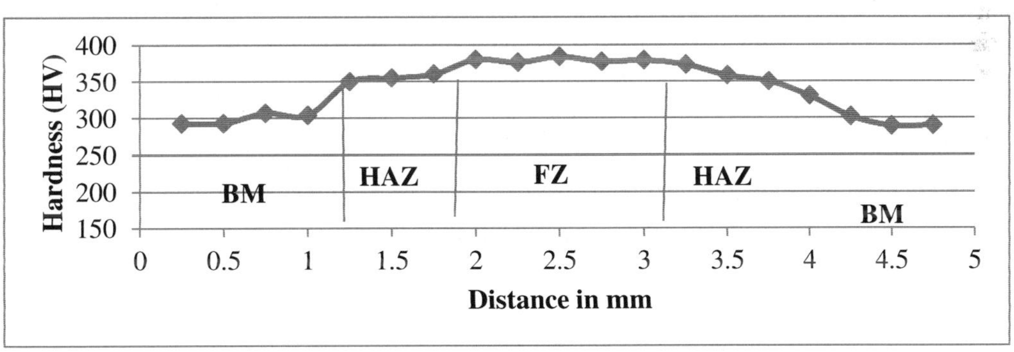

Figure4. Micro-hardness plot for across the weld zone.

3.2 Microstructure

Three different regions are found in the weld zone Microstructure, fusion zone, heat affected zone (HAZ) and base metal is shown in fig 5. Optical images of three zones are also shown in figure 6. Fusion zone shows equiaxed dendritic structure with ferrite and martensitic structure. HAZ can be divided into two regions. 1st region near fusion zone which cools slower because of adjacent hot fusion zone forms coarse grain HAZ and 2nd region is near base metal where fine grain HAZ is formed because of faster cooling rate due cool adjacent base metal [7]. HAZ zone shows presence of ferrite and fine perlite lamellae and parent material showing soft ferrite and perlite microstructure.

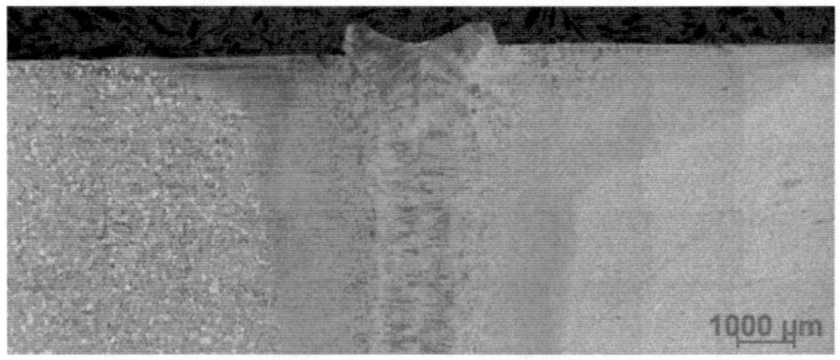

Figure5. Electron beam welding microstructure.

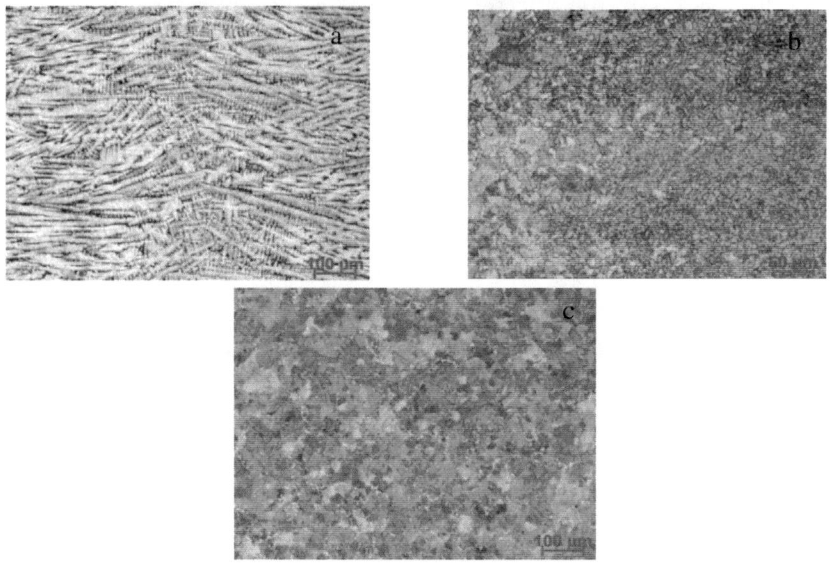

Figure6. Optical images of a) Fusion Zone b) HAZ zone c) Base Material.

3.3 Transverse Tensile strength

Transverse tensile strength results are shown in table 3. In all three samples tensile test failure location is in base metal (as shown in fig. 7) so we can say that weld material is stronger than parent metal. 18% increase in the tensile strength of material was found after electron beam welding. Better properties can be explained by microstructure results showing fine dendritic microstructure in the weld zone which is result of higher cooling rate in welding

Table 3.Transverse tensile test result

Specimen No.	1	2	3
Ultimate tensile strength (MPa)	871	875	904
Base Metal strength (MPa)	747		

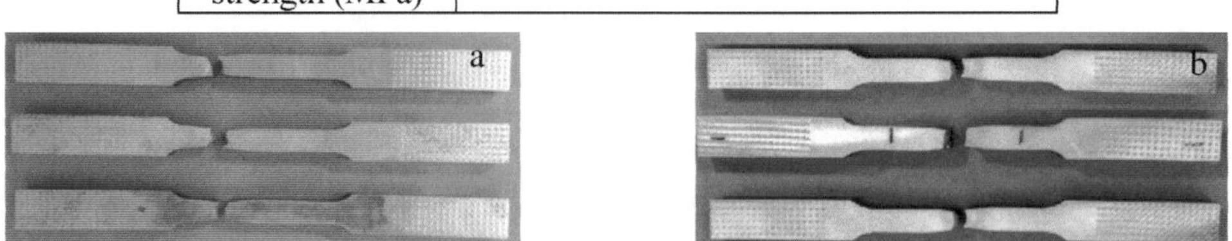

Figure7. Tensile test failure location in a) EBW b) Parent Material.

IOP Conf. Series: Materials Science and Engineering **229** (2017) 012028 doi:10.1088/1757-899X/229/1/012028

3.4 Impact Strength

Impact strength of welded and parent material is shown in table 4. Toughness of the weld joint reduced significantly. Since EBW uses no filler material so the composition of weld metal is same as base metal and weld metal is cast structure which formed with high cooling rate. So there is loss in toughness of material in as welded condition [8].

Table4. Impact strength result.

Specimen No.	1	2	3
Impact strength (J)	11.2	12.5	11.8
Base Metal Impact strength (J)	48		

4. Conclusion

Electron beam welding can be used in high depth application in automotive industry. Electron beam welding study carried out on SAE 5137H steel shows following weld characteristics.

- Average 360-378 HV hardness is found in the fusion zone is higher than parent material this is due to fusion zone has fine dendritic microstructure.
- Electron beam welding has higher tensile strength than parent metal, because of less heat input and low HAZ.
- As compared to parent material weld is very low toughness as weld metal is like a cast structure and formed by high cooling rate.

References

[1] Gokul Ananth M., B.Sathish Babu, P.Chakravarthy, K.Jayakumar,A.Manickavasagam, N.V.S Arunprakash, K.M. Gopalakrishnan, "Experimental Investigations on Electron Beam Welding of Austenetic/Ferritic Stainless Steel for Space Applications", *IJRMET* Vol. **3**, Issue 2, May - Oct 2013

[2] Weglowski M. st., S. Błacha, A. Phillips, "Electron beam welding- techniques and trends- Review", *Vacuum* **130** (2016), 72-92

[3] Benjamin Joseph, D. Katherasan P. Sathiya and C. V. Srinivasa Murthy, "Weld metal characterization of 316L(N) austenitic stainless steel by electron beam welding process", *International Journal of Engineering, Science and Technology,* Vol. **4**, No. 2, 2012, pp. 169-176

[4] Dietrich W., "Investigation Into Electron Beam Welding of Heavy Sections", AWS 59th Annual Meeting held in New Orleans, Louisiana,during April 3-7, 1978

[5] Takashi I., Masahiro O., Yukio T., Yukio T., Koij T., Kunio K., Rikio C., Seij I., "Development of heavy steel plates with excellent electron beam weldability " , Nippon steel technical report No. **58**, July 1993

[6] Srivastava B. K., Dr. Tewari S. P., Jyoti P., " Review on effect of preheating and or post weld heat treatment on mechanical behaviour of ferrous metal", *International Journal of Engineering Science and Technology* Vol. **2**(4), 2010, 625-631

[7] Błacha S., M.St. Węglowski, S. Dymek, M. Kopuściański, "Microstructural Characterization and Mechanical Properties of Electron Beam Welded Joint of High Strength Steel Grade S690QL", Arch. Metall. Mater., Vol. **61** (2016), No 2B, p. 1193–1200

[8] Sivaramakrishnan N., K. S. Raja and K. Prasad Rao, "Effect of Post-weld Heat Treatment on Weld Metal Impact Toughness of a Semi-Austenitic PH Stainless Steel", *Welding Research Supplement*, 200-s-207-s, august 1994

IOP Conf. Series: Materials Science and Engineering **229** (2017) 012029 doi:10.1088/1757-899X/229/1/012029

Copolymerization of Glycolide and ε-Caprolactone Using 12-Aminolauric Acid Modified Montmorillonite

HAV Gallos[1] and LQ Reyes[1,*]

[1]School of Chemical Engineering and Chemistry, Mapua University, Manila, Philippines

*lqreyes03@gmail.com

Abstract. Poly(glycolide-co-ε-caprolactone) (PGLYCL) nanocomposites were prepared by copolymerization glycolide (GLY) and ε-caprolactone (ε-CL) in the presence of varying loadings 12-aminolauric acid (12-ALA)-modified montmorillonite. Copolymerization was successfully achieved based on the increase in polymer molecular weight after the reaction determined by gel permeation chromatography (GPC). The amount of the poly(glycolide) block and poly(ε-caprolactone) block units in the copolymer, identified by proton nuclear magnetic resonance (^{1}H-NMR) spectroscopy, suggested that the increase in organo-clay loading cause a reduction GLYL: ε-CLL ratio. The arrangement of the monomers in the polymer products was elucidated to have an ABA triblock structure, where PCL block is the central block and the PGLY is found at both end of the copolymer. The presence of intercalated and exfoliated silicates in the nanocomposites were observed by x-ray diffraction (XRD) analysis. The biocompatibility of the nanocomposites with NCTC 292 mouse normal fibroblast was high relative to untreated cell cultures using tetrazolium bromide (MTT)-dye reduction assay.

1.Introduction

Synthetic biodegradable polymers are designed to perform first their function and then later on be degraded by the body without exhibiting any biocompatibility and excretion issues. Because of this along with their tunable properties, they have gained wide interest not only in the fields of drug delivery applications but also in smart biomaterials, medical devices and tissue engineering applications [1]. Biodegradable polymers commonly used as biomaterials can be prepared from a major class of synthetic polymers, the poly(α-hydroxy acids) or commonly known as polyesters. One example of a polymer biomaterial is polyglycolide (PGLY); which is produced from ring-opening polymerization of glycolide (GLY). PGLY is insoluble in most organic solvent due to its high degree of molecular arrangement, making it difficult to process, unlike the copolymers of PLGY such as the popular poly(lactide-co-glycolide) (PLGA). Poly(caprolactone) (PCL) is known to be useful as polymer for long-term controlled drug delivery and implantable devices because of its relatively low rate of degradation [1]. The copolymerization of GLY with other cyclic esters were previously exhibited the the research group of Díaz-celorio [2] where they synthesized a copolymer of poly(glycolide-co-methylene carbonate) copolymers via ring opening polymerization (ROP) using organometallic stannous (II) octoate ($SnOct_2$) as the catalyst were conducted by some research group . While $SnOct_2$ is the most popular catalyst for this reaction, few studies on using alternative eco-

Content from this work may be used under the terms of the Creative Commons Attribution 3.0 licence. Any further distribution of this work must maintain attribution to the author(s) and the title of the work, journal citation and DOI.
Published under licence by IOP Publishing Ltd

friendly catalysts have been carried out [3]. Montmorillonite clays have been found to be effective catalysts for many organic reactions, including polymerization reactions [4]. Thus, they also contribute a great deal in the shift towards making clays as the most promising/ alternative green catalyst [5]. The present study explored the ability of the organo-modified clay to catalyze a ring-opening polymerization of GLY with ε-CL. The functional groups for aliphatic polyesters were determined by Fourier Transform Infrared (FTIR) and Proton Nuclear Magnetic Resonance (^{1}H-NMR) spectrometry. In addition to that, from the ^{1}H-NMR spectra, the terminal polymer chain was determined through end group analysis and the relative amount of GLY and ε-CL in the samples were calculated. Though the relationship of organo-modified clay loading and molecular weight of the polymers are not yet established, the present study also aimed to relate these two factors using Gel Permeation Chromatography (GPC). Finally, the extent of polymer intercalation within the clay silicate layers was also evaluated using X-ray Diffraction (XRD).

2.Methodology

2.1. In situ intercalative polymerization of glycolide and caprolactone with 12-aminolauric acid

Na-montmorillonite (Nanofill 116) with 1.16 meq/g cation exchange capacity (CEC) was loaded with 90% 12-aminolauric acid (12-ALA) to produce an organically modified clay by exchanging the the Na+ counter ions of montmorillonite with 12-ALA in aqueous media. Then, polyester-clay nanocomposites were prepared by *in situ* intercalative polymerization of cyclic esters, namely ε-CL and GLY, with AMMT. Briefly, 50:50 wt % of GLY:ε-CL, with varying loading of AMMT (1, 3, 5 10 and 15 wt. %), were placed in a rection flask. The mixture was reacted at 150°C for 24 hours with continuous Nitrogen gas purging. The products were dissolved in chloroform and then precipitated in methanol. As a control, polyesters, from each monomer, were synthesized using the conventional $SnOct_2$ catalyst with 1, 4-butanediol (BDO) as the initiator.

2.2 Molecular characterization of the AMMT-catalyzed polyesters

The molecules weight properties of the polyester products were determined by gel permeation chromatography (GPC) with polystyrene standards using a Shimadzu LC-20AD Prominence liquid chromatograph equipped with a Shimadzu RID-10A refractive index detector (RID). Samples were eluted with tetrahydrofuran (THF) through a TSK-GEL $G3000H_{HR}$ at a flow rate of 1 mL/min. The function groups of the copolymer products were determined using Shimadzu IR Prestige-21 Fourier Transform Infrared (FTIR) spectrophotometer and samples weree scanned at a rate or 4 cm^{-1} for 32 times ranging from 4000 to 600 cm^{-1}. To elucidate the molecular structure of the polyester products and their actual GLYL and ε-CLL content, ^{1}H-NMR spectroscopy was performed using a Varian 500MHz spectrometer. The polyester products were dissolved deuterated chloroform ($CDCl_3$) with trimethylsilane (TMS). The order of oligo-GLY and oligo-ε-CL in the copolymer was evaluated by based on the presence methylene group adjacent to hydroxyl (4.67ppm or 3.62ppm) and carboxyl group (4.79ppm or 2.21ppm). The actual amount of GLYL and ε-CLL was derived from the proton NMR spectra using the following equation (Equation 1):

$$\%mol = \frac{A_{4.82ppm}/n_{4.82ppm}}{A_{4.82ppm}/n_{4.82ppm}+A_{1.66}/n_{1.66}} x100\% \quad (1)$$

where $A_{4.82}$ and $n_{4.82}$ is the peak integration and the number of methylene protons giving rise to the chemical shift characteristic of GLYL and the $A_{1.66}$ and $n_{1.66}$ is the peak integration and the number of methylene protons giving rise to the chemical shift of ε-CLL. To determine the formation of the nanocoposites, x-ray diffraction analysis using a Shimadzu XRD-700 x-ray diffractometer. The x-ray diffractograms were taken at an accelerating voltage of 40kV using an x-ray source was CuKα1 (1.540598 Å) at a rate of 2.0°/min from 3° up to 90°. The *in vitro* cytotoxicity of the clay-catalyzed polyesters was performed using indirect contact MTT-dye reduction assay. Mouse normal fibroblast (NCTC 292, ATCC® TIB-67™) were incubated in MEM supplemented with 10%v/v horse serum, 1%v/v non-essential amino acids, 2mM L-glutamine, 1mM sodium pyruvate and 2%v/v of 7.5%

sodium bicarbonate and 1%v/v antibiotic-antimycotic. A 5.0×10^3 cells/mL were seeded into 96-well plates and then incubated for 24 hours with the polymer extracts.

3.Results and Discussion

3.1. Synthesis of polyester-clay nanocomposites by in situ intercalative polymerization

Poly(glycolide-co-ε-caprolactone) PGLYCL with different loading amount of the organo-clays (1, 3, 5%, 10%, and 15% by weight) were prepared via *in situ* intercalative ring-opening polymerization as shown in Figure 1.

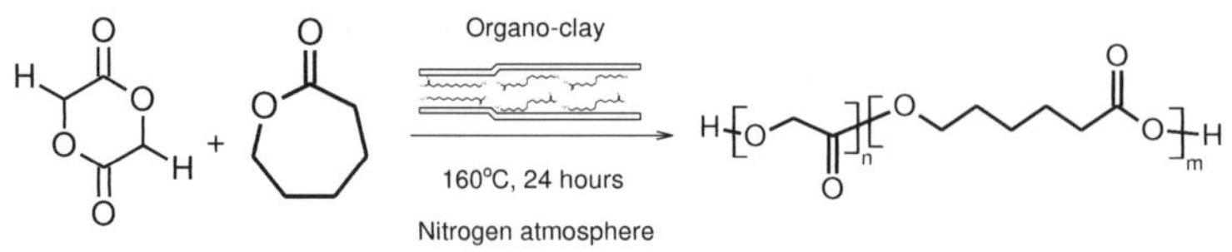

Figure 1. In situ intercalative polymerization of GLY and ε-CL using 12-ALA modified MMT.

3.2. Molecular weight analysi of the copolyesters

The molecular weight properties of the $SnOct_2$-catalyzed PGLYCL and PGLYCL nanocomposites are shown in Table 1. The molecular weight of PGLYCL nanocomposites was observed to be increase with increasing clay loading. The presence of the clay's silicate layers during polymerization may have obstructed the diffusion of the monomers toward the growing polymer chains of the polymer chains. This might also be one of the reasons why molecular weights of PGLYCL polymers in the nanocomposites are much lower than that of $SnOct_2$-catalyzed PGLYCL. The sensitivity of the polymerization reaction to the presence of water molecules should also be considered. While the clay is organically modified, there are still traces of water present in the structure as observed.

Table 1. Molecular weight properties of PGLYCL-$SnOct_2$ and PGLYCL nanocomposites.

Sample	%Recovery	M_n	M_w	PDI
PGLYCL-$SnOct_2$	96%	7691	12884	1.6752
PGLYCL-1AMMT	94%	4269	4288	1.0044
PGLYCL-3AMMT	95%	3938	3956	1.0046
PGLYCL-5AMMT	92%	1462	3104	2.1231
PGLYCL-10AMMT	94%	1683	3017	1.7926
PGLYCL-15AMMT	93%	1499	3076	2.0520

3.3. Structure elucidation of the copolymers

The representative ^{1}H-NMR spectra of the $SnOct_2$ and clay-catalyzed polyesters are shown in Figure 2. Chemical shifts observed in the PGLYCL nanocomposites are all consistent with the peaks observed in $SnOct_2$-catalyzed PGLYCL. Signals in the range of δ = 1 ppm to 2.5 ppm are all attributed to methylene protons of PCL while peaks within δ = 4 ppm to 5 ppm are attributed mainly to PGLY methylene protons. The chemical shifts of $SnOct_2$-catalyzed PGLYCL and clay-catalyzed PGLYCL with 1% and 5% wt organo-clay exhibited intense and narrow peaks. End group analysis of the PGLYCL products was conducted to determine the order of the monomers in the polymer chain. The chemical shift of a methylene protons neighboring the hydroxyl and carboxylic acid group for PGLY blocks were found at 4.69 ppm and 4.29 ppm, respectively. The PCL blocks are found in the central part of the molecular structure. In addition, the actual amount of PGLY and PCL content in the copolymer were calculated based on the relative peak intensity that was characteristic for each monomer. Assuming the complete conversion of monomers during polymerization, the expected mole percent of GLY would be 52%, 51%, 52% and 51% for PGLYCL with 1, 5, 15 wt % the organo-clay and PGLYCL-$SnOct_2$, respectively. However, copolymer analyses on the representative peaks of GLYL at δ = 4.82 ppm and ε-CLL at δ = 1.66 ppm varied from the theoretical values. The calculated

IOP Conf. Series: Materials Science and Engineering **229** (2017) 012029 doi:10.1088/1757-899X/229/1/012029

mole percent of GLYL for PGLYCL with 1, 5, 15 wt % clay and PGLYCL-$SnOct_2$ are 45.57, 40.99, 22.20 and 44.22 %, respectively. There is a decrease in GLYL content with increasing organo-clay content which could be attributed to the initial phase of the monomers. At room temperature, GLY appears to be a white powder while ε-CL is an oily liquid. During the initial stage of polymerization, the more mobile ε-CLs react with one another while GLY independently melts. After melting, only then GLY react with other GLY molecules or with ε-CL monomers. The resulting polymer products were soft and/or appeared like hard waxes. It should also be noted that during the purification stage, the PGLYCL materials were soluble in $CHCl_3$, but there were insoluble white polymer precipitates attributed to the PGLY homopolymer or PGLY-rich polymer. And so, during ^{1}H-NMR analyses, it was possible that what were analyzed are only those copolymers that in soluble in $CHCl_3$ while the PGLY-rich polymers were removed or filtered out from the solution. The use of 1 to 5 wt. % of organo-clay allowed the copolymerization of the monomers with relatively similar GLYL and ε-CL composition in PGLYCL-$SnOct_2$. The PGLYCL with 15 wt. % of clay may have a surplus of active initiation sites where GLY and ε-CL polymerized independently, such that their individual polymers did not grow in the same chain.

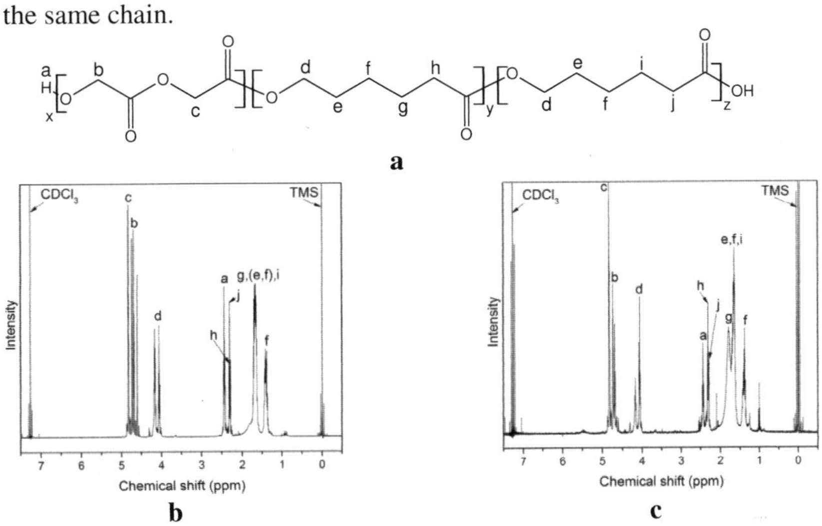

Figure 2. Proposed molecular structure of (a) PGLYCL and the ^{1}H-NMR spectra of the (b) PGLYCL-$SnOct_2$ and (c) PGLYCL-5AMMT.

3.4. Infrared investigation of ester bond formation after polymerization

The infrared spectra of the GLY, ε-CL, $SnOct_2$-catalyzed PGLYCL and PGLYCL nanocomposites are shown in Figure 3. A typical cyclic ester has a sharp of C=O stretching vibration at 1715cm^{-1}; although the exact position depends on the type of carbonyl structure. For ε-CL, the C=O stretch is observed at 1741cm^{-1} (Figure 3.3a) while it is 1751cm^{-1} for GLY (Figure 3.3b). For the $SnOct_2$-catalyzed PGLYCL and the PGLYCL nanocomposites, the C=O stretch of an aliphatic ester is found at 1735cm^{-1}, consistent with all polymer products. This shows that polymerization of GLY with ε-CL has been successful as evidenced by the formation of ester linkages. The peaks observed at 2947 to 2950 cm^{-1} and 2862 to 2870cm^{-1} are attributed to asymmetric and symmetric stretching of CH_2 groups of the hydrocarbon chains. The characteristic spectral peaks associated to OAMMT, found at 3600 cm^{-1} and 995cm^{-1} were not detected in the spectra of PGLYCL nanocomposites which is probably due to the small amount of the organo-clay in the polymer matrix that is not detectable by FTIR instrument.

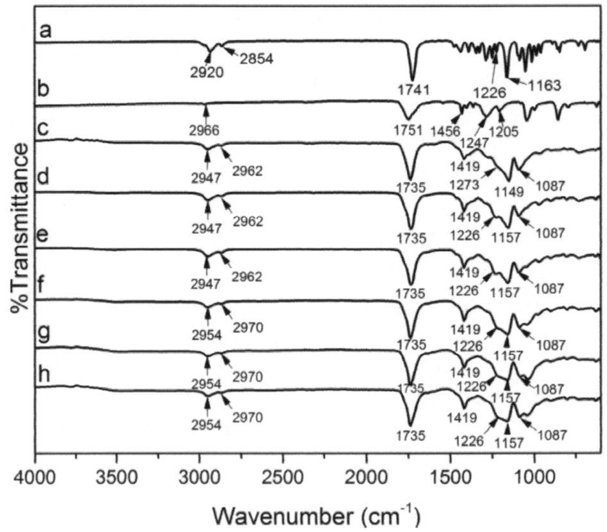

Figure 3. FTIR spectra of PGCL synthesized using (a) ε-CL, (b) GLY, (c) PGLYCL-$SnOct_2$ and (d) PGLYCL-1AMMT, (e) PGLYCL-3AMMT, (f) PGLYCL-5AMMT, (g) PGLYCL-10AMMT and (h) PGLYCL-15AMMT.

3.5. Formation of polyester-clay nanocomposites

The extent of the clay's silicate dispersion and delamination within the polymer matrix was assessed using XRD. The x-ray difrractogram of the organo-clay, PGLYCL-$SnOct_2$ and PGLYCL nanocomsposites are shown in Figure 4. First, organo-modification of Na^+-MMT with 12-ALA was done prior to polymerization to achieve greater compatibility between the hydrophilic clay and hydrophobic polymer. A shift of peak at 2θ = 4.85° relative to the 2θ position of Na^+-MMT found at 7.12°, shifting of peak towards lower 2θ is a result of an increase in the basal spacing at (001) plane of organo-clay's silicate galleries from 1.24 nm (Na+-MMT) to 1.82 nm (AMMT). The (001) peak were absent for PGLYCL-1MMT and PGLYCL-5AMMT, suggesting that the clay silicates are dispersed in the polymer matrix. The diffraction peak of the clay's (001) plane in the PGLYCL-10AMMT nanocomposite (Figure 3.4e) is 5.1° that corresponds to 1.73 nm and PGLYCL-15AMMT nanocomposite (Figure 3.4f) exhibited the (001) peak at 5.15° that corresponds to and 1.7 nm. Again, the d_{001}-spacing of the unmodified Na^+-MMT is 1.25 nm. Assuming that the surfactant in the clay undergoes Hoffman elimination (Carrasco et al., 2011) [6] during polymerization, there is a 0.48 nm increase for PGLYCL-10AMMT and 0.46 nm increase for PGLYCL-10AMMT of the clay's d_{001}-spacing due to polymer intercalation. The polymer chains that are intercalated in the clay layers may adopt a coexisting (in transition from one phase to another) monolayered to a bilayered structure.

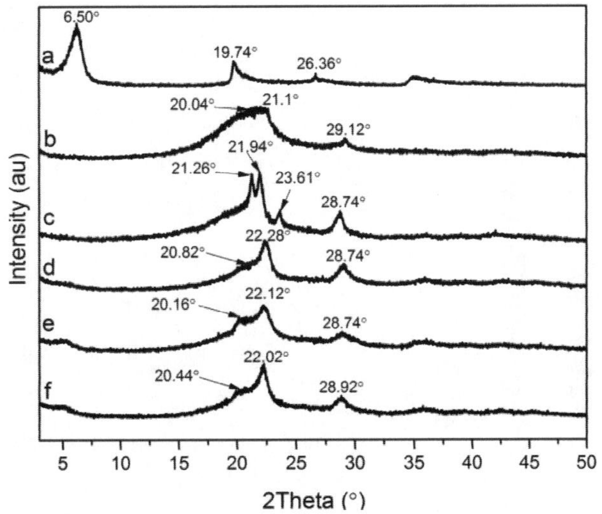

Figure 4. X-ray diffractograms of (a) organo-clay, (b) PGLYCL-$SnOct_2$, (c) PGLYCL-1AMMT, (d) PGLYCL-5AMMT, (e) PGLYCL-10AMMT and (f) PGLYCL-15AMMT.

3.6. Biocompatibility of the polyester-clay nanocomposites

The biocompatibility of the PGLYCL-$SnOct_2$ and PGLYCL-AMMT to NCTC 929 cells was determined by *in vitro* non-contact MTT assay. The cell viability and also the mitochondrial integrity, upon exposure to nanocomposite extract, were measured based on the reduction of the MTT-dye into purple formazan. The cytotoxicity profiles of the PGLYCL samples at 100% extraction media concentration are shown in Figure 5. Similar to the PCL and PGCLY samples, the PGLYCL material exhibited high cell viability that are more than 80%. The presence of clay within the polymer matrix did not affect the biocompatibility of the nanocomposites, and even at 15% clay loading, the % cell survival is around 95%. Representative photographs of viable cells after exposure to the polymer and nanocomposite extract are shown Figure 6. The inherent biocompatibility of the copolymers is again due to their biodegradability, allowing them to be processed by the cellular machinery. The doxorubicin (positive control) resulted to 17% cell survival. The photograph of viable cells in culture media (negative control) and non-viable cells after exposure to doxorubicin, a cytotoxic compound (positive control) are shown in Figure 3.5.

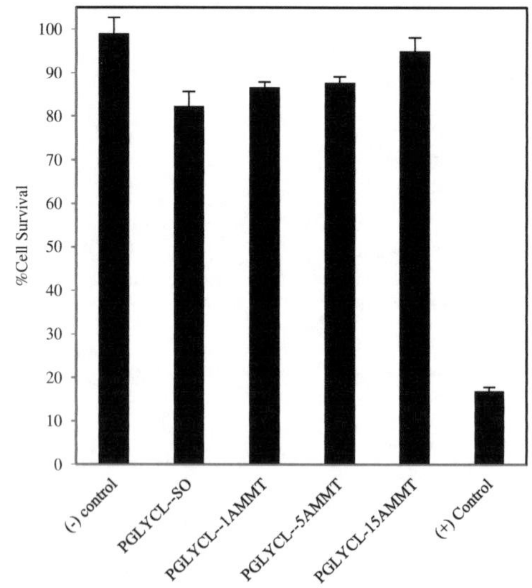

Figure 5. Cytotoxicity of the $SnOct_2$-catayzed and clay-catalyzed PGLYCL materials. The negative (-) control represents the incubation of cells in cell culture media. The positive control (+) represents the incubation of cells with doxorubicin.

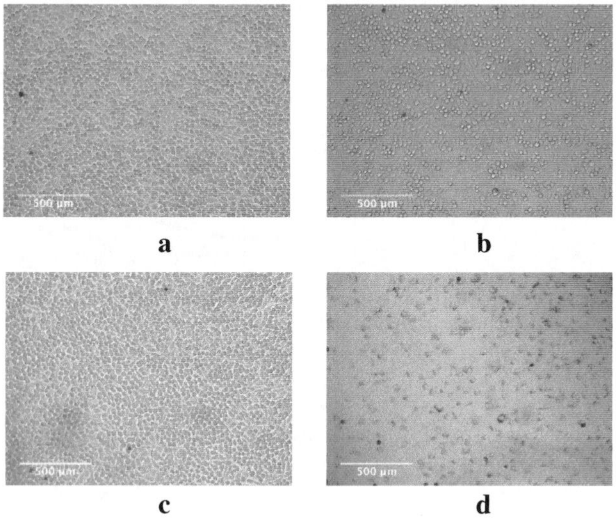

Figure 6. Optical Micrographs of cells exposed to (a) PGLYCL-$SnOct_2$ and (b) PGLYCL-5AMMT, (c) Culture media and (d) Doxorubicin (positive control) (Magnification: 100X).

4.Conclusion

The copolymerization of GLY and ε-CL was successfully synthesized by ring-opening polymerization using AMMT as catalyst. The molecular structure of the polymer products was observed to have a blocks of GLY (A) and ε-CLL (B) units in ABA order. The present study demonstrated the catalytic activity of the organically modified clay to polymerize the cyclic esters but with relatively low molecular weight. The polymerization, dispersion of the clay catalyst was achieved, forming polymer-clay nanocomposites. The composite materials with exfoliated/ intercalated structures were observed. Interestingly, the nanocomposites also exhibited high cellular biocompatibility with NCTC mouse normal fibroblast, which shows the possibility of the materials find applications as a biomedical polymer.

References

[1] J. Middleton and D. Ph, "Tailoring of Poly (lactide-co-glycolide) to Control Their Properties for Medical Device and Pharmaceutical Applications."

[2] E. Díaz-Celorio, L. Franco, Y. Márquez, A. Rodríguez-Galán, and J. Puiggalí, "Thermal degradation studies on homopolymers and copolymers based on trimethylene carbonate and glycolide units," *Thermochim. Acta*, vol. 528, pp. 23–31, 2012.

[3] L. S. Nair and C. T. Laurencin, "Biodegradable polymers as biomaterials," *Prog. Polym. Sci.*, vol. 32, no. 8–9, pp. 762–798, 2007.

[4] N. Kaur and D. Kishore, "Montmorillonite: An efficient, heterogeneous and green catalyst for organic synthesis," *J. Chem. Pharm. Res.*, vol. 4, no. 2, pp. 991–1015, 2012.

[5] D. E. Kherroub, M. Belbachir, and S. Lamouri, "Cationic Ring Opening Polymerization of Glycolide Catalysed by a Montmorillonite Clay Catalyst," *Bull. Chem. React. Eng. Catal.*, vol. 9, no. 1, pp. 74–80, 2014.

[6] F. Carrasco, J. Gámez-Pérez, O. O. Santana, and M. L. Maspoch, "Processing of poly(lactic acid)/organomontmorillonite nanocomposites: Microstructure, thermal stability and kinetics of the thermal decomposition," *Chem. Eng. J.*, vol. 178, pp. 451–460, 2011.

IOP Conf. Series: Materials Science and Engineering **229** (2017) 012030 doi:10.1088/1757-899X/229/1/012030

Polyaniline – Carrageenan - Polyvinyl Alcohol Composite Material Synthesized Via Interfacial Polymerization, its Morphological Characteristics and Enhanced Solubility in Water

R C K Montalbo[1] and M C Marquez[2,*]

[1]Institute of Chemistry, University of the Philippines-Diliman, Diliman, Quezon City, Philippines, 1101
[2]School of Chemical Engineering and Chemistry, Mapua University, Muralla St., Manila, Philippines, 1002

*menandro.marquez@gmail.com

Abstract. In recent years, conducting polyaniline (PAni) has been a popular interest of research in the field of conducting polymers due to its relatively low cost, ease of production, good conductivity, and environmental stability. Many studies however, have focused on improving its short-comings such as its limited processability and solubility in common solvents. In this study, PAni, soluble in water was produced via interfacial polymerization with chloroform as the organic solvent. Poly(vinyl alcohol) (PVA) and kappa(κ), iota(ι) and lambda(λ) - carrageenan (κ CGN, ι CGN, λ CGN) were added to the aqueous layer to stabilize PAni in the medium. FTIR and UV-Vis absorption spectra of the solutions as well as the fabricated film confirmed the existence of PAni emeraldine salt (PAni-ES). FTIR spectrum also confirmed the peaks corresponding to the interaction of PAni with the CGNs. Moreover, PVA-CGN played a very large role on the stability of the PAni nanofibers integrated on the PVA-CGN matrix. The morphologies of the products were further investigated using SEM and TEM. Polymer electrolyte for supercapacitor or an interfacial layer for organic solar cell is being targeted as potential application of the synthesized water soluble PAni.

1. Introduction

Since the discovery of intrinsically conducting polymers (ICPs) in the 1970's, the field has been an attractive area of research. Tens of thousands of published works on ICPs have already been produced. The study of ICPs has been popular due to their conductivities in the semiconductor range. Applications in electronics, energy, membranes, and sensors have also greatly contributed to the increase in studies of ICPs[1].

One of the most studied ICPs is polyaniline (PAni). Unlike other ICPs, PAni can exist in three forms depending on the degree of oxidation. Each degree of oxidation has different property and a characteristic color. PAni has attracted much attention due to these behaviors and due to its good conductivity, good environmental stability, relatively low cost, and facile synthesis. Similar to many other ICPs however, PAni is limited by its poor processability and solubility in common solvents [2].

Previous studies have shown that PAni can be stabilized by blending with thermoplastic polymers such poly(vinyl alcohol) (PVA)[3], poly(ethylene oxide) (PEO)[4], and poly(ethyl acrylate) (PEA)[5].

Content from this work may be used under the terms of the Creative Commons Attribution 3.0 licence. Any further distribution of this work must maintain attribution to the author(s) and the title of the work, journal citation and DOI.
Published under licence by IOP Publishing Ltd

It has also been proposed that in the presence of high molecular weight anionic polysaccharides such as carrageenan[6] and chitosan[7], ionic interactions between PAni and the water-soluble polysaccharide improve the dispersibility of the PAni-polysaccharide composite in water. By interfacial synthesis, it has been observed that PAni nanofibers can be dispersed in water without the use of templates and retain good electrochemical performance [8].

This study was undertaken to improve the solubility of PAni in water by using relatively inexpensive materials and simple methods of synthesis. In this study, water-soluble PAni was synthesized by interfacial polymerization. The organic layer consisted of aniline dissolved in chloroform. Poly(vinyl alcohol) (PVA) and carrageenan (CGN) dissolved in 1.0 M HCl comprised the aqueous layer to provide a stable matrix for the PAni polymerization during interfacial synthesis. The primary objective was to graft the hydrophilic emeraldine salt form of PAni onto the water-soluble PVA-CGN matrix by ionic interactions between PAni and CGN.

2. Methodology

2.1 Materials
Ammonium persulfate (APS, 98%), ethanol (EtOH, 99.9%), and PVA (99-99.8%) were purchased from J.T. Baker. Chloroform (99.8%) was purchased from RCI Labscan. Aniline was purchased from Sigma Aldrich. HCl (36-38%) was purchased from UniChem

2.2 Interfacial Polymerization of PAni into a CGN-PVA Matrix
In the interfacial synthesis of PAni, the molar ratio between aniline and APS was kept at 4:1 with equal volumes of aqueous and organic phase. In order to maximize interactions between PAni and CGN, ιCGN was used instead of κ -CGN.

In the synthesis described in this section, 0.15 g of ι-CGN was used in order to maintain the same molar concentration of repeating CGN units. The interfacial synthesis of PAni was studied with PVA only, ι-CGN only, with PVA and ι-CGN, and without PVA or ι-CGN. The aqueous solutions were prepared by dissolving the components in 30 mL of deionized water at 90°C with vigorous stirring for 1 hour. After dissolution of ι-CGN and PVA, the solution was allowed to cool prior to addition of 0.9 g of APS with sonication. It was observed that APS reacts with CGN on heating. Allowing the solution to cool first prevents the reaction from occurring. Organic phase was prepared by dissolving aniline in 30 mL of chloroform under sonication. The organic and aqueous solutions were allowed to cool in an ice bath before the aqueous phase was carefully poured onto the organic phase. The setups were kept under refrigeration without agitation for 18 hours. The setups were vacuum filtered with Whatman 40 filter paper then subsequently washed with EtOH, HCl, and deionized water. The precipitates were then dried in a vacuum oven at 60°C for 24 hours. Due to the small particle size of the PAni from Int-PAni-CGN-PVA, no precipitates were recovered from the solution. Instead, 5 mL of the filtrate was cast on a petri dish and dried in a vacuum oven at 60°C for 24 hours.

3. Results and Discussion

3.1 Fourier Transform Infrared Spectroscopy

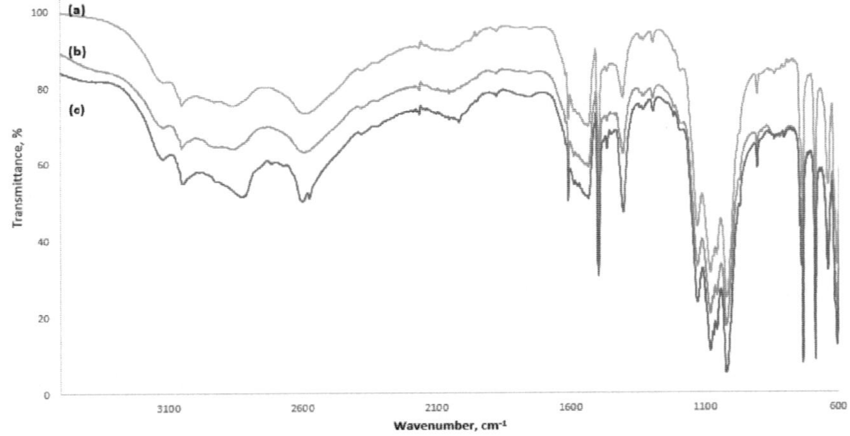

Figure 1. IR Spectra of (a) Int-PAni-κCGN-PVA-Cast, (b) Int-PAni-íCGN-PVA-Cast, and (c) Int-PAni-λCGN-PVA-Cast

IR spectra of Int-PAni-κCGN-PVA-Cast, Int-PAni-íCGN-PVA-Cast, Int-PAni-λCGN-PVA-Cast display similar peaks at the same positions as shown in Figure 1. All characteristic bands of PAni and the CGNs are observed. From the IR data, all three components interact similarly regardless of the type of CGN used. The spectra of the types of CGNs display similar characteristic peaks and they only differ by their of SO_4 peaks. κ-CGN has one peak corresponding to the 4-sulfate galactose functional group, í-CGN has two peaks corresponding to its 4-sulfate galactose and 2-sulfate-3,6-anhydrogalactose, and λ-CGN has two peaks corresponding to its 2-sulfate galactose and 6-sulfate galactose. All the SO_4 peaks occur between 800 cm^{-1} to 850 cm^{-1}. In the composites however, no sulfate bands are detectable since they overlap with the out-of-plane C-H deformation on the 1,4-disubstituted benzene ring of PAni which are found within the same range, but with greater intensities. The broad O-H band of CGN also significantly decreased in intensity in the composites. Relative to the other bands, the bands of the $N\text{-}H^{*+}$ stretches, the polaron band, in the C-H plane of the composites are relatively prominent. Longer bond length is associated with greater band intensities, suggesting interactions between the PAni $N\text{-}H^{*+}$ and the CGNs. The summary of the identified peaks and their corresponding interpretations is presented in Table 1.

Table 1. Position of characteristic FTIR peaks of (a) Int-PAni-κCGN-PVA, (b) Int-PAni-íCGN-PVA, and (c) Int-PAni-λCGN-PVA

	Bands	a	b	c
CGN	O-H stretch of water recrystallization	1680	1660	1650
	Sulfate symmetric and asymmetric vibrations	1375	1376	1374
	Galactose C-O and C-O-C vibrations	917, 1041	912, 1042	913, 1043
PAni	Quinoid deformation	1563	1567	1560
	Benzenoid deformation	1505	1505	1505
	C-N stretch	1312	1304	1316
	N-H+ stretch in C-H plane	1138	1134	1140
	Out-of-plane C-H deformation on 1,4-disubstituted benzene ring	821	823	818
	C-N+ stretch	1235	1242	1235
Composites	O-H and N-H Interactions	2600	2588	2599

IOP Conf. Series: Materials Science and Engineering **229** (2017) 012030 doi:10.1088/1757-899X/229/1/012030

3.2 Scanning Electron Microscopy

The SEM images of the samples showed varying degrees of homogeneity among the samples as shown in Figure 2. Int-PAni-κCGN-PVA, had a smooth morphology, the surface appears to be relatively uniform and smooth. Smooth surfaces are characteristic of blends with good homogeneity. Int-PAni-ίCGN-PVA had fibrous morphology but the fibers fused together as the film dried. Fusion of one-dimensional structures is characteristic of the formation of three-dimensional PAni structures. It was observed that granules formed on top of the fibrous surface. PVA inherently forms smooth flakes while PAni forms granules in an inhomogeneous blend. It can then be said that Int-PAni-λCGN-PVA cast had poor homogeneity compared to the other two samples due to the separation of PAni from the matrix. The image produced was expected since only κ-CGN and ί-CGN are able to form fibers in their pure form. The λ-CGN cannot form fibers due to the steric hindrance induced by its three sulfate groups.

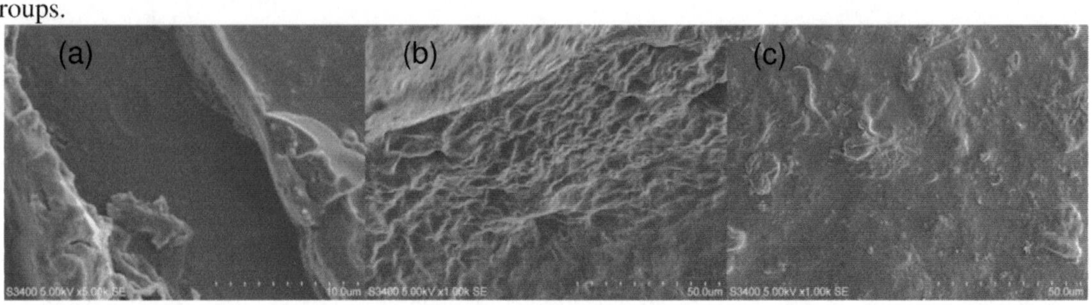

Figure 2. SEM Images directly casted from an aqueous layer containing (a) κCGN, (b) ίCGN, and (c) λCGN

Shown in Figure 3 are the SEM images of the composite film after soaking it in ethanol.

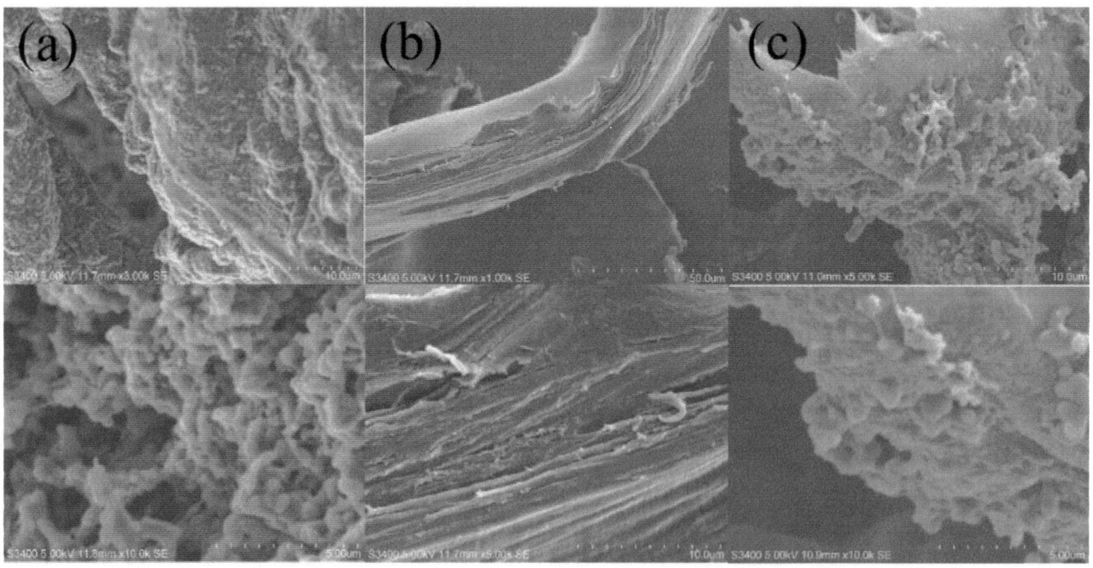

Figure 3. SEM Images After Ethanol Soaking of (a) Int-PAni-κCGN-PVA at x3.0k (top) and x10.0k (bottom), (b) Int-PAni-ίCGN-PVA at x1.0k (top) and x5.0k (bottom), and (c) Int-PAni-λCGN-PVA at x5.0k (top) and x10.0k (bottom)

The SEM image of Int-PAni-κCGN-PVA showed a granular morphology. The smooth flakes associated with PVA were not visible at the surface of the film. In Int-PAni-κCGN-PVA, the PAni agglomerated on the surface of the PVA fibers. The SEM image of Int-PAni-κCGN-PVA, Figure 3a showed a granular morphology. The smooth flakes associated with PVA were not visible at the surface of the film. In Int-PAni-κCGN-PVA, the PAni agglomerated on the surface of the PVA fibers. The SEM image of Int-PAni-λCGN-PVA, Figure 3c showed two distinct morphologies, similar to Int-PAni-λCGN-PVA-Cast. The smooth surface of the PVA is clearly seen underneath the PAni granules. Additionally, the sizes of the granules were less uniform than those of Int-PAni-κCGN-PVA. The SEM images of both interfacially polymerized samples containing λ-CGN indicate that it is not suitable for obtaining homogeneous composites with PAni and PVA. The most distinct morphology had been obtained from Int-PAni-ίCGN-PVA, Figure 3b. The image shows the cross-section of the film and part of the film's surface at the upper portion of the image. The edges of the fibers appeared clearly in the cross-section, while the surface of the film appeared smooth and void of any nanostructures. During drying therefore, the individual fibers aligned parallel to each other and self-assembled into planar sheets. The planar sheets then stacked on top of each other to produce the film.

3.3 UV-Vis and Solubility

Scraping 20 mg off from the films and dissolving in 30 mL of deionized water with vigorous stirring produced green solutions that remained green after filtration. The absorbance bands shown in Figure 4 at 429 nm, 424 nm, and 440 nm, confirmed the presence of the benzenoid transition in the Int-PAni-κCGN-PVA, Int-PAni-ίCGN-PVA, and Int-PAni-λCGN-PVA, respectively. Bands at 860 nm, 825 nm, and 864 nm confirm the presence of the polaron transition in the Int-PAni-κCGN-PVA, Int-PAni-ίCGN-PVA, and Int-PAni-λCGN-PVA, respectively. The presence of both transitions confirmed that the solutions of the dissolved films contain PAni emeraldine salt, the conducting form of PAni. Shown in the insets in Figure 4 are the photographs of the composite material that was dissolved in water. It has been observed that the particles are uniformly dispersed in the solvents and that a clear green solution was observed. This uniform dispersability indicates that the synthesized composite is higly soluble in water. However, this research only guarantees the solubility of the composite material to 0.67mg/mL as tested.

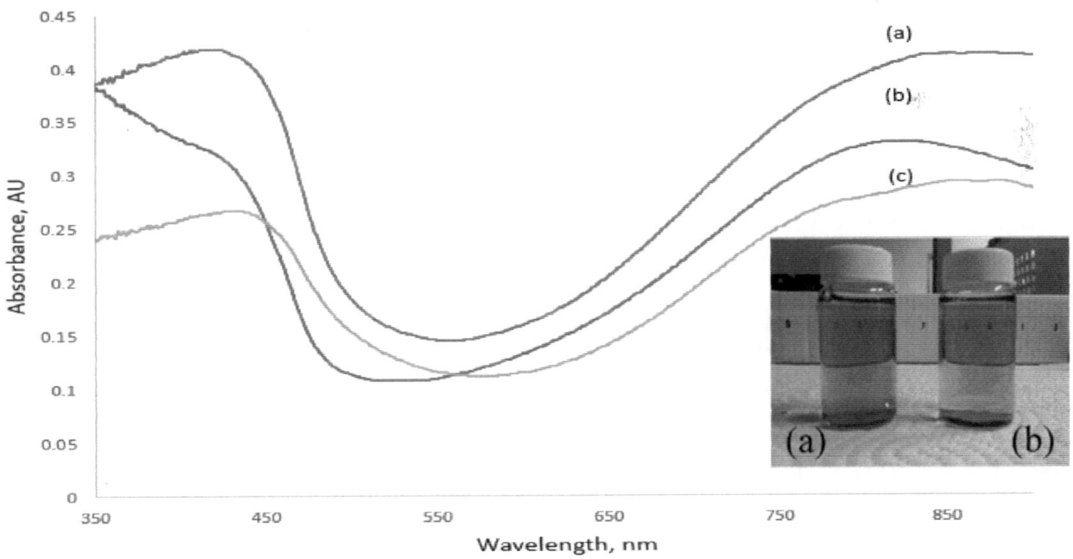

Figure 4. Absorbance Spectra (a) Int-PAni-κCGN-PVA, (b) Int-PAni-ίCGN-PVA, and (c) Int-PAni-λCGN-PVA , insets dissolved Solutions of (a) Int-PAni-κCGN-PVA, (b) Int-PAni-λCGN-PVA

4. Conclusion

In this study, PAni was synthesized by interfacial polymerization into a CGN-PVA matrix. It was found that excessive amounts of CGN or PVA lead to the gelation of the solution. In the case of the most suitable type of CGN to be used, it was observed that ί-CGN, being intermediate of κ-CGN and λ-CGN, imparted better properties when added to the aqueous matrix. κ-CGN produced composites with different morphologies when drop-casted or soaked in EtOH. Due to its inability to produce fibers, λ-CGN produced composites with poor homogeneity. In both the drop-casted and EtOH-soaked films, composites containing ί-CGN were produced possessing the morphologies of aligned nanofibers that underwent self-assembly into larger structures. The high solubility of the synthesized composite material in water could pave way to the development of thin films and eventually functional devices without the use of toxic solvents.

Acknowledgment

The authors would like to thank the Philippine Council for Industry, Energy and Emerging Technology Research and Development - Department of Science and Technology (PCIEERD-DOST) of the Philippine Government for the financial support they have extended for this study.

References

[1] Berrada, K.,Quillard, S.,Louarn, G., Lefrant, S. (1995) Polyaniline and Substituted Polyanilines: A Comparative Study of the Raman Spectra of Leucoemeraldine, Emeraldine and Pernigranaline. Synthetic Metals, 69, 201-204.

[2] Bhadra, S., Singha, N. K.,Khastgir, D. (2008) Semiconductive composites from ethylene 1-octene copolymer and polyaniline coated nylon 6: Studies on mechanical, thermal, processability, electrical, and EMI shielding properties. Polymer Engineering & Science, 48 (5), 995-1006.

[3] Subrahmanyam, A. R., Geetha, V., Kumar, A., Alakanandana, A.,Kumar, J. S. (2012) Mechanical and Electrical Conductivity Studies of PANI-PVA and PANI-PEO Blends. International Journal of Material Science, 2 (1), 27-30.

[4] Arenas, M. C., Sánchez, G., Martínez-Álvarez, O. (2014) Electrical and morphological properties of polyaniline–polyvinyl alcohol. Composites: Part B, 56, 857-861.

[5] Castillo-Ortega, M. M., Dal Castillo-Castro, T., Ibarra-Bracamontes, I., Nuño-Donlucas, S. M.; Puig, J. E., Herrea-Franco, P. J. (2007) Urea sensing film prepared by extrusion from DBSA-doped polyaniline-poly(styrene-co-potassium acrylate) in a poly(n-butyl methacrylate) matrix. Sensors and Actuators B: Chemical, 125 (2), 538-543.

[6] Vega-Rios, A., Olmedo-Martinez, J. L., Farías-Mancilla, B., Hernández-Escobar, C. A., Zaragoza-Contreras, E. A.(2014) Synthesis and electrical properties of polyaniline/iota carrageenan biocomposites. Carbohydrate Polymers, 110, 78-86.

[7] Yavuz, A. G., Uygun, A., Bhethanabotla, V. R. (2009) Substituted polyaniline/chitosan composites: Synthesis and characterization. Carbohydrate Polymers, 75, 448-453.

[8] Huang, J., Kaner, R. B. (2004)A General Chemical Route to Polyaniline Nanofibers. Journal of the American Chemical Society, 126 (3), 851-855.

IOP Conf. Series: Materials Science and Engineering **229** (2017) 012031 doi:10.1088/1757-899X/229/1/012031

Influence of the brittle behavior of work materials on polishing characteristics

Satoshi Sakamoto[1], Masaya Gemma[1], Keitoku Hayashi[1], Yasuo Kondo[2], Kenji Yamaguchi[3], Takao Yakou[1] and Susumu Arakawa[4]

[1]Yokohama National University, 79-2 Tokiwadai, Hodogaya-ku, Yokohama, Japan
[2]Yamagata University, 4-3-16 Jonan, Yonezawa, Japan
[3]Yonago National College of Technology, 4448 Hikona-cho, Yonago, Japan
[4]Kurashiki Boring Kiko Co., Ltd., 2-4-20 Matsue, Kurashiki, Japan

*Corresponding E-mail: sakamoto-satoshi-tv@ynu.ac.jp

Abstract. Diamond electrodeposited wire tools are frequently used to cut thin wafers from hard and brittle materials. However, microcracks sometimes appear during the slicing process. The appearance of microcracks adversely affects slicing efficiency and slicing accuracy. In this study, we examine the influence of brittle behavior on the polishing characteristics such as polishing depth and tool wear. This is the first step toward investigating the influence of the brittle behavior of work materials on slicing characteristics. Ceramics such as alumina, silicon carbide, and zirconia are used as work materials. Even with the same degree of hardness, we found that the polishing depth values were greater for materials exhibiting brittle behavior. In the polishing of high-hardness materials, abrasive grains were badly damaged during the initial stages of polishing. Damage to the abrasive paper was less in wet polishing as compared with dry polishing. Moreover, wet polishing had a greater polishing depth than dry polishing. The polishing characteristics of the brittle materials were similar to the grooving characteristics produced using wire tools; however, both these characteristics depend on the brittle behavior of the work materials. Therefore, by performing simple polishing tests, estimating the state of grooving or slicing using wire tools is possible.

1. Introduction

Multi-wire sawing is an excellent slicing method for electronic or optical parts made from hard and brittle materials, such as silicon ingots, magnetic materials, ceramics, and sapphires. In recent years, multi-wire saws using diamond electrodeposited wire tools have gained acceptance as a mainstream slicing method for large-diameter ingots or high-hardness materials [1, 2]. However, microcracks may occur during the slicing of hard and brittle materials. Such microcracks negatively influence the slicing efficiency and the slicing accuracy. In addition, they prevent wafer thinning. Therefore, it is important to clarify the relation between the brittle behavior of the work materials and the processing characteristics.

In our previous studies, we have investigated multi-wire sawing to achieve precise and very efficient slicing of hard and brittle materials [3, 4]. This study aims to examine the influence of the brittle behavior of the work materials on slicing characteristics such as the generation mechanism of the sliced surface and wear characteristics of the wire tool. In this report, we examine the influence of brittle behavior on the polishing characteristics. This is the first step toward investigating the influence

Content from this work may be used under the terms of the Creative Commons Attribution 3.0 licence. Any further distribution of this work must maintain attribution to the author(s) and the title of the work, journal citation and DOI.
Published under licence by IOP Publishing Ltd

of the brittle behavior of the work materials on the slicing characteristics. Therefore, we performed loading-unloading tests and polishing experiments for various ceramics.

2. Experimental procedure

Figure 1 shows the schematic of a nanoindenter used in this study and summarizes the conditions for the loading–unloading tests (see table 1). Loading–unloading tests were performed using a triangular pyramid diamond indenter. The measuring force was 500 mN, and the loading rate was 0.7 mN/s. The holding time was 10 s.

Figure 2 shows the schematic of the reciprocating abrasion tester used for the experiment. Table 2 summarizes the conditions for the polishing experiments and materials used. The distance of the reciprocating movement was 100 mm, and the polishing time was a maximum of 30 min. The table speed (i.e., the reciprocating speed) was 1212 mm/min. The polishing load applied to the work materials was 26.7 N. Ceramic plates such as alumina (Al_2O_3), silicon carbide (SiC), and zirconia (ZrO_2) were used as work materials. The cross-sections of the work materials were rectangles with dimensions of 10 mm × 3 mm. In the previous study, we had clarified that Al_2O_3 and ZrO_2, which we used as work materials, have almost the same hardness. SiC has a greater hardness than both Al_2O_3 and ZrO_2 [5]. In wet polishing, a glycol-based water-soluble coolant was used as the working fluid. The wet polishing experiments were conducted in a liquid medium.

A scanning electron microscope (SEM) image of the unused diamond abrasive paper is shown in figure 3. The abrasive paper was made up of diamond grains fixed on a copper plate via nickel plating. The particle size of the diamond grains was approximately 30 μm.

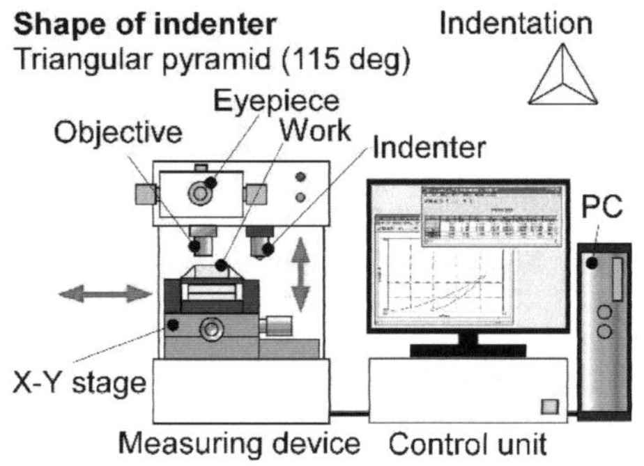

Figure 1. Schematic of the nanoindenter.

Table 1. Experimental conditions for loading-unloading test.

Indenter	
Shape	Triangular pyramid
Ridgeline angle	115 [°]
Indenter	500 [mN]
Indenter	0.7 [mN/s]
Indenter	10 [s]

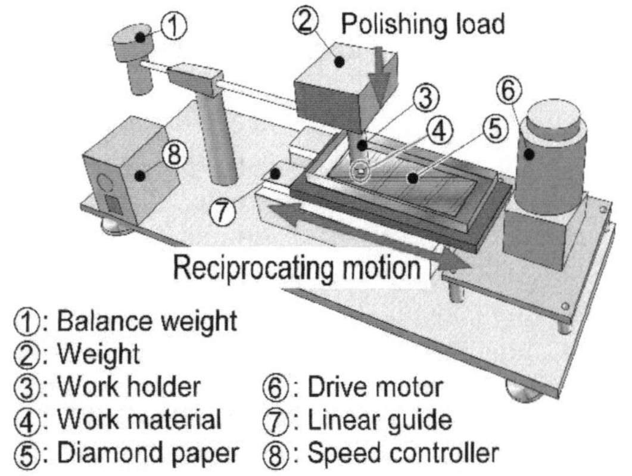

Figure 2. Schematic of the abrasion experimental setup.

Table 2. Conditions for polishing experiments and materials used.

Abrasive paper	
Abrasives	Diamond grain
Particle size	30 [μm]
Base material	Nickel-plated copper plate
Work material	Al_2O_3, SiC, ZrO_2
Cross section	Rectangle (3 × 10 [mm^2])
Polishing speed	1212 [mm/min]
Polishing load	26.7 [N]
Polishing time	30 [min]
Moving distance	100 [mm]

IOP Conf. Series: Materials Science and Engineering **229** (2017) 012031 doi:10.1088/1757-899X/229/1/012031

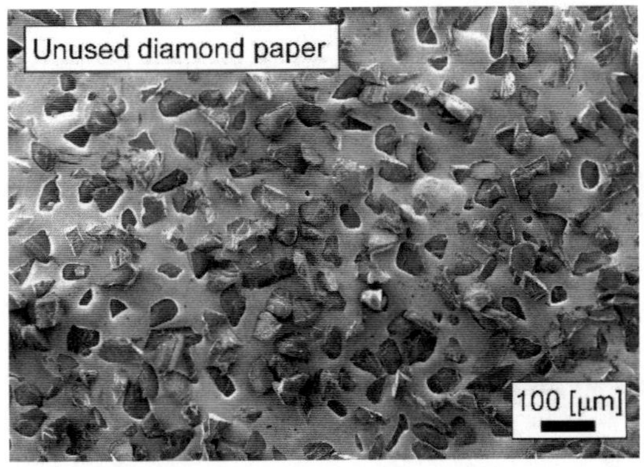

Figure 3. SEM image of the unused diamond abrasive paper

3. Experimental results and discussion

3.1. Results of loading-unloading test

Results of the loading–unloading test are shown in figure 4. For Al_2O_3 and SiC, we confirmed the presence of multiple microfractures that preceded the indenter. However, no microfractures were observed in the loading–unloading test performed for ZrO_2. This result proved that Al_2O_3 and SiC demonstrate more brittle behavior than ZrO_2.

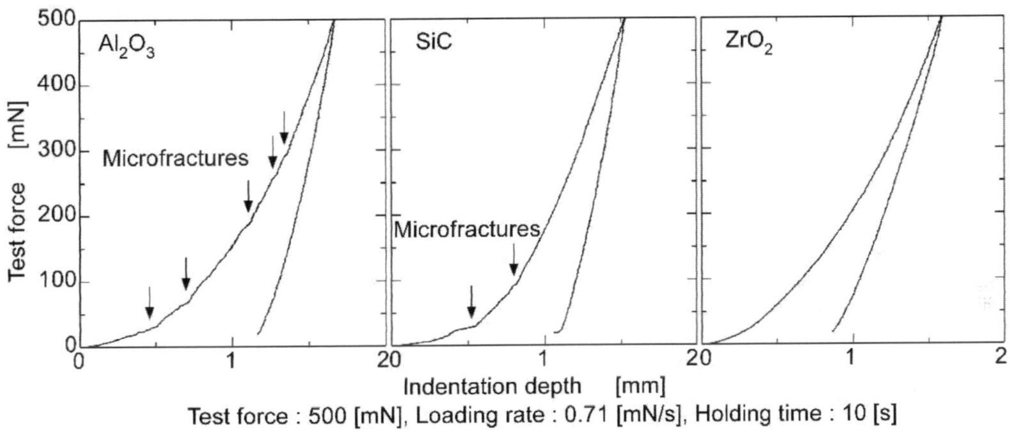

Figure 4. Results of loading-unloading test.

3.2. Influence on polishing depth

Figure 5 shows the relation between the polishing time and the polishing depth. This is the result of dry polishing. The polishing depth of each material increases with the passage of polishing time. The polishing depth of Al_2O_3 is greater than ZrO_2 of similar hardness. The brittle behavior of Al_2O_3 is remarkable compared with other work materials, which is one of the reasons for the increase of the polishing depth (see figure 4). In addition, SiC with high hardness has a slightly greater polishing depth than ZrO_2. SiC has high hardness, and ZrO_2 has high toughness; therefore, we did not observe any remarkable increase in the polishing depth accompanying the polishing time.

Figure 6 shows the relation between the polishing time and the polishing depth in wet polishing. The polishing depth of wet polishing is much greater than that of dry polishing. As compared with dry polishing, the polishing depth of each material increased linearly with the passage of polishing time (see figure 5). This is because the loading (or clogging) of the diamond abrasive paper was suppressed. In particular, remarkable effects are obtained with Al_2O_3 and ZrO_2. Diamond grains of the abrasive papers were badly damaged during SiC polishing because SiC has high hardness. Therefore, the effect of suppressing the loading is considered to be small.

Figure 7 shows the polishing depth per unit time in dry polishing, and figure 8 shows the results of wet polishing. In wet polishing, the polishing depth per unit time is much greater than that of dry polishing. The polishing depth of the work materials decreased with the passage of polishing time. However, in wet polishing of Al_2O_3 and ZrO_2, it was possible to maintain the polishing depth per unit time. We obtained these results because the loading (clogging) of the diamond abrasive paper was suppressed. Although SiC has high hardness, it showed brittle behavior similar to Al_2O_3; therefore, the polishing depth at the initial stage of polishing was high. However, the polishing depth per unit time of SiC greatly decreased with the lapse of polishing time. This is because polishing of SiC having high hardness caused severe damage to the abrasive paper at the initial stages of polishing.

These results obtained are similar to the trends observed in groove processing using a diamond electrodeposited wire tool [5, 6].

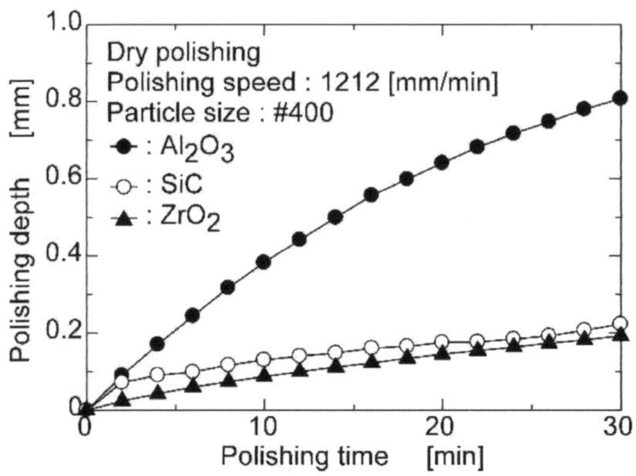

Figure 5. Relation between polishing time and polishing depth in dry polishing.

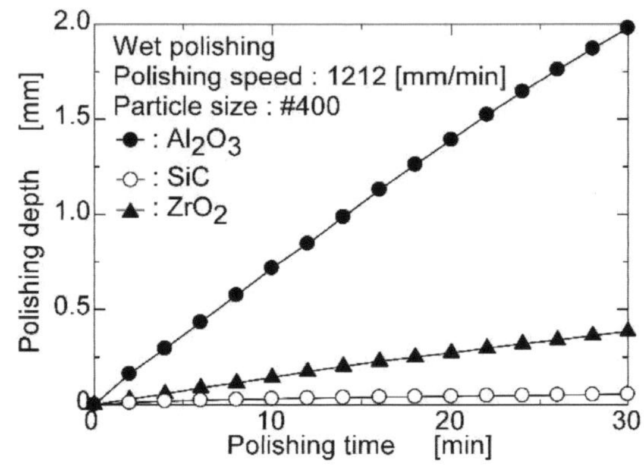

Figure 6. Relation between polishing time and polishing depth in wet polishing.

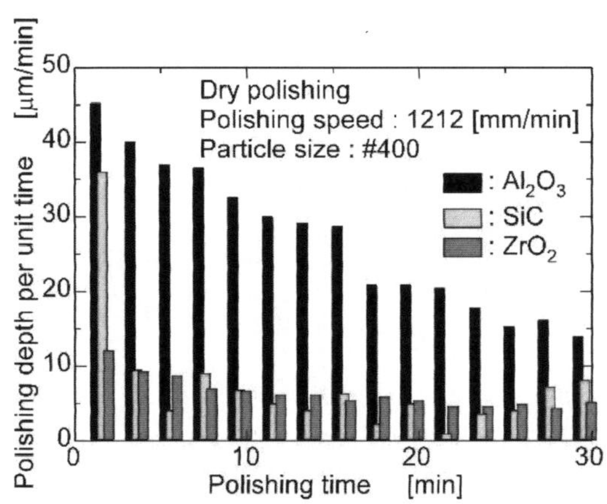

Figure 7. Relation between polishing time and polishing depth per unit time in dry polishing.

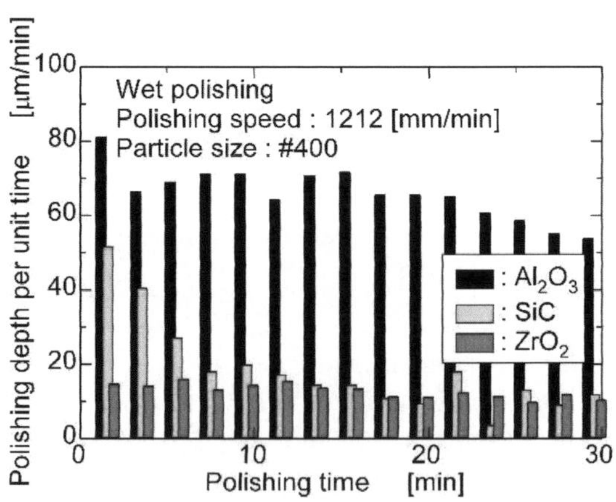

Figure 8. Relation between polishing time and polishing depth per unit time in wet polishing.

3.3. Influence on polished surface

Figure 9 shows images of polished surfaces for dry polishing. The polishing time was 30 min. The properties of the polished surfaces of Al_2O_3 and SiC are quite similar. However, scratch marks due to the diamond abrasive grains of the abrasive paper were clearly visible on the polished surface of ZrO_2. This is probably because ZrO_2 has greater toughness than Al_2O_3 and SiC. The wet polished surface was almost the same as the dry polished surface, and scratch marks were clearly visible only on the polished surface of ZrO_2.

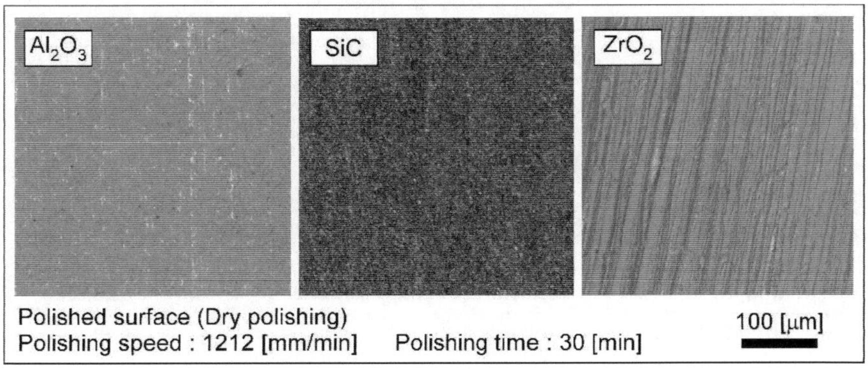

Figure 9. Examples of results of load-unloading test.

3.4. Influence on diamond abrasive paper after polishing

Figure 10 shows the images of the diamond abrasive papers after dry polishing for 30 min. All the diamond abrasive papers were severely damaged in the dry polishing experiments. In addition, loading occurred in all the abrasive papers. In the polishing of Al_2O_3, the diamond abrasive grains were worn out; however, the diamond abrasive grains could be clearly observed on the abrasive paper. In the polishing of ZrO_2, diamond abrasive grains could be observed despite severe loading. However, it was impossible to clearly confirm the presence of diamond abrasive grains on the abrasive paper after the polishing of SiC.

Figure 11 shows the images of the diamond abrasive papers after wet polishing for 30 min. As with dry polishing, all of these abrasive papers were damaged. Particularly, after the polishing of SiC, the diamond abrasive grains in the abrasive paper were worn out and flattened. However, the presence of diamond abrasive grains could be confirmed on all abrasive papers. In wet polishing, if the lubrication is improved and loading is suppressed, damage to the abrasive paper will be reduced.

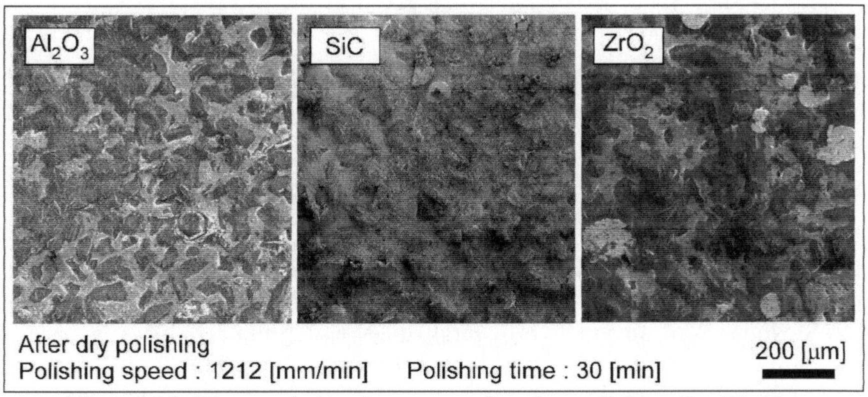

Figure 10. Diamond abrasive paper after dry polishing for 30 min.

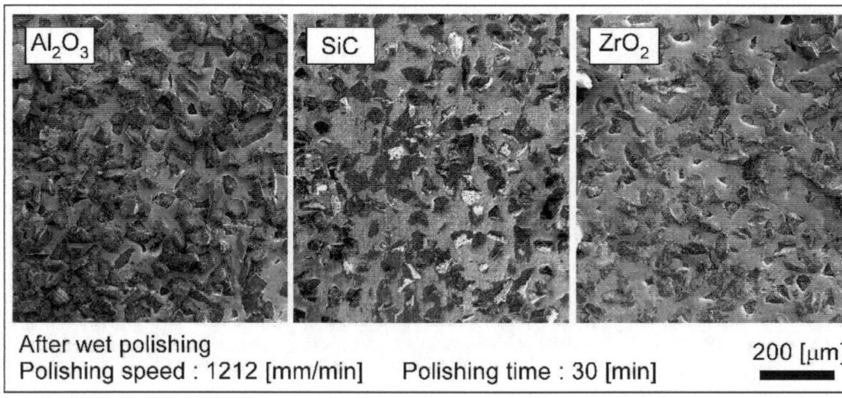

Figure 11. Diamond abrasive paper after wet polishing for 30 min.

These results are similar to the trends observed in our previous experiments wherein the wear characteristics of a wire tool were observed during grooving using a diamond electrodeposited wire tool [5, 6]. In other words, it became clear that it is possible to estimate the machining state when using a wire tool by performing simple polishing tests.

4. Conclusions

We experimentally investigated the influence of the brittle behavior of work materials on polishing characteristics. The main conclusions obtained from this study are as follows. The hardness and brittle behavior of the work materials affect the polishing characteristics. Even with materials having the same degree of hardness, the polishing depth of materials that exhibit brittle behavior will be greater. In the polishing of high-hardness materials, the abrasive grains were badly damaged in the early stages of polishing. Damage to the abrasive paper was less in wet polishing as compared to dry polishing, and the polishing depth in wet polishing was greater than dry polishing. The results of this study confirm the results obtained by grooving using a diamond electrodeposited wire tool. Therefore, it is possible to estimate the state of grooving or slicing using fine wire tools by performing simple polishing tests.

Acknowledgments

We thank Mr. Ryuichi Iida (Tokyo Gakugei University) and Instrumental Analysis Center of Yokohama National University for their support and cooperation. This work was supported by JSPS KAKENHI Grant Number JP12345678.

References

[1] H Wu 2016 *Precis. Eng.* **43** 1
[2] Ishikawa K et al. 1995 *High Efficiency and High Precise Slicing of Hard and Brittle Materials* (Tokyo: Industrial Publishing & Consulting, Inc.)
[3] Sakamoto S, Kondo Y, Yamaguchi K, Murakami N and Akita N 2006 *Proc. 8th Int. Conf. on Progress of Machining Technol. (Matsue)* (Hiroshima: Technical Committee of Machining of Difficult-to-cut Materials of the Japan Society for Precision Engineering) p 397
[4] Sakamoto S, Yamaguchi M, Kondo Y, Yamaguchi K and Nomura J A 2014 *Key Eng. Mater.* **625** 597
[5] Sakamoto S, Gemma M, Hayashi K, Kondo Y, Yamaguchi K, Yamaguchi M and Fujita T 2016 *Key Eng. Mater.* **703** 17
[6] Sakamoto S, Hayashi K, Gemma M, Kondo Y, Yamaguchi K, Yamaguchi M and Fujita T 2016 *Key Eng. Mater.* **719** 132

IOP Conf. Series: Materials Science and Engineering **229** (2017) 012032 doi:10.1088/1757-899X/229/1/012032

Feature Extraction from Sensor Data Streams for Optimizing Grinding Condition

Y Kondo[1,*], Y Higashimoto[1], S Sakamoto[2], T Fujita[3] and K Yamaguchi[3]

[1] Graduate School of Science & Engineering, Yamagata University, 4-3-16, Jhonan, Yonezawa, 992-8510, Japan
[2] Yokohama National University, 19-1 Tokiwadai, Yokohama, 240-8501, Japan
[3] Yonago National College of Technology, 4448 Hikona, Yonago, 683-8502, Japan

*kondo@yz.yamagata-u.ac.jp; http://eastbook.yz.yamagata-u.ac.jp/

Abstract. A visualization method for time-series sensing data was designed to optimize grinding condition. The fluctuation pattern of time-series data streams can be visualized as a white and black pattern by utilizing the spindle power change rate average. The designed visualization method was applied to a condition monitoring in lapping operation. The relation between the fallout abrasive grain content and lapping behaviour was experimentally examined. In the lapping with grinding fluid containing no fallout abrasive, the spindle power decreased in a monotone manner with lapping time, while in the lapping with fallout abrasive, the spindle power decreased with lapping time up to 20s of lapping and then tended to converge on a constant value. The spindle power change rate average displayed as a white and black pattern reproduced the changes of spindle power very well. The appearance probability of white or black pattern has a strong relation with the fallout abrasive content and the designed data processing scheme could make possible to predict the grinding fluid condition from the easy-handling grinding test.

1. Introduction

The condition monitoring is one of the most important requirements for the automated and unattended manufacturing systems. The typical monitoring technique operates according to the vibrations, force, power consumption and acoustic emission because these are influenced by the cutting tool state and the material removal process conditions [1],[2]. Most of the promising methods have low detection sensitivity without a variety of statistical processing because the sensing data often show the large dispersion even in the data obtained from the machining under the same condition.

The authors have been studying tool condition monitoring using spindle power consumption [3]. The spindle power consumption is a fairly accurate measure of the deterioration of tool condition and its evolution is a picture of the evolution of cutting force. However, the motor current always fluctuates 1 to 30% due to the temperature rise, magnetic loss and so on [4]. This means that an alternative feature extraction method is required to monitor the tool condition with a high precision from the time-series sensing data of spindle power consumption.

In this study, we will design a visualization method for spindle power consumption change in grinding operation to understandably display the fluctuation pattern of time-series sensing data. The applicability of designed visualization method to the condition monitoring in grinding operations was experimentally examined through the lapping test of SUS304.

Content from this work may be used under the terms of the Creative Commons Attribution 3.0 licence. Any further distribution of this work must maintain attribution to the author(s) and the title of the work, journal citation and DOI.
Published under licence by IOP Publishing Ltd

2. Feature extraction from time-series data streams

Figure 1 shows the condition monitoring algorithm for grinding and lapping operations. For example, 4 time-series sensing data were supposed to be collected during grinding operations. Series 1 & 2 and series 3 & 4 show the very similar fluctuation pattern but have a different absolute value with each other. The similarity of fluctuation pattern of time-series data was selected as the feature of time-series data stream for detecting the state of grinding and lapping operation, because the motor current always fluctuates 1 to 30% due to the temperature rise, magnetic loss and so on.

To discuss the similarity of fluctuation pattern, spindle power change rate average was utilized. At 1st, the time series data was segmented at fixed time intervals and the spindle power change rate average, which represents the magnitude and the changing direction of spindle power, was calculated for each segment. The calculated spindle power change rate average would have so much uncertainty since the most time-series sensing data consists of long-term data (trend component) and period data (noise component). To eliminate the noise component, the spindle power change rate average with the absolute value less than the predefined threshold α was removed. The residual spindle power change rate average is displayed as a white and black pattern in which the black pattern shows the segment with the absolute value more than α and the white pattern shows the segment with the absolute value less than α. The completed white and black pattern corresponds to a simplified expression of time-series sensing data, spindle power consumption, in grinding and lapping operations.

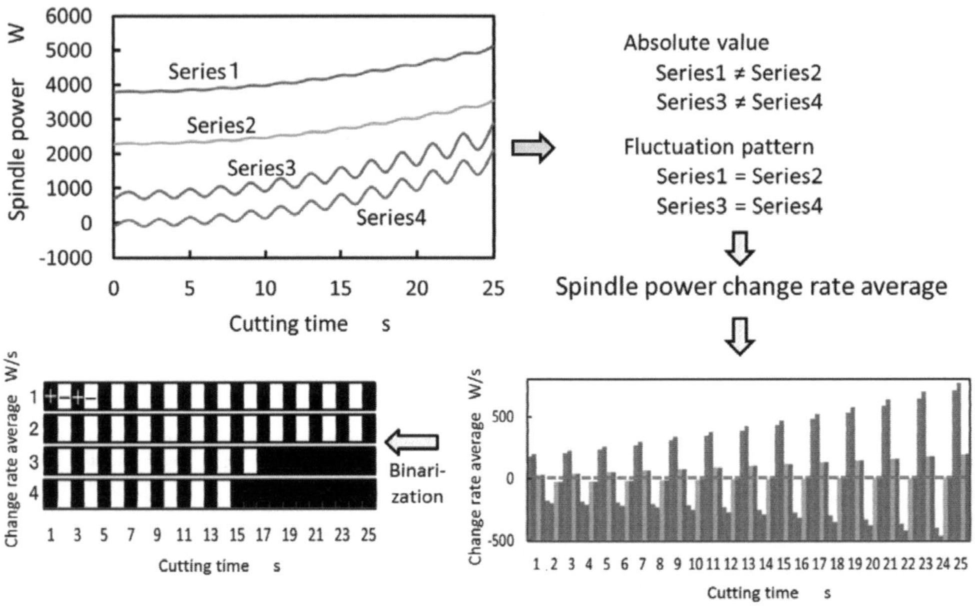

Figure 1. Condition monitoring algorithm for grinding and lapping operations

3. Experimental method

We examined the applicability of designed method to predict the contamination condition of grinding fluid, through the lapping operation of AISI304. Figure 2 shows the schematic drawing of lapping apparatus. Three pieces of work materials (AISI304) are lapped at a time with #1000 SiC abrasive paper under the condition of 2.1 [m/s] of peripheral speed of lapping disk (φ200mm). During the lapping operation, the coolant is supplied on the abrasive paper at 20 cm^3/min of pouring rate. A simulated grinding fluid, which is 5% water-soluble grinding fluid containing 0, 0.001, 0.005, 0.01 or 0.05 wt% of #800 GC abrasive grain, was prepared. The GC abrasive grain is simulating the fallout abrasive grains released from a grinding wheel into coolant. The power consumed P [kW] by the electric motor of automatic polisher was measured at 0.2s interval using a power meter (HIOKI 3330). As the rule, the power consumed is given by

$$Ft \cdot V = 1000 \times P \tag{1}$$

where Ft [N] is the tangential force on abrasive and V [m/s] is the peripheral speed of abrasive [5]. Once the power consumed by electrical motor had been measured, the tangential force Ft can be calculated easily.

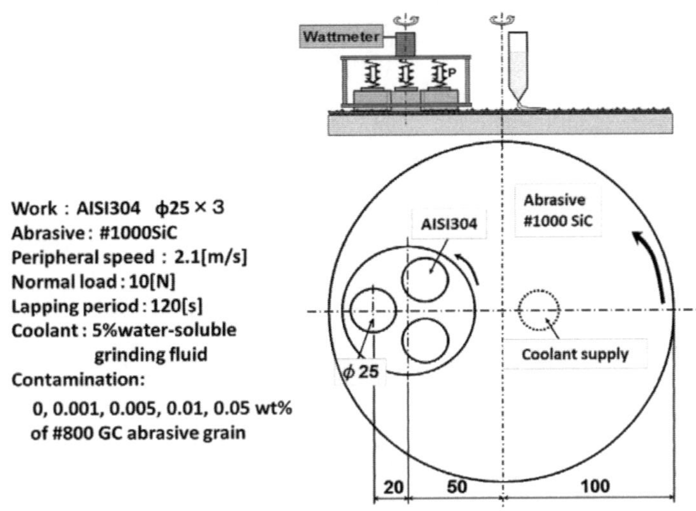

Figure 2. Schematic drawing of lapping apparatus and lapping conditions

4. Results and discussions

4.1 Spindle power change in lapping operation

Figure 3 summarizes the spindle power changes when the SUS304 was lapped for 120s using a simulated grinding fluid containing 0, 0.001, 0.005, 0.01 or 0.05 wt% of #800 GC abrasive. In the lapping with grinding fluid containing no #800 GC abrasive (virgin fluid), the spindle power decreased in a monotone manner with lapping time. While in the lapping with grinding fluid containing #800 GC abrasive, the spindle power decreased with lapping time up to 20s of lapping and then tended to converge on a constant value. When looking at the period (noise) component, the amplitudes of noise component tends to increase with the #800 abrasive grain content. This indicates that the long-term (trend) component in spindle power change has a stronger relation with the abrasive grain content rather than the short-term (noise) component.

4.2 Feature extraction based on power change rate average

The time-series spindle power data was divided into small segment at interval of 0.6s and then the spindle power change rate average was calculated for each segment. The interval of 0.6s was determined from the preliminary experiments. Figure 4 summarizes the change of spindle power change rate average when the SUS304 was lapped for 120s using a simulated grinding fluid containing 0, 0.001, 0.005, 0.01 or 0.05 wt% of #800 GC abrasive. Positive and negative values alternately appear in the spindle power change rate average stream, independent of the abrasive content. This means that it is difficult to define the feature of time-series spindle power data from the change rate average stream because the fluctuation width of period (noise) component is equal or larger than that of long-term (trend) component in spindle power change.

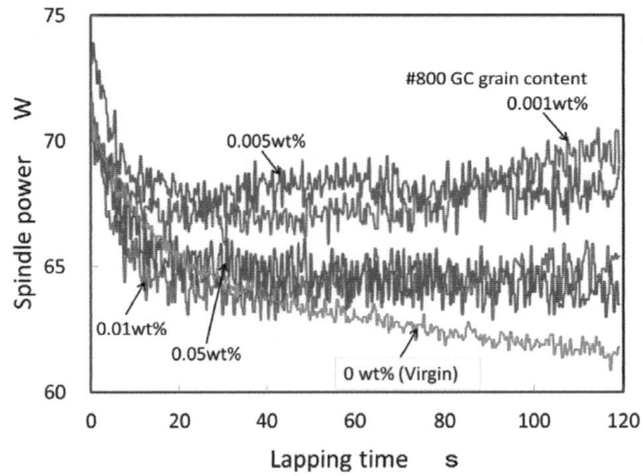

Figure 3. Spindle power change in lapping operation

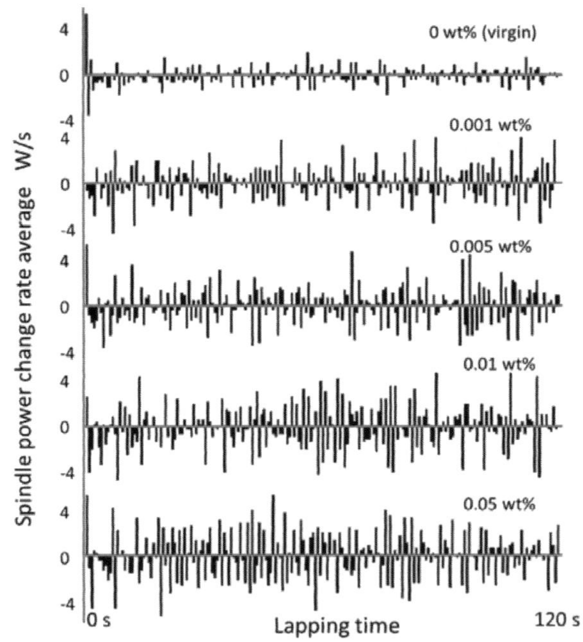

Figure 4. Change of spindle power change rate average

To extract the trend component from spindle power change, elimination of period component was attempted. Figure 5 shows the distribution curve of spindle power change rate average in the lapping with virgin fluid. The distribution has a normal distribution with the standard deviation of σ= 0.63. If the predefined threshold α sets the standard deviation σ= 0.63, the spindle power change rate average with the absolute value more or less than σ is the top 15.87% or the lower 15.87%. In this study, we defined that the top 15.87%, the middle 15.87 ~ 84.13% and the lower 15.87% of spindle power change rate average have a certainly rising spindle power slope interval, the slope interval with a large cumulative error and a certainly falling spindle power slope interval, respectively.

Figure 6 shows the spindle power change rate average displayed as a white and black pattern in which the black pattern shows the segment with the certainly rising spindle power slope interval and the white pattern shows the segment with a large cumulative error or a certainly falling spindle power slope interval. The completed white and black pattern reproduces the changes of spindle power showed in Fig.4. This shows that the fluctuation pattern of spindle power can be visualized by processing the time-series data according to the processing scheme of Table 1.

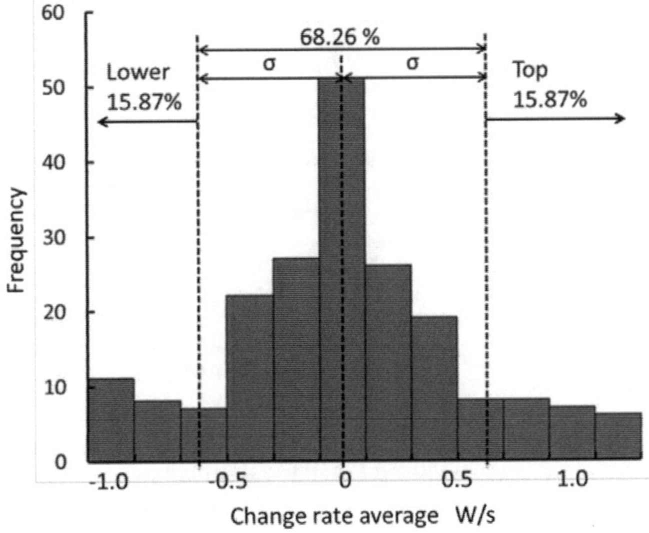

Figure 5. Distribution curve of spindle power change rate average

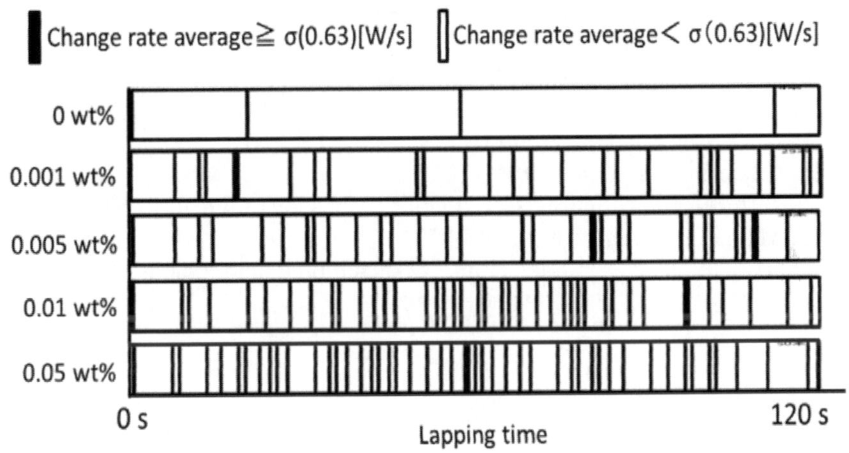

Figure 6. Spindle power change rate average displayed as a white and black pattern.

5. Conclusion

We designed a visualization method for spindle power consumption change in grinding operation to understandably display the fluctuation pattern of time-series sensing data. The fluctuation pattern of time-series data streams can be visualized as a white and black pattern by utilizing the spindle power change rate average. The designed visualization method was applied to a condition monitoring in lapping operation. The spindle power change rate average displayed as a white and black pattern reproduced the changes of spindle power very well. The appearance probability of white or black pattern has a strong relation with the fallout abrasive content and the designed data processing scheme could make possible to predict the grinding fluid condition from the easy-handling grinding test.

Table 1. Data processing scheme

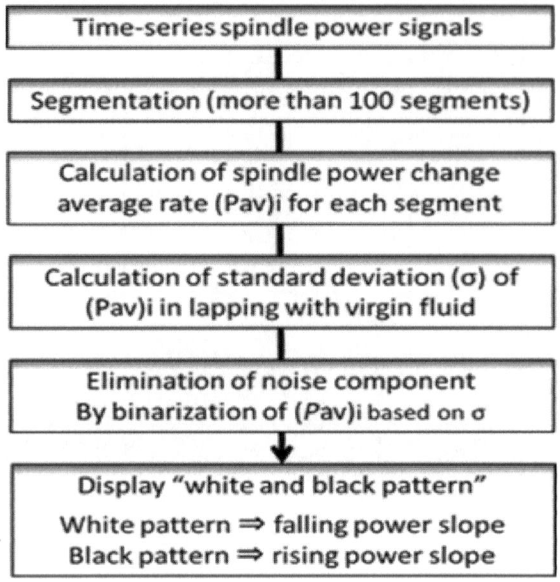

References

[1] Tonshof, H.K. et al: Development and Trends in Monitoring and Control of Machining Process, Ann. CIRP, 37-2, (1988), 611-622.

[2] S.Madhave Reddy, A.Chennakesava Reddy: Latest Development in Condition Monitoring of Machining Operations, J. of Applied Science, 12(10), 938-946, (2012).

[3] Y.Kondo et al. :A Supervisory System in Mechanical Process Leading to Electric Powe Saving, Key Eng. Materials, vol.624, (2014), 123-128.

[4] Ketteler G. : Influence on Monitoring Systems which Affect their Reliability, 2nd Int. workshop of Intell. Manuf. Systems, (1999), 768-777.

[5] Boothroyd, G., Knight W., Fundamentals of machining and machine tool, Marcel Dekker, (1989).

IOP Conf. Series: Materials Science and Engineering **229** (2017) 012033 doi:10.1088/1757-899X/229/1/012033

Experimental Study of Tool Wear and Grinding Forces During BK-7 Glass Micro-grinding with Modified PCD Tool

A Pratap[1], P Sahoo[2], K Patra[3*] and A A Dyakonov[4]

[1, 2, 3] Indian Institute of Technology Patna, Bihta, Patna-801103, Bihar, India
[4] South Ural State University, 76, Lenin Avenue, Chelyabinsk, 454080, Russian Federation

* kpatra@iitp.ac.in

Abstract. This study focuses on the improvement in grinding performance of BK-7 glass using polycrystalline diamond micro-tool. Micro-tools are modified using wire EDM and performance of modified tools is compared with that of as received tool. Tool wear of different types of tools are observed. To quantify the tool wear, a method based on weight loss of tool is introduced in this study. Modified tools significantly reduce tool wear in comparison to the normal tool. Grinding forces increase with machining time due to tool wear. However, modified tools produce lesser forces thus can improve life of the PCD micro-grinding tool.

1. Introduction

Glass is a difficult to machine material which possesses very high hardness, no conductivity, high corrosion resistance and extremely fine aesthetic properties for which it is widely used in micro fluidics, optical lenses and beam splitting lenses [1, 2]. Glass and other hard and brittle materials are never easy to machine as they offer high resistance to tool and create large machining forces which results in rapid tool wear. In the recent concluded decades, mostly unconventional processes such as laser micromachining [3], electro discharge machining [4] and ultrasonic machining [5] were used to create micro-features in hard to machine material surfaces because they are non contact type machining processes where cutting tool does not come in direct contact with the workpiece and hence no problem of tool wear. However, these processes are also not one stop solution for the machining of all types of brittle and hard materials as some processes like EDM can machine only conductive materials whereas laser machining produces sever heat affected zone and surface damage. Due to above discussed limitations of unconventional processes; development of some mechanical micromachining process becomes inevitable. Due to high brittleness of glass, crack formation is very common during machining. To avoid brittle cracks during machining, it is desirable to achieve ductile mode failure in brittle materials. In the recent times, micro-grinding process emerged as a solution for the micro-machining of hard and brittle materials where diamond and cubic boron nitride (CBN) abrasive tools were used [6, 7]. Micro-grinding can be done at very small depth of cuts so that ductile mode material removal can be maintained. While choosing depth of cut, it is also to be considered that depth of cut should not be less than critical depth of cut otherwise more ploughing will take place without actual chip formation [8]. Two types of micro-grinding tools have been normally applied so far, one is coated and another is solid. Diamond coated tools were used to produce high quality surfaces in optical glasses [4, 9]. On the other hand, different shaped polycrystalline diamond solid tools e. g. square shape, D-shape, circular shape were applied for micro-grinding of BK-7 glass to produce different levels of surface finish [5]. Cavity or groove at the tool bottom was also observed as

Content from this work may be used under the terms of the Creative Commons Attribution 3.0 licence. Any further distribution of this work must maintain attribution to the author(s) and the title of the work, journal citation and DOI.
Published under licence by IOP Publishing Ltd

an effective technique to remove the micro-chips from the tool-workpiece interaction area during micro-grinding of quartz glass [10]. For the reduction of burrs and micro-cracks, coating of the workpiece surface with different types of resins was done before performing micro-grinding. Coating of the surface produces a restraining effect on the crack propagation [11]. Recently, ultrasonic vibration assisted micro-grinding of silica glass has been performed to reduce the grinding forces and to improve the surface quality. Normal and tangential forces also declined by at least 50% because of abrasive-work impact effect and interrupted cutting [12, 13].

Chip removal from the grinding area is a key factor for producing smooth surface and reducing grinding forces along with tool wear. As discussed above, different types of tool modification strategies were adopted to remove the chips effectively from the grinding zone and to prevent lodging of the tool. This study is focused on reduction of tool wear through tool modification. Two types of modified PCD micro-grinding tools are prepared through micro-WEDM. One tool has one groove on its end face while another one has two perpendicular grooves on its end face. Then, performance of three tools e. g. tool without any groove (NG), tool with one groove (1G) and tool with two grooves (2G) are observed in the micro-grinding of BK-7 glass based on grinding forces and tool wear.

2. Experiments

2.1. Experimental setup

All micro-grinding experiments were carried out on CNC hybrid micro machine tool- DT110 (Mikrotools Pte. Ltd, Singapore) having programmable multi axis controller (PMAC). Maximum allowable bed movements in x, y, z directions are 200, 100 and 100 mm, respectively. Position accuracy of the machine is 1μm/ 100 mm. Range of spindle speed is 100 rpm to 60000 rpm and it uses AC servo motor controlled angular contact bearing for high precision rotation. The micro-grinding setup for grinding of BK-7 glass using different shaped PCD micro-grinding tools is shown in figure 1.

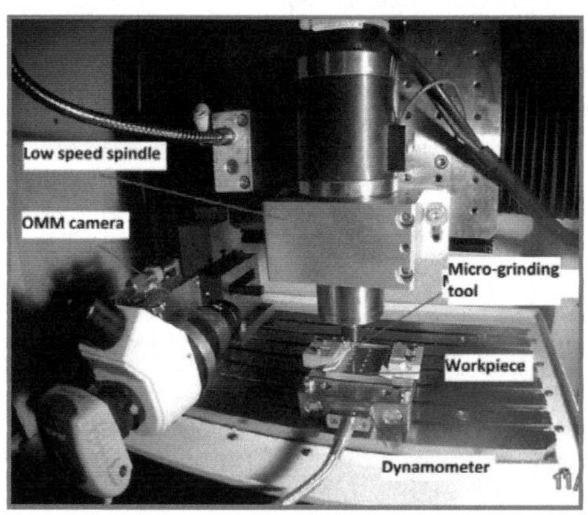

Figure 1. Experimental setup of micro-grinding.

2.2. Tool modification

Micro-tool used in this study is made up of polycrystalline diamond tip brazed on a 3 mm diameter tungsten carbide shank. Tip diameter of the tool is approximately 0.70 mm and it has included taper angel of 45°. Micro-grooves on the end face of the tools are fabricated through micro-WEDM using setup as shown in figure 2(a). Zinc coated brass wire of diameter approximately 0.07 mm is used as sacrificial electrode during WEDM. Operating parameters of micro-WEDM are given in figure 2(b). Three types of tools are used in this study. First tool has flat end face, second tool has one groove on the end face while third has two perpendicular grooves on the end face as shown in figure 2(b). Width and depth of the groove made is approximately 0.09 mm and 0.05 mm, respectively. Machining time for fabricating groves on the tool end face was captured using stopwatch. Machining time for tool type

1G is 32 minutes while for tool type 2G is 55 minutes. Machining time for tool type 2G is not exactly twice of tool type 1G, because for tool type 2G, when groove no 2 is fabricated, it has to remove less material as some of the material is already removed during fabrication of groove 1.

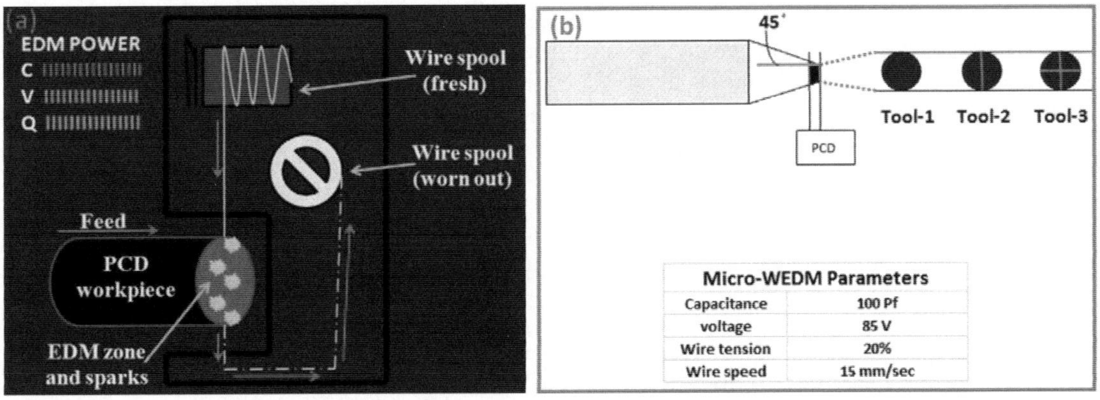

Figure 2. (a) micro-WEDM setup for tool modification; (b) Types of tool and micro-WEDM parameters

2.3. Micro-grinding experiments and measurements

After preparation of 3 types of tool, micro-grinding experiments are performed. Grinding parameters [14], conditions and different measurements adopted are presented in Table 1. Tool-workpiece orientation, feed direction and depth of cut direction during experiments are depicted in figure 3. Small pockets of 1.5 mm length are machined on BK-7 glass.

Table 1. Micro-grinding conditions and measurements adopted

Experimental conditions		Measurements adopted	
Speed	2000 rpm	Before experiments	WEDM machining time for tool NG, 1G and 2G
Feed rate	25 μm/min		Tool profile and weight measurements
Depth of cut	5 μm	During experiments	Grinding force
Lubrication	Dry	After experiments	Grinding surface morphology inspection using optical inspection
Environmental temp.	25 °C		Surface roughness measurement Tool inspection using microscope Tool wear

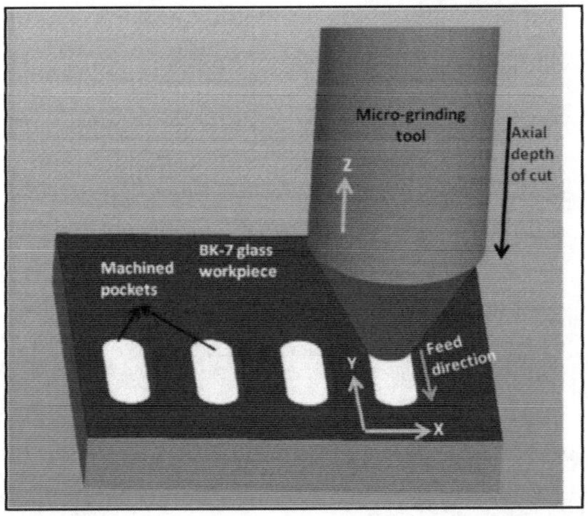

Figure 3. Tool-workpiece orientation and machining directions during machining

IOP Conf. Series: Materials Science and Engineering **229** (2017) 012033 doi:10.1088/1757-899X/229/1/012033

3. Results and discussions

3.1. Tool wear study

Significant tool wear is one of the few negative aspects of micro-grinding process. But it is very difficult to characterize the tool wear phenomena in micro-grinding because of multiple and random cutting edges. In this study, tool wear pattern of three types of tools (NG, 1G and 2G) are observed through microscopic images of the end face. Weight loss of the tools during grinding is also taken as a measure of tool wear. Tool wear were analyzed after grinding of 4.5 mm length. Microscopic images of the tool end face taken at different magnification level can be seen in figure 4. Enlarged views of the craters and pitting marks are shown using 50X magnification. End face morphology of tool type NG after grinding is shown in figure 4(a). Big craters as well as lot of pitting marks can be observed on tool face. Length and width of the crater are measured and found to be 85.34 μm×57.04 μm. Wear pattern is also non-uniform when going from centre to periphery. For tool type 1G, edge craters developed are not as severe as compared to that of tool type NG but are very significant to damage the edges of the machined pockets. Size of the cracks is around 58.58 μm×49.79 μm. End face of tool type 2G after grinding is shown in figure 4(c). Craters developed are not as deeper or wider as of tool types NG and 1G because two micro-grooves on the face provides enough space for chip removal from grinding zone. Crater size is found to be 30.11 μm×14.38 μm.

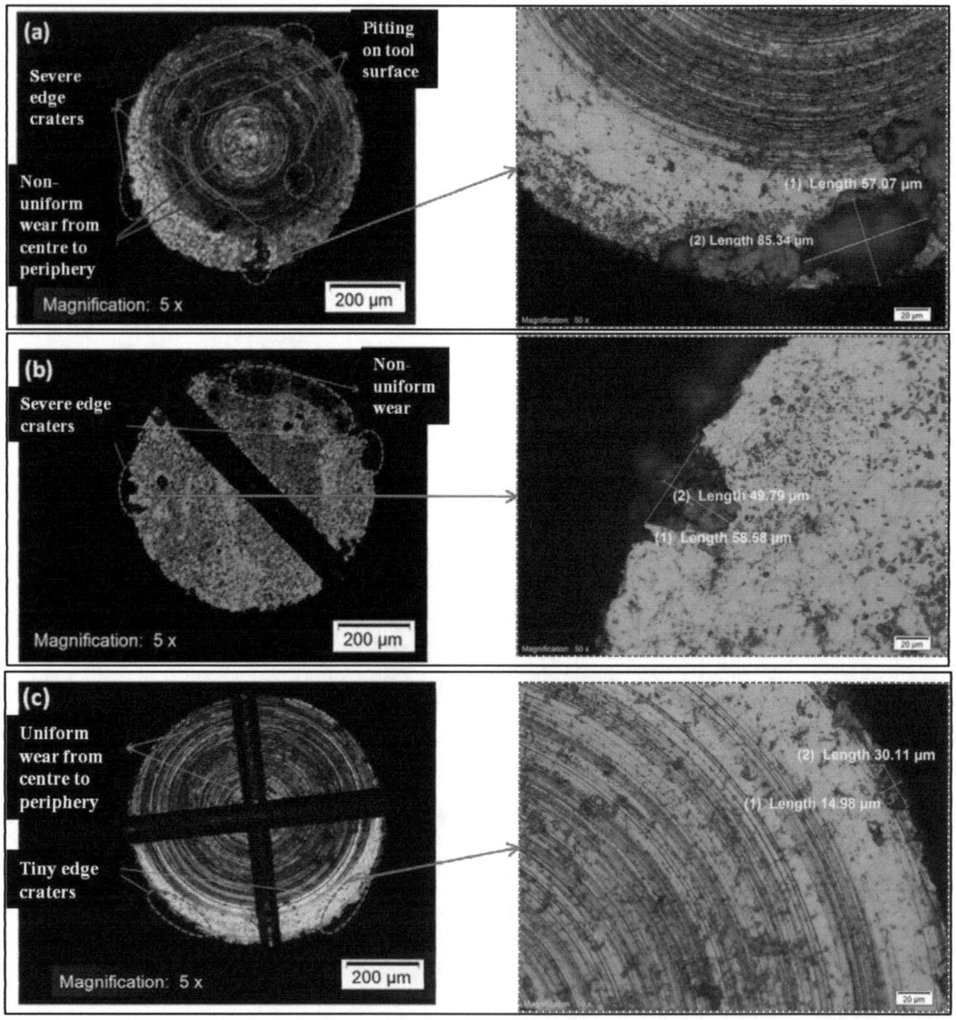

Figure 4. Worn out end faces of tool types (a) NG, (b) 1G, (c) 2G

Amount of tool wear is approximately quantified through weight loss measurement. Highly precise micro weighing machine (Make- Mettler Toledo, Resolution- 1 μg) is used to measure the weight of the tool. Tools were thoroughly cleaned in ultrasonic cleaner before measuring weight so as to remove any adhered microchips on the tool face. Weight of the tool before and after micro-grinding is measured and difference in the weight is the approximate amount of tool wear. Since grinding length is only 4.5 mm, expected amount of tool wear is very less. As shown in figure 5, weight losses of tool types NG, 1G and 2G during micro-grinding are 29 μg, 19 μg and 16 μg, respectively. It is clear that amount of tool wear is least for tool type 2G because of less contact area between tool and workpiece as well as reduced rubbing of microchips on tool surface. However, cleaning of the tool is carried out before measuring weight of the tool; still some microchips may be there on the tool surface which leads to error in the weight measurement of the tool as weight of these microchips are neglected. This problem should be taken care of in the future studies.

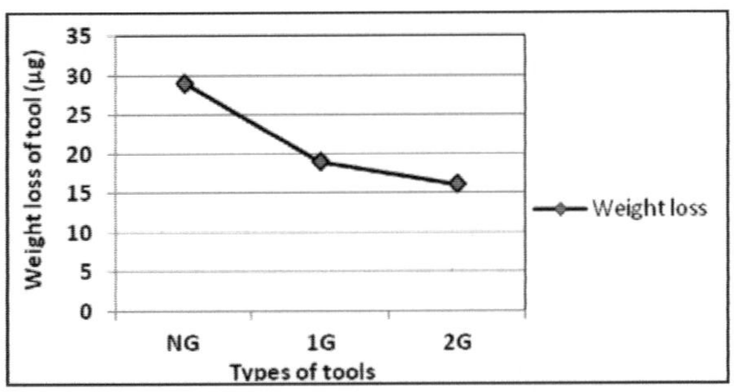

Figure 5. Weight loss of the tools (amount of tool wear) during micro-grinding.

3.2. Variations of grinding forces and ground surface morphology with time

Variations of grinding forces F_x, F_y and F_z with respect to time are shown in figures 6 (a), (b) and (c), respectively. Increase in machining time causes tool wear and hence increase in grinding forces in all the directions. Grinding forces are observed for three types of tool at the machining times of 0, 60, 120 and 180 minutes. From the figures, it can be observed that grinding force is generally increasing with machining time except some exceptions. At the machining time of 60 minute, there is a sudden rise in F_x and F_y values for tool type 2G. This is due to clogging of chips or re-sharpening of tool for any particular case [15]. Tool re-sharpening at any instant may lead to sharp cutting edges, which in turn increases the chip load and grinding forces. Overall, grinding forces has an increasing tendency with machining time and grinding forces for tool type 2G is always lesser as compared to tool types NG and 1G. Surface of the pockets at different machining times is shown in figure 7. Initially, brittle fractures and damage at the edges and surface of the pockets are more due to very sharp edges of the tool. After some time, surface obtained is smoother and damage free as compared to initial surface under balanced stage of the tool. But the pocket surface again starts deteriorating after some time because of the progressive tool wear.

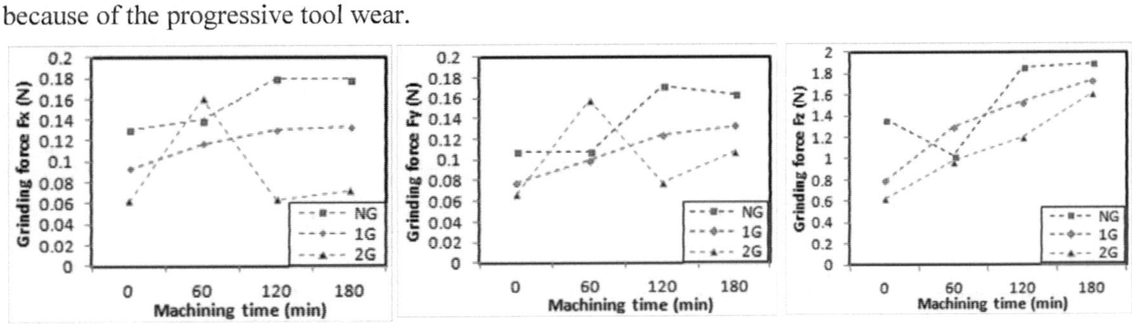

(a) (b) (c)

Figure 6. Variation of cutting force with time for all tools (a) F_x (b) F_y (c) F_z

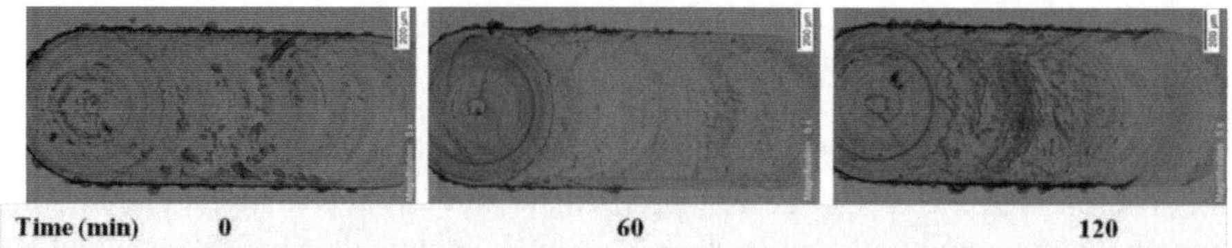

Figure 7. Pocket surface obtained at different machining times for 2G tool.

4. Conclusions

Following conclusions can be made from this study

- Wire-EDM is an efficient and precise technique to machine PCD tools which is very difficult to machine with conventional machining processes.
- Modified tools are able to flush away the microchips from the grinding zone.
- It is observed from the microscopic images of the tool that modified tool suffers less wear as compared to that of normal tool during micro-grinding of BK-7 glass.
- Weight loss measurement technique is used to quantify the amount of tool wear where amount of tool wear is minimum for tool type 2G.
- Grinding forces first decrease with time due to rounding of sharp edges and then increase due to progressive wear of cutting edges.
- Effects of tool wear can also be visualized from the grinding surface morphology.

Acknowledgement

This article is a part of the bilateral research project «Modeling and analysis of high speed hybrid micromachining»:

1. The reported study was partially supported by DST, Govt. of India, research project No. INT/RUS/RFBR/P-226.
2. The reported study was partially supported by RFBR, research project No. 15-58-45017 ИНД_a.

References

[1] Chena ST and Lin SJ, 2011 *Journal of Materials Processing Technology* **211** 1581– 89.
[2] Perveen A, San WYand Rahman M, 2012 *International Journal of Advanced Manufacturing Technology* **61** 101–115.
[3] Dubey AK and Yadava V, 2008 *International Journal of Machine Tools & Manufacture* **48** 609–628.
[4] Liao YS, Chen ST, Lin CS and Chuang TJ, 2005 *Journal of Micromech. Microeng* **15** 1547–55.
[5] Thoe TB, Aspinwall DK and Wise MLH, 1998 *Int. J. Mach. Tools Manufacturing* **38** 239–255.
[6] Aurich JC, Carrella M and Walk M, 2015 *CIRP Annals - Manufacturing Technology* **64** 325–328.
[7] Pratap A, Patra K and Dyakonov AA, 2016 *Procedia Engineering* **150** 969 – 974.
[8] Anand RS and Patra K, 2014 *Machining Science and Technology*, **18:3** 323-347.
[9] Gabler J and Pleger S, 2010 *International Journal of Machine Tools & Manufacture* **50** 420–424.
[10] Chen ST and Jiang ZH, 2015 *Journal of Materials Processing Technology* **216** 206–215.
[11] Cheng J, Wang C, Wen X and Gong Y, 2014 *International Journal of Machine Tools & Manufacture* **85** 36–48.
[12] Hua ZJ, Yan Z, Qiang TF, Shuo Z and Shen GL, 2015 *International Journal of Advanced Manufacturing Technology* **78** 1893–1904.
[13] Hua ZJ, Yan Z, Shuo Z, Qiang TF, shen GL and zhen DR, 2014 *Hindawi Publishing Corporation: Shock and Vibration.*
[14] Perveen A, San WY and Rahman M, 2011 *Int. J. Abrasive Technology* **4:4** 304–324.
[15] Perveen A, Jahan MP, Wong YS and Rahman M, 2011 In *Advanced Materials Research Trans Tech Publications* **264** 1085-90.

Mechanical, Thermal and Acoustic Properties of Open-pore Phenolic Multi-structured Cryogel

Rui Yao[1,2,*], Zhengjun Yao[1,2], Jintang Zhou[1,2], Peijiang Liu[1,2] and Yiming Lei[1,2]

[1]College of Materials and Technology, Nanjing University of Aeronautics and Astronautics, Nanjing 211100, China
[2]Jiangsu Key Laboratory of Advanced Structural Materials and Application Technology, Jiangsu, China

*277462297@qq.com

Abstract: Open-pore phenolic cryogel acoustic multi-structured plates (OCMPs) were prepared via modified sol gel polymerization and freeze-dried methods. The pore morphology, mechanical, thermal and acoustic properties of the cryogels were investigated. From the experimental results, the cryogels exhibited a porous sandwich microstructure: A nano-micron double-pore structure was observed in the core layer of the plates, and nanosized pores were observed in the inner part of the micron pores. In addtion, compared with cryogel plates with uniform-pore (OCPs), the OCMPs had lower thermal conductivities. What's more, the compressive and tensile strength of the OCMPs were much higher than those of OCPs. Finally, the OCMPs exhibited superior acoustic performances (20% solid content OCMPs performed the best) as compared with those of OCPs. Moreover, the sound insulation value and sound absorption bandwidth of OCMPs exhibited an improvement of approximately 3 and 2 times as compared with those of OCPs, respectively.

1. Introduction

Since the bonding strength, rate of carbon residue and chemical stability of phenolic resin are high, meanwhile, it has low smoke and toxic. Phenolic resin can be used as adhesive, fire insulation and anti-corrosion materials [1-3]. Among them, phenolic foams with light weight, good heat resistant and low thermal conductivity exhibit good application prospects on the fields of thermal insulating materials [4,5]. Nevertheless, the applications of traditional phenolic foam materials are limited because of their large brittleness and high powder dropping rate[6]. While phenolic cryogels which are prepared by freeze-drying method has declined the brittleness and powder dropping rate. This is mainly caused by their three-dimensional reticular structure [7,8]. In addition, the solid contents of the materials is low, which has badly blocked the heat conduction in the thermal transfer process. Moreover, due to their nano-sized holes, thermal convection and radiation are severely weakened and obstructed. The phenolic cryogel materials exhibit better thermal insulation performance under above effects [9-12].

Therefore, compared with phenolic traditional phenolic foam materials, cryogel materials own higher implementation value. In addition, the nano-sized holes of cryogel materials are mostly even with hole area percentages more than 95%. This big amount of holes offer sufficient space for sound to

Content from this work may be used under the terms of the Creative Commons Attribution 3.0 licence. Any further distribution of this work must maintain attribution to the author(s) and the title of the work, journal citation and DOI.
Published under licence by IOP Publishing Ltd

shuttle and rub with the cryogel skeleton. Hence, the sound absorption ability of phenolic cryogels is excellent [13]. However, attributed to the uniform pore structures of the cryogels, the sound absorption spectra of these materials is narrow. At certain frequencies, the pores with similar size exhibit good sound absorption performance, while the absorption performance emerge significantly decreases at other frequencies [14]. To solve this issue, OCMPs were designed and prepared. The OCMPs were composed of three layers: the top and bottom of the layers owned a pyknotic structure with sectional tiny pores, while the interlayer exhibited an multi-pores structrue which owned micro-pores together with irregularly dispersed nano-pores [15,16]. Attributed to their broad size distribution, this mult-structure lead to the increase of the sound absorption spectrum width. Therefore, these materials can be used as sound absorption materials in the fields of building, vehicle, and aerospace, such as auto interior decoration, household ceiling, submarine stealthy materials and so on[17,18].

2. Experimental details

2.1. Preparation of phenolic cryogels

Phenolic cryogels were prepared by a three-step method. Phenolic hydrogels were made from resorcinol (R) and formaldehyde (F) [19], then acetic acid was used to dispose the hydrogels. After freeze-drying, phenolic cryogels were obtained [20].

First, water, sodium carbonate, formaldehyde, resorcinol were mixed uniformly (molar ratio, R: F=1: 2, Na_2CO_3: R=1: 150) and made into hydrogel plates under certain conditions (80 °C, 1-2 h, 50KHz ultrasound). Second, the hydrogels were put into the 2wt% acetic acid solution disposing for 5 days, then they were immersed into 5wt% acetic acid solution (1 day) for further enhancement. Third, through freeze-drying at -50°C by lyophilizer for 1 day, open-pore phenolic cryogel acoustic multi-structured plates (OCMPs) were obtained. In addition, cryogel plates with uniform-pore (OCPs) were obtained after the hydrogel plates were disposed by 5wt% acetic acid solution and 99wt% tert butyl alcohol for 5 and 7 days, respectively

For OCMPs samples A1 to A4, the solid contents of the hydrogels were 10wt%, 15wt%, 20wt% and 25wt%. Meanwhile, one OCP sample B1 were prepared with 20wt % solid contents.

2.2. Microstructures and Characterizations.

Field Emission Scanning Electron Microscopy (FESEM) was measured through S-4800 ice launch scanner (Hitachi, Japan). X-ray photo electron spectroscopy (XPS) was obtained using Thermo ESCALAB 250XI multifunctional imaging electron spectrometer (Thermo, USA). Pore analysis was tested by JW-BK132F specific surface area and pore size analyzer (Jingwei, China). Fourier Transform Infrared Spectroscopy (FTIR) was measured by PerkinElmer Spectrum 100 FT-IR Spectrometer (PerkinElmer, America). X-ray Diffraction (XRD) was measured by Bruker D8 Advance Eco (Bruker, Germany).

2.3. Mechanical testing

A CMT5105 electronic universal testing machine was used to test the composites' compression and tensile performances. To produce an average and variance, five replicates were applied in the mechanical tests.

2.4. Thermal testing

Perkin-Elmer Pyrisl calorimeter and DSC-2 thermal analyzer were used to proceed TG-DSC analysis. The temperatures were ranged from 30 to 850°C with 10°C/min heat increasing rate under N_2 atmosphere.

Thermal conductivities of the plates were measured by a Hot Disk TechmaxTPS1500 thermal meter (Sweden).

2.5. Acoustic testing

Impedance Measurement Tube for ASTM Type 4206A, Pules Multi-analyzer system Type 3560 and Power Amplifier Type 2716 (B&K, Denmark) were used to evaluate transmission loss and absorption coefficient.

3. Results and discussion

3.1. Pore mophology and size analysis of the aerogels.

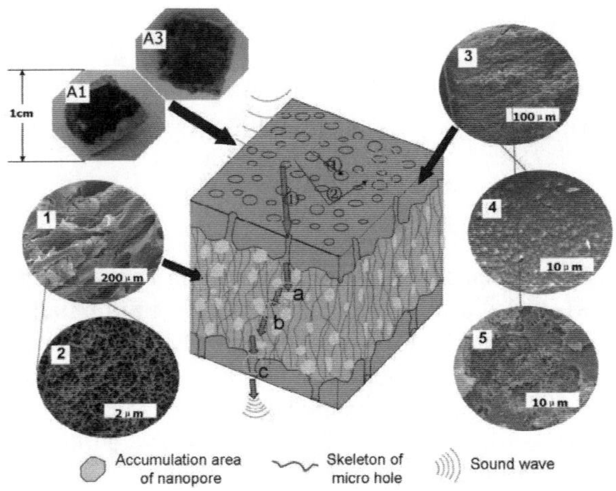

Figure 1. The photographs of the actual products A1 and A3, as well as the SEM images of A3 and the simulated diagram and sound transmission mechanism of OCMP samples.

The photographs of the actual samples A1 and A3, as well as the SEM images of sample A3 and the simulated diagram of OCMPs' sound transmission mechanism are illustrated in Figure 1. Seen from the photographs, the boundary walls of sample A3 were thinner than those of A1. The OCMPs exhibited a porous sandwich structure, which can be viewed from the SEM images. Through further observation, the specific structure of the cryogels were as follows. The morphology of the boundary walls were displayed in the SEM images '3, 4, and 5'. They showed a poriferous surface and a relatively dense structure compared with the center section. While the structure of the interlayer contained micro-holes with randomly dispersed nano-pores on their inner part, which was exhibited in the SEM images '1 and 2' [21]. The formation mechanism of the sandwich microstructure is explained through the simulated diagram. The SEM image '2' has roughly assumed the skeletons of the hydrogels, however, when freeze-dried, the skeletons of the hydrogels are partly destroyed by the surface tension of water under pumpdown action, which emerges from the center to the surface. As a consequence, thick boundary walls are formed by the agglomerated materials which spread from the cracked skeleton. Meanwhile, the nano-pores has fused together, which lead to the development of micro holes in most part of the interlayer region, however, there still maintain nano-pores in other areas. The skeleton of A3 which own higher solid content is stronger than A1, hence, it's more difficult for A3 to destroy its interlayer by the surface tension of water than A1. In addition, the boundary walls will become thnner with the decrease of the cracked skeleton materials. Eventually, with the help of the images of the actual samples and the microstructure, a simulated diagram of OCMPs' sound transmission mechanism was constructed in Figure 1, and the mechanism will be elaborated below.

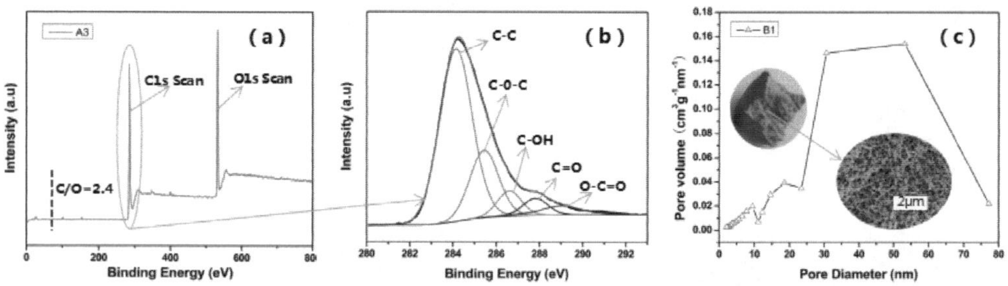

Figure 2. XPS measurements of A3, as well as the morphology, macroscopic pattern, and pore size desorption of sample B1.

XPS curves of A3, as well as the morphology, photograph, and pore size desorption (BJH method) of sample B1 are displayed in Figure 2. In Figure 2a, around 286 and 533 eV, two peaks appeared, which is attributed to C and O elements, respectively. The atomic ratio of carbon and oxygen (C/O ratio) was 2.4, and it's in accordance with the range of theoretical ratio which was between 2 and 4 according to chemical dehydration-condensation reaction (R:F=0.5), in the other hand, this result stated that the cryogels had complex chemical bonds. In addition, observed from the peaks caused by C–O–C and C=O groups in Figure 2b, the same conclusion was obtained. Simultaneously, there exised characteristic peaks attributed to C–C, C–OH and O-C=O groups, while the peak of C–C groups also contained methylene bonds [22]. Seen from Figure 2c, the pore diameters of sample B1 were mostly ranged from 30 nm to 50 nm. In addition, the nano-pores of sample B1 exhibited almost the same size as that of sample A3. OCMPs has similar nano-pores structure as OCPs, albeit they have quite different macroscopic patterns.

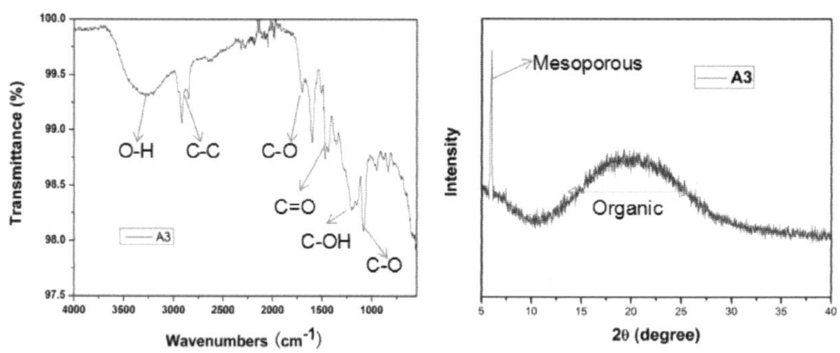

Figure 3. FTIR and XRD curves of A3.

Seen from the FTIR image (left image of Figure 3), RF cryogels show the similar groups as XPS image. Viewed from the peaks of FTIR image, there existed O-H, C-C, C-O, C=O and C-OH groups in the cryogels. Through above analysis, it's obvious that RF cryogels were obtained successfully.

In XRD curves (right image of Figure 3), RF cryogels emerge broad peaks, which is attributed to that they are organic materials. In addition, there existed poignant peaks on the curve of the cryogels at the small angle, indicating that there were mesoporous inside these cryogels, which was consistent with the result of the Figure 2c.

3.2. Mechanical and thermal properties of the aerogels

IOP Conf. Series: Materials Science and Engineering **229** (2017) 012034 doi:10.1088/1757-899X/229/1/012034

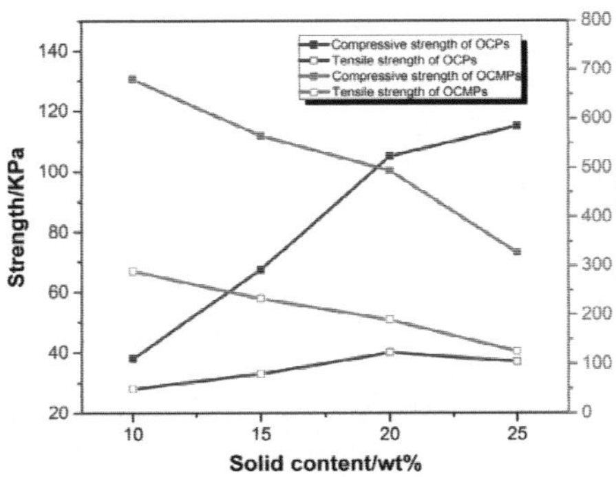

Figure 4. Compressive and tensile strength of OCPs and OCMPs.

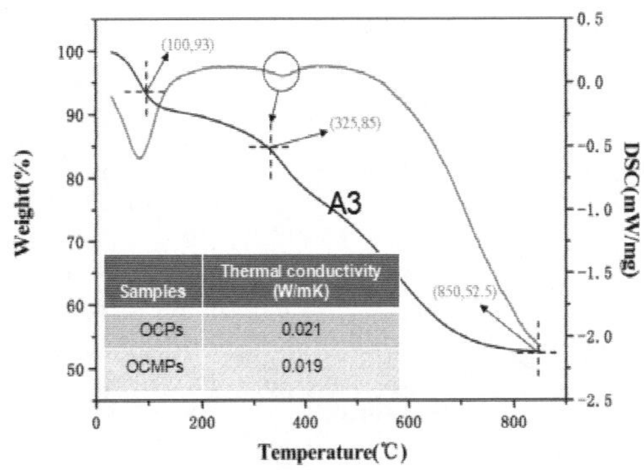

Figure 5. TG-DSC curves of A3 as well as thermal conductivity of OCPs and OCMPs

The compressive and tensile strength of the cryogels are shown in Figure 4. When the solid content reaches 10 wt.%, the compression and tensile strength of OCMPs reached their maximum. After that, they began to decrease significantly. While OCPs exhibited the opposite results. They were the minimum at the 10 wt.% solid content, then gradually increased. For OCPs, since they were uniform structure, the increase of the solid content will surely increase the structure strength, lead to the increase of their compressive and tensile strength. But for OCMPs, the mainly bearing part was the boundary walls, they became thin with the increase of solid content concluded from above results. So their strength will surely decrease. In addition, compared with the strength of OCPs, the compressive strength and tensile strength of OCMPs are much higher, which is also caused by the boundary walls.

Figure 5 displays the TG-DSC curves of sample A3. Three key points of the sample were marked in the TG-DSC curves. Since the specific surface area and porosity of cryogels were large, the water vapor in the air was absorbed into the pores. When the temperature reached 100 °C, the absorbed water was evaporated which made 7% weight lost of the sample, and this was the first key point. After that, the second key point will appear at 325 °C with 15% weight lost of the sample. At this temperature, the sample will be decomposed acutely. Accordingly, in the DSC curve, an exothermic peak will emerge. Finally, the third point was found at the temperature of 850 °C, and the weight lost of the sample become invariant at 47.5%, and this stability was caused by the C element (The groups of H and O elements were completely broken and wiped off, while C element maintained under higher heat resistance). Generally speaking, the cryogel materials had high thermal resistance, and they will become carbon cryogels with higher thermal resistance.

The thermal conductivities of cryogel samples under normal temperature are tested and showed in the table of Figure 5. Viewed from the table, the thermal conductivity of OCMPs was higher than that of OCPs, which was mainly attributed to the block action of the close-grained boundary walls to the thermal convection.

3.3. Acoustic properties of the aerogels as well as sound absorption and insulation mechanism

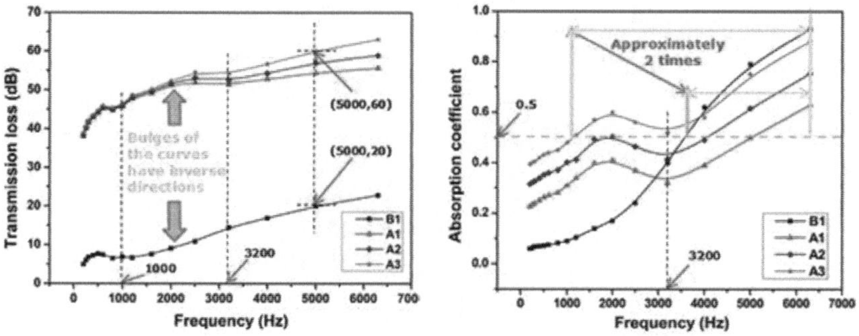

Figure 6. The transmission loss and sound absorption coefficient of OCMP and B1 samples.

Figure 6 shows the sound performances of OCMPs and B1 samples. Sample A4 was excluded from the test for its agglomeration. The transmission loss of B1 increased quickly at first with the increase of frequency, then the growth speed lowed down. For sound absorption coefficient, it was the same trend. In addition, the sound performances of OCMPs exhibited similar trends as those of B1. Samples A3 and B1 have same ingredient and solid content, while they have a biggest difference which reflects in their structures. B1 has an uniform nanopores structure, while A3 has solid boundary walls at two surfaces with inner micro pores and nanopores. The solid boundary walls which obstruct the sound waves make the TL value of A3 much higher than that of B1. But for sound absorption coefficient, the boundary walls make less sound waves be consumed by the inner pores which in turn leads to the decline of sound absorption. At the same time, the boundary walls have through-holes which reduce barrier effects and enhance consumption of sound waves. In view of the comprehensive function of above two aspects, these two samples present similar sound absorption coefficients.

Nevertheless, from 1000 Hz to 3000 Hz, there occurred a significant difference: Compared with other frequencies, the sound performances showed great improvement. In general, compared the transmission loss of A3 with that of B1 at 5000 Hz, the improvement was almost 3 times. Meanwhile, compared the samples A1, A2 and A3, it is concluded that the sound absorption coefficient will increasingly improve with the increase of solid content. The primary sound insulation mechanism is elaborated in the simulated diagram of Figure 1: As shown in Figure 1, a sound wave transmits into the surface and interior of OCMPs, it will become three waves: absorbed, reflected, and transmitted wave. The boundary walls is the significant factor to form the reflected wave. After reflection, the remained wave is running along path 1. Then, the sound will rub with the medium (Figure 1"a,b,c"), the sound wave which is consumed here is the absorbed wave. After absorption and reflection of the materials, The sound wave that survives is the transmitted wave. Under this sound insulation mechanism, the sandwich structured OCMPs guarantee sound waves to spread and consume with the surface and interior of the materials adequately. In addition, when the absorption coefficient exceeds 0.5, the absorption frequency spectrum of OCMPs and OCPs have improved about 2 times. The solid content of sample A3 is the highest compared with other samples, so its skeleton structure is compact which can absorb more sound energy. This will surely make sample A3 exhibit the best sound performance. However, with the increase of solid content, the boundary walls will become thinner, which will decline the transmission loss in some degree [23].Through above analysis, the 20% solid content OCMPs have presented the best sound performance.

4. Conclusions

In this study, a type of a sandwich-like multi-structure was designed for obtaining better acoustic performance. Afterwards, the OCMPs were observed to have been manufactured into this type of structure through microscopic morphology. In addtion, compared with cryogel plates with uniform-pore (OCPs), the OCMPs had lower thermal conductivities. What's more, the compressive and tensile strength of the OCMPs were much higher than those of OCPs. Finally, The OCMPs exhibited superior acoustic performance (20% solid content OCMPs performed the best) as compared with that of OCPs. Moreover, the sound insulation value and sound absorption bandwidth of OCMPs exhibited an improvement of approximately 3 and 2 times as compared with those of OCPs, respectively.

Acknowledgements
This work is supported by the Opening Project of Jiangsu Key Laboratory of Advanced Structural Materials and Application Technology (PAPD). as well as National Natural Science Foundation of China (Funding NO. 51672129)

References
[1] A. Tadaharu, K. Takao and H. Masahiro, International Journal of Adhesion and Adhesives , 53 , 53 (2015).
[2] X.Y Sui and Z.Z Wang, Polymers for Advanced Technologies , 24 , 593 (2013).
[3] K. Guo, X.G. Li, Z.W. Yue, X.G. Wang, F.J. Yan, X.M. Wang, Jianmei Gao and Zhibin Fan, Polymer composites , 36 , 1346 (2015).
[4] Y. Wang, S.J. Wang, C. Bian, Y.H. Zhong and X.L. Jing, Polymer Degradation and Stability , 111 , 239 (2015).
[5] L.P. Bian, J.Y. Xiao, J.C. Zeng, S.L. Xing, C.P. Yin and A.Q. Jia, Materials & Design , 54 , 230 (2014).
[6] U.A. Amran, S. Zakaria, C.H. Chia, S.N.S. Jaafar and R. Roslan, Industrial Crops and Products, 72 , 54 (2015).
[7] R. Zhang, Y.G. Lu, L. Zhan, X.Y. Liang, G.P. Wu and L.C. Ling, Carbon , 41 , 1660 (2003).
[8] L.I. Grishechko, G. Amaral-Labat, A. Szczurek, V. Fierro, B.N. Kuznetsov and A. Celzard, Microporous and Mesoporous Materials, 168 , 19 (2013).
[9] T.A. El-Brolossy, S.S. Ibrahim and E.A. Alkhudhayr, Polymer Composites, 36 , 1242 (2014).
[10] E. Cuce, P.M. Cuce, C.J. Wood and S.B. Riffat, Renewable and Sustainable Energy Reviews, 34 , 273 (2014).
[11] R.J. Goldstein, E.R.G. Eckert, W.E. Ibele, S.V. Patankar, T.W. Simon, T.H. Kuehn, P.J. Strykowski, K.K. Tamma, A. Bar-Cohen, J.V.R. Heberlein, J.H. Davidson, J. Bischof, F.A. Kulacki, U. Kortshagen, S. Garrick and V. Srinivasan, International Journal of Heat and Mass Transfer, 48 , 819 (2005).
[12] E.M. Sparrow, S.S. Kang and W. Chuck, International Journal of Heat and Mass Transfer, 30 , 1237 (1987).
[13] G.P. Wu, J.B. Yang, D.P. Wang, R. Xu, K. Amine, C.X. Lu, Mater Lett, 115 , 1 (2014).
[14] G.Q. Tang, Z.G. Jiang, X.F. Li, H.B. Zhang, D. Aravind, Carbon, 77 , 592 (2014).
[15] W.F. Chen, S.R. Li, C.H. Chen, L.F. Yan, Adv Mater, 23 , 5679 (2011).
[16] X. Xu, H. Li, Q.Q. Zhang, H. Hu, Z.B. Zhao, J.H. Li, et al, ACS Nano, 9 , 3969 (2015).
[17] J.Z. Feng, J. Feng, Y.G. Jiang, C.R. Zhang, Mater Lett, 65 , 3454 (2011).
[18] R. Baetens, B.P. Jelle, A. Gustavsen, Energ Buildings, 43 , 761 (2011).
[19] H. Reuter, Adv Mater, 3 , 568 (1991).
[20] F. He, C. Sui, X.D. He, M.W. Li, Mater Lett, 152 , 9 (2015).
[21] T. Zhu, K.Y. Teng, J. Shi, L. Chen, Z.W. X, Compos Sci Technol, 123 , 276 (2016).
[22] Q.Y. Peng, Y.B. Li, X.D. He, X.C. Gui, Y.Y. Shang, Adv Mater, 26 , 3241 (2014).
[23] P.S. Allan, A. Ahmadnia, R. Withnall, Polym Test, 31 , 312 (2012).

IOP Conf. Series: Materials Science and Engineering **229** (2017) 012035 doi:10.1088/1757-899X/229/1/012035

Fracture Behavior of Dielectric Elastomer under Pure Shear Loading

D Ahmad[1] and K Patra[2*]

[1]Department of Mechanical Engineering, Indian Institute of Technology, Patna 801103, Bihar.
[2]Department of Mechanical Engineering, Indian Institute of Technology, Patna 801103, Bihar.

[*]Corresponding Author: kpatra@iitp.ac.in

Abstract. Dielectric elastomer has become a very important material for many emerging applications areas like optics, micro fluidics, sensors, actuators and energy harvesting. However, these elastomer components are prone to fracture or catastrophic failure because of defects likes notches, flaws, and fatigue crack, impurities which occur during production or during service. To make better use of this material, it is important to investigate fracture characteristics under different operating conditions. This study experimentally investigated the effects of notch length and strain rate on the fracture toughness, failure stretch and failure stress of acrylic elastomer under pure shear deformation mode. It is observed that failure stretch depends on notch length and independent of strain rate, but failure stress decreases with increasing notch length and increases with increasing strain rate. It is also found that fracture toughness is independent of notch lengths. However, fracture toughness is found to increase with strain rate.

1. Introduction

Now a day's dielectric elastomer is used as an actuator material since it undergoes large deformation when stimulated electrically [1–4]. Recent study also reveals that dielectric elastomer can also be an important engineering material for sensors and energy harvesting applications [5–7]. However, the high stretch ability can fall markedly when the material contains notches/flaws, which leads to inhomogeneous deformation and fracture of the material [8].

Despite the wide scale recognition, very few researchers have investigated the effect of notch and other process parameters in the fracture behavior of dielectric elastomer. Kaltseis et al [9] conducted test on natural rubber and VHB 4910 acrylic rubber with notch in the sample and found that the fracture energy at 1/s is almost double for natural rubber as compared to VHB 4910. Chen [10] determined fracture characteristics of natural rubber under dynamic loading. Loung et al [11] also conducted tests on natural rubber and damage mechanisms during tensile test were described. A fracture criterion (T) was explained and scanning electron microscope was used to evaluate damage in the specimens at the beginning of the tearing. Schmidt et al [8] investigated rupture of pristine VHB 4910 membranes by performing uniaxial tension, pure shear, and membrane inflation tests at relatively low stretch-rates. They found that the failure stretch to be around 9 and essentially independent of the

Content from this work may be used under the terms of the Creative Commons Attribution 3.0 licence. Any further distribution of this work must maintain attribution to the author(s) and the title of the work, journal citation and DOI.
Published under licence by IOP Publishing Ltd

IOP Conf. Series: Materials Science and Engineering **229** (2017) 012035 doi:10.1088/1757-899X/229/1/012035

imposed state of deformation. They investigated the rupture behavior of the VHB but did not investigate the effect of notch in the sample. Phar et al [12] also experimentally investigated rupture characteristics of VHB 4905 as a function of stretch-rate and sample height and they experimentally showed that the introduction of a pre cut into this material dramatically reduces the failure stretch. Furthermore, the failure stretch was found to decrease with increase of sample height, and the fracture energy was found to increase with stretch-rate but remained unaffected by change in sample height.

To understand the fracture behavior of elastomers, energy based approach is adopted because it involves energy balance for the whole specimen rather than requiring the knowledge of complex stress and strain fields at the crack tip which is difficult to understand for highly nonlinear kind of materials like elastomers. Recently, for materials having large stretch ability, pure shear test were used to calculate fracture energy by different researchers [13, 14].

Marano et al [14] applied J-integral method to calculate fracture toughness at crack onset, for filled and unfilled natural rubbers. However, the effects of different notch length and fracture toughness of the elastomer samples were not considered. The present work considered the effects of notch length and strain rate on fracture toughness at crack onset, failure stress and failure stretch of the dielectric-elastomer, VHB 4910.

2. Experimental Specifications

Acrylic elastomer (VHB 4910) is used in the present investigation which is commercially available in the roll form. The thickness of the material is 1 mm. The mass density of the elastomer is 960 kg/m^3 and operating temperature ranges from -10 ^oC to 90 ^oC. All the testing was performed at a room temperature of 23 ^oC. VHB 4910 specimens were tested under pure shear conditions using a computer controlled electro-magnetic universal testing machine (UTM) of Zwick/Roell, Model Z050 with 2.5KN load cell at ambient temperature. Test Xpert-II software is interfaced with UTM through a computer to control the test environment.

A pure shear test fixture has been developed and fitted in the UTM with one end fixed to the lower stationary grip and other end connected with 2.5 kN load cell. Figure 1 (a) shows a complete set up. Closer view of a precut sample fixed to pure shear test fixture is shown in figure 1 (b). The width of the test fixture is 320 mm and height of the sample is taken to be 17 mm. The width is almost 19 times more than the height taken. Hence even at a very large deformation of the sample, the contraction in the width direction is very small as compared to elongation in the stretching direction. Therefore material deformation state can be considered as pure shear state [8]. The sample is properly fixed in the grips and tightened with bolts so that slippage does not occur at larger stretching. Different lengths of notches (32mm, 64mm, 106mm, and 160mm) are introduced by using a scissor in the vertical edge of the sample.

Propagation of crack during testing of notched sample is shown in figure 2 (a) to (d). At the starting of the test the height H of the sample is 17mm as shown in the figure 2 (a) with a notch length of 64 mm. 2 (d). The notch tip experiences a large blunting before reaching a critical value as shown in figure 2 (b) and then propagates along the width of the sample perpendicular to the movement of crosshead as shown in figure 2 (c) [15]. Ultimately, the sample tears into two parts at the end of the test as shown in the figure 2 (d). Virgin sample just before failure is also shown in figure 2 (e) and brittle failure of virgin sample is shown in figure 2 (f).

For a notch of 64mm length, the test is conducted at 0.1/min,1/min,10/min and 60/min strain rates in order to understand the behavior of material at different strain rates. Higher strain rate test cannot be done on the machine because of its limitation. The fracture energy and fracture toughness during each test is calculated.

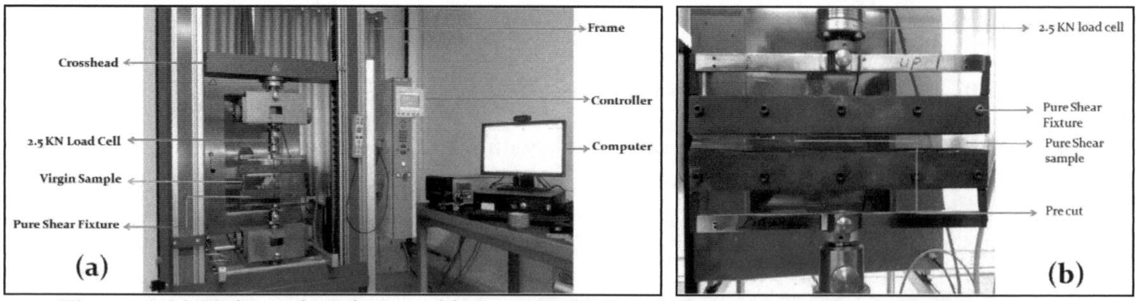

Figure 1 (a) Universal testing machine set up (b) Specially developed pure shear fixture

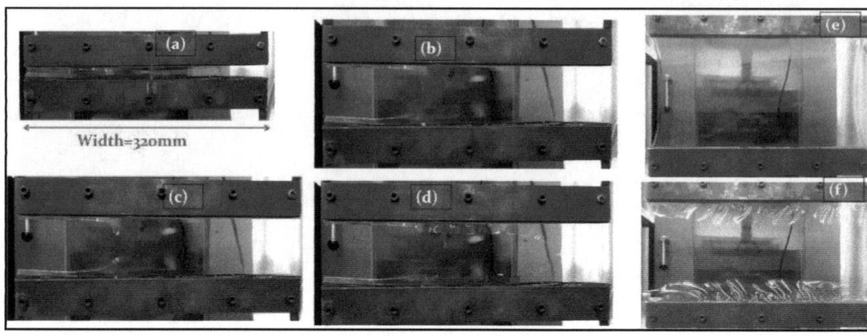

Figure 2 (a) Initial 64mm length of notch in un-deformed state (b) Blunting of crack tip just before crack propagation (c) Initiation of crack propagation (d) End of crack propagation (e) Virgin sample without notch prior to fracture (f) Brittle fracture of virgin sample without external notch.

3. Results and Discussions

3.1. Effects of notch length on fracture toughness and fracture energy.

We have taken four different notch lengths (32mm, 64mm, 106mm, and 160mm) which is 10%, 20%, 33% and 50% of width (=320mm) and conducted three repeatability tests for each sample at a strain rate of 10/min (0.166/s). The stress-stretch curve is plotted in a single graph to compare the behaviour as shown in figure 3. Furthermore, fracture toughness (J_c) and fracture energy (T_c) are calculated for all the tests conducted as given in Table 1.

Fracture toughness can be obtained by using the following equation [14]

$$J_c = U_c / B(W - a_0) \tag{1}$$

where, J_c = Fracture toughness of the sample.

B = Thickness of the sample.

U_c = Area under force–stretch curve.

W = Width of the specimen.

a_0 = Length of the notch.

Fracture energy at the onset of crack propagation can be calculated by using the following equation [14]

$$T_c = w * H \tag{2}$$

where, T_c = Critical Fracture Energy.

w = Strain energy density in an un-notched specimen at the same strain at which fracture onset occurs in the notched specimen.

H = Height of the sample

From table 1, fracture energy for the virgin sample is quite high (28.26 kJ/m^2) as compared to that of notched samples (3-3.5 kJ/m^2). In Figure 3, the fracture toughness J_c and fracture energy T_c have been plotted with mean values and standard deviation and it is seen that both are in good agreement. Hence in the current work we have taken the fracture toughness, J_c as fracture criteria because this criterion is generally used for viscoelastic material [14]. From table 1 and figure 4, it is clearly demonstrated that the fracture toughness J_c, for 32 mm, 64 mm, 106 mm and 160 mm notch lengths is almost constant. Dragoni et al [16] also demonstrated that fracture toughness is a material property and is independent of introduced notch length and the value of fracture toughness is around 7 kJ/m^{2}for natural rubber.

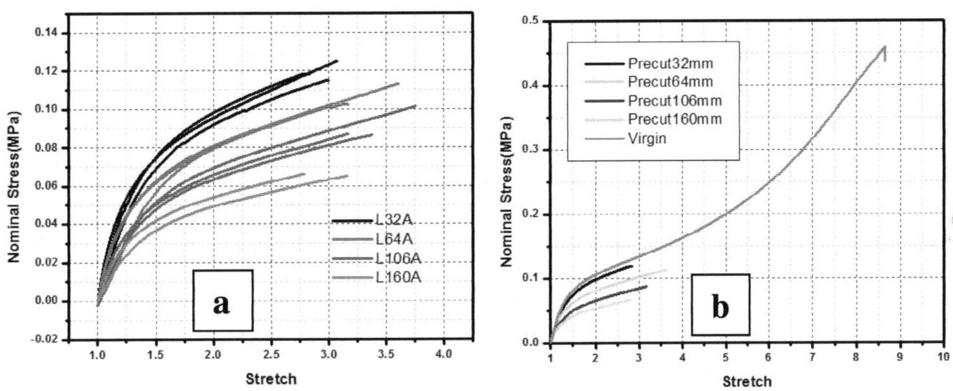

Figure 3. Variation of stress-stretch curve for (a) different notch lengths at a strain rate of 10 per min (b) virgin sample and notched sample.

Table 1. Variation of average fracture toughness, fracture energy, failure stretch and failure stress of notched and virgin sample at 10 per min.

Notch Length (mm)	Fracture Energy,T_c (kJ/m^2)	Fracture Toughness,J_c (kJ/m^2)	Failure Stretch	Failure Stress (MPa)
32	3.22	3.20	2.97	0.1193
64	3.85	3.53	3.23	0.104
106	4.33	3.44	3.43	0.0914
160	3.23	3.22	2.98	0.064
Virgin	28.26		8.62	0.451

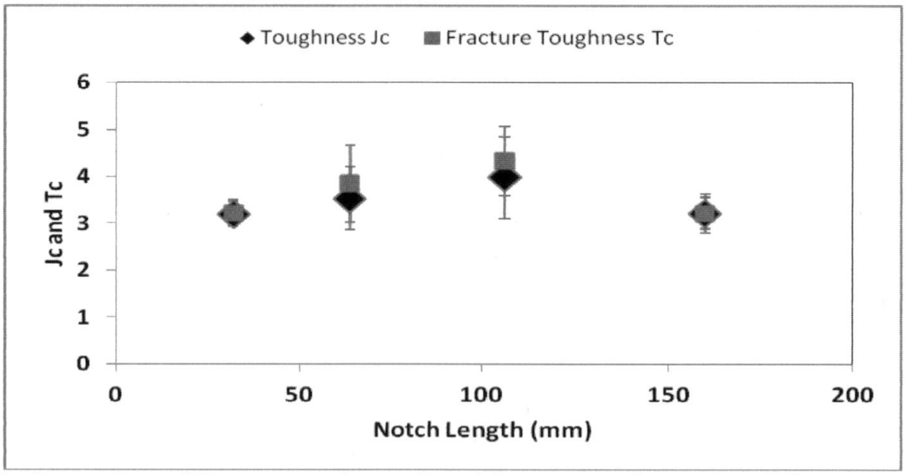

Figure 4. Comparison of Fracture toughness, J_c and fracture toughness T_c at different notch length

3.2. Effect of notch length on failure stretch and failure stress

Stress-stretch graph is plotted for virgin and notched samples to visualize the area under the curve (figure 3.b). From the figure 3 (b), it is seen that with the increase in lengths of notch, the stress at a particular stretch will always be lesser. For 32 mm notch, failure stress is 0.12 MPa while for 160 mm notch length failure stress decreases down to 0.064 MPa as shown in figure 5(a). This is because with increasing lengths of notch, lesser width is available for stretching. Therefore, work done to rupture (area under the stress-stretch curve) keeps on decreasing with notch length (see in figure 3.a). Hence force needed to rupture the material will be less that results in lower stress [10, 17].

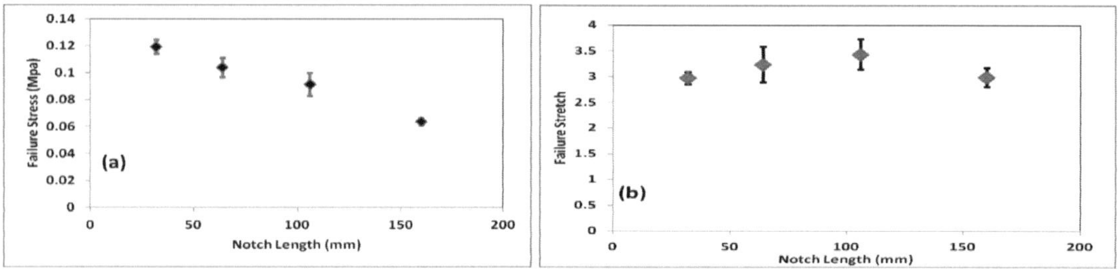

Figure 5. Variation of (a) failure stress and (b) failure stretch with notch length.

From figure 5(b), it is demonstrated that failure stretch for notched samples are almost constant for different notch lengths in the range of 3 to 3.5. This is because failure stretch depends on fracture toughness. Fracture toughness of the sample is almost constant. Failure stretch also does not vary with notch length.

3.3. Effect of strain rate on fracture toughnes (Jc,), failure stretch (λ_f) and failure stress.

Calculated fracture toughness for 64 mm long notched sample (20% of width) at varying strain rate of 0.1/min, 1/min, 10/min and 60/min are shown in figure 6(a). Experiments at higher strain rate above 60/min cannot be done because of the limitation of machine. For strain rate of 60/min, the fracture toughness, Jc obtained is in well agreement with Kaltseis et al [9]. The values of fracture energy for strain rate of 0.1/ min, 1/min, 10/min, 60/min is 1.33 kJ/m^2, 1.81kJ/m^2, 3.16 kJ/m^2 and 4.71 kJ/m^2 respectively which shows increasing trend. In figure 6(a), the effect of strain rate on fracture energy of different earlier research works are compared. Fracture energy values from the works of Chen [10] and Kaltseis et al [9] and Pharr et al [12] are compared with the present work. At increasing strain rate,

IOP Conf. Series: Materials Science and Engineering **229** (2017) 012035 doi:10.1088/1757-899X/229/1/012035

fracture energy of natural rubber is showing increasing trend [10]. At a strain rate of 1/s, the fracture energy of the present work is in good agreement with Kaltseis work as shown in figure 6 (a).

Phar et al [12] also investigated the effect of strain rate on fracture energy and the current investigation follows the same trend as shown in the figure 6 (a). Fracture energy increases with strain rates because at higher strain rates there is little relaxation [12]. From figure 5 (b), we see that failure stretch with strain rate is almost constant (λ_f =2.6) for notched sample. This is because the failure stretch λ_f is directly proportional to the number of monomers present in the sample [17].With increasing strain rate failure stretch remains constant as number of monomers is same for all strain rates. From figure 6 (c), it is demonstrated that failure stress also increases from 0.05 MPa to 0.14 MPa with strain rate for a 64 mm notched sample. This is because the viscoelastic material gets less time for stress relaxation with strain rate and material stiffening occurs. Therefore, more stress is required to fail the material [12].

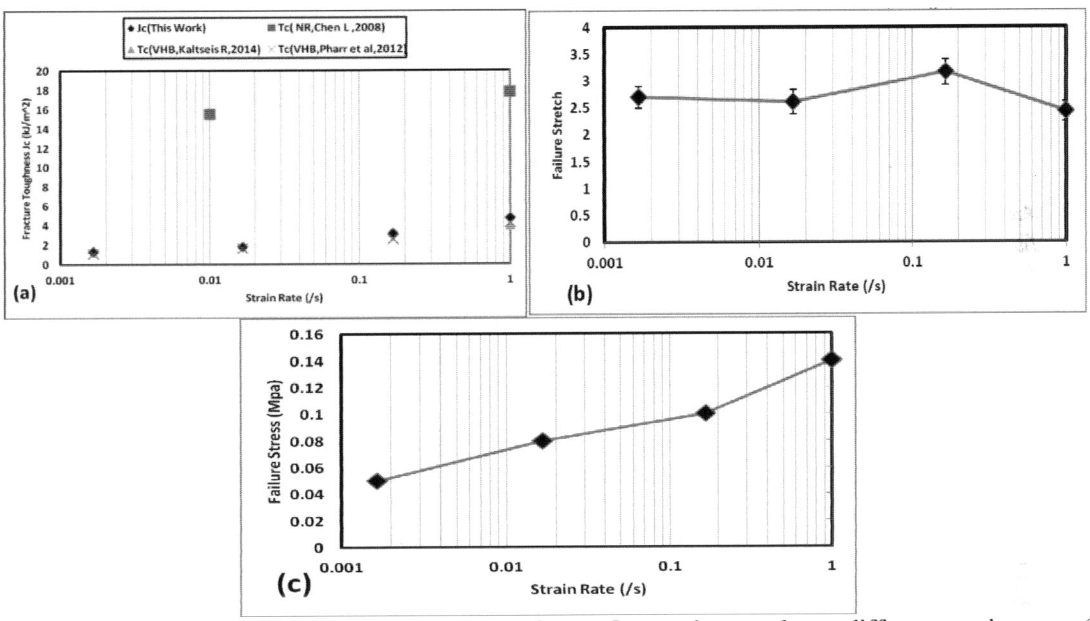

Figure 6. (a) Comparison of mean fracture toughness from other works at different strain rates (b) Variation of mean failure stretch (c) mean failure stress with strain rate. (Notch Length=64mm)

4. Conclusion

The present wok successfully investigated the failure characteristics of VHB 4910 under different notch lengths. It is observed that failure stretch does not depends on notch length and strain rate, but failure stress decreases with increasing notch length and increases with increasing strain rate. It is also found that fracture toughness is independent of notch lengths. However, fracture toughness is found to increase with strain rate. These results are very important to design dielectric elastomer based products where initial flaw size is expected to be large.

Acknowledgement

The reported study was partially supported by DST, Govt. of India, research project No. INT/SIN//P-03.

References

[1] Wissler M and Mazza E 2007 Mechanical behaviour of an acrylic elastomer used in dielectric elastomer actuators *Sensors Actuators A* **134** 494–504

[2] Wissler M and Mazza E 2007 Electromechanical coupling in dielectric elastomer actuators *Sensors Actuators, A Phys.* **138** 384–93
[3] Pelrine R, Kornbluh R, Pei Q and Joseph J 2000 High-Speed Electrically Actuated Elastomers with Strain Greater Than 100%. *Science* **287** 836–9
[4] Moscardo M, Zhao X, Suo Z and Lapusta Y 2008 On designing dielectric elastomer actuators *J. Appl. Phys.* **104** 1–7
[5] Czech B, van Kessel R, Bauer P, Ferreira J A and Wattez A 2010 Energy harvesting using dielectric elastomers *Proc. 14th Int. Power Electron. Motion Control Conf. EPE-PEMC 2010* 18–23
[6] Chiba S, Waki M, Wada T, Hirakawa Y, Masuda K and Ikoma T 2013 Consistent ocean wave energy harvesting using electroactive polymer (dielectric elastomer) artificial muscle generators *Appl. Energy* **104** 497–502
[7] McKay T G, O'Brien B M, Calius E P and Anderson I A 2011 Soft generators using dielectric elastomers *Appl. Phys. Lett.* **98** 1–4
[8] Schmidt A, Rothemund P and Mazza E 2012 Multiaxial deformation and failure of acrylic elastomer membranes *Sensors Actuators, A Phys.* **174** 133–8
[9] Kaltseis R, Keplinger C, Adrian Koh S J, Baumgartner R, Goh Y F, Ng W H, Kogler A, Tröls A, Foo C C, Suo Z and Bauer S 2014 Natural rubber for sustainable high-power electrical energy generation *RSC Adv.* **4** 27905–13
[10] Chen L 2008 Tear Energy of Natural Rubber Under Dynamic Loading, Master Thesis,University of Acron, Ohio
[11] Luong R, Isac N and Bayraktar E 2007 Damage initiation mechanism in rubber sheet composites during the static loading **28** 19–26
[12] Pharr M, Sun J Y and Suo Z 2012 Rupture of a highly stretchable acrylic dielectric elastomer *J. Appl. Phys.* **111**
[13] Elmukashfi E and Kroon M 2012 Numerical analysis of dynamic crack propagation in rubber *Int. J. Fract.* **177** 163–78
[14] Marano C, Boggio M, Cazzoni E and Rink M 2014 Fracture Phenomenology and Toughness of Filled Natural Rubber Compounds Via the Pure Shear Test Specimen *Rubber Chem. Technol.* **87** 501–15
[15] Rivlin R S and Thomas a. G 1953 Rupture of rubber. I. Characteristic energy for tearing *J. Polym. Sci.* **10** 291–318
[16] Dragoni E and Medri G 1988 Fracture toughness evaluation of natural rubber *Theor. Appl. Fract. Mech.* **10** 79–83
[17] Gent A N 2012 *Rubber Chemistry and Technology* (Munich: Hanser Publishers)

IOP Conf. Series: Materials Science and Engineering **229** (2017) 012036 doi:10.1088/1757-899X/229/1/012036

Thermal Processes of the Isolating Materials in Problems of Nondestructive Diagnostics of the Main and Distributive Power Stations

M N Dubyago* and N K Poluyanovich

Southern Federal University, Russia

*w_m88@mail.ru

Abstract. Methods of calculation of thermal processes of the isolating materials are considered, in problems of nondestructive diagnostics of the main and distributive power stations. Theoretical justification of calculation of thermal processes of the multilayered isolating structure was given. On the basis of the analysis of results of operation, service life, physical processes of "ageing" and use of diagnostic parameters of the isolating structure characteristics of resistance of the equipment to refusal are offered. Time of refusal and a residual service life allow to predict using of dependences of refusal resistance.

1. Introduction

Temperature is one of the major factors defining wear and destruction of the insulating materials, power stations. If temperature becomes nearer to the most admissible, begins process of intensive thermal wear of the insulating material, its thermal aging. As a result can occur thermal breakdown of the insulating material. Factors of environment and electromagnetic processes complicate a task of the analysis of the thermal modes, from the point of view of heating engineers, on some orders.

2. Model of thermal processes

Let's consider a power cable as a system of the following thermal homogeneous bodies, Figure 1.

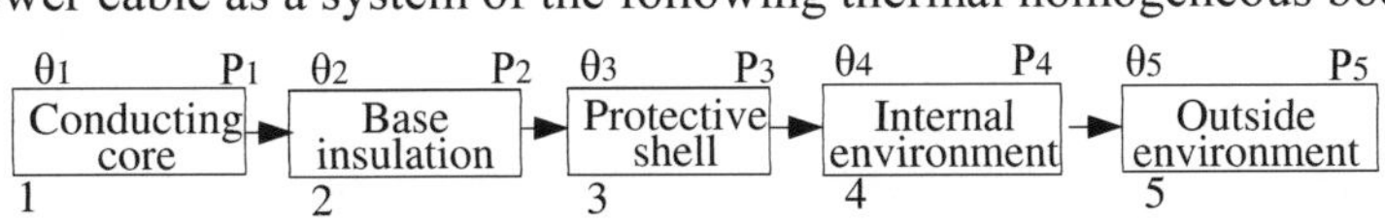

Figure 1. Structural diagram of thermal processes.

The mathematical model of thermal processes of the insulating material [1,2], doesn't consider a number of factors of the insulating materials influencing heating – dielectric losses, superficial effect in veins, losses in the screen, a way of screen grounding. In the work is considered the mathematical model of thermal processes of the power cable line (PCL) which isn't using empirical coefficients and dependences, but allowing a message calculation of temperature of the most heated point of insulation of a cable with a sufficient accuracy. Write down the system of the differential equations for thermal processes of single-core PCL considering specific factors where: θ_1-θ_5 – temperatures of uniform bodies, figure 2, °C; C_1-C_4 – thermal resistance of the corresponding uniform bodies; P_1-P_4 – losses of active power, respectively, in a current carrying vein, insulation of a vein, the screen and a protective cover.

Content from this work may be used under the terms of the Creative Commons Attribution 3.0 licence. Any further distribution of this work must maintain attribution to the author(s) and the title of the work, journal citation and DOI.
Published under licence by IOP Publishing Ltd

$$\begin{gathered} C_1 \cdot d\theta_1/dt + (\theta_1 - \theta_2)/R_1 = P_1; \\ C_2 \cdot d\theta_2/dt + (\theta_2 - \theta_1)/R_1 + (\theta_2 - \theta_3)/R_2 = P_2; \\ C_3 \cdot d\theta_3/dt + (\theta_3 - \theta_2)/R_2 + (\theta_3 - \theta_4)/R_2 = P_3; \\ C_4 \cdot d\theta_4/dt + (\theta_2 - \theta_1)/R_3 + (\theta_4 - \theta_5)/R_4 = P_4. \end{gathered} \tag{1}$$

On the basis of system (1) is synthesized the thermal equivalent circuit of the isolating material, Figure 1.

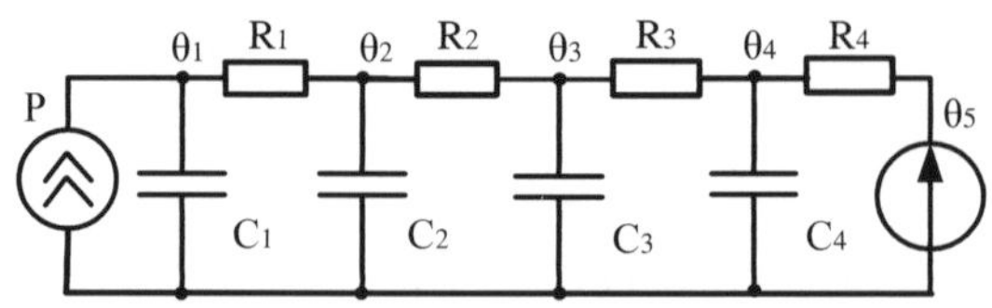

Figure 2. Thermal equivalent circuit.

We can note that θ_2 temperature of the most heated point of isolation of a cable (Figure 2). Thermal capacities of uniform bodies (Figure 3), the APV power cable of Pu of - 1x30/25-10 C1-C3 can be found on formulas:

$$\begin{gathered} C_1 = c_1 \cdot \beta_1 \cdot (\pi \cdot {d_1}^2/4) \cdot l; \\ C_2 = c_2 \cdot \beta_2 \cdot (\pi/4) \cdot (d_2^2 - d_1^2) \cdot l; \\ C_3 = c_3 \cdot \beta_3 \cdot (\pi/4) \cdot (d_3^2 - d_2^2) \cdot l; \\ C_4 = c_4 \cdot \beta_4 \cdot (\pi/4) \cdot (d_5^2 - d_4^2) \cdot l. \end{gathered} \tag{2}$$

Thermal resistance of uniform bodies of PCL R1-R3 are calculated by the formulas:

$$R_1 = 1/(2 \cdot \gamma_1 \cdot \pi \cdot l); \tag{3}$$

$$R_2 = (\ln(d_2/d_1))/(2 \cdot \gamma_2 \cdot \pi \cdot l); \tag{4}$$

$$R_3 = (\ln(d_3/d_2))/(2 \cdot \gamma_3 \cdot \pi \cdot l); \tag{5}$$

$$R_4 = (\ln(d_5/d_4))/(2 \cdot \gamma_4 \cdot \pi \cdot l). \tag{6}$$

Heat conductivity of metals considerably exceeds heat conductivity of dielectrics therefore the ratio is fair: $R_1 \ll R_2$; $R_1 \ll R_3$.

Losses of power, taking into account dependence of resistance of a vein of a cable on temperature, are equal

$$P = I^2 \cdot R_{el} = I^2 \cdot \rho 20 \cdot (1 + \alpha \cdot (\theta_1 - 20)) \cdot ((4 \cdot l)/(\pi \cdot d_1^2)), \tag{7}$$

where I – the current proceeding on a vein of a cable (A): R_{el} – cable vein resistance length of l (Ohm): ρ20 – the specific resistance of material at a temperature θ1 = 20 °C (Om·m); α – the temperature coefficient of resistance (TCR) of material (1/°C).

Losses in a cable vein taking into account the phenomenon of superficial effect which is shown in increase in resistance of the conductor on alternating current of R_ω in comparison with resistance on a direct current of R_0, are equal:

$$P_1 = I_1^2 \left(\frac{R_\omega}{R_0}\right) \rho 20\big(1 + \alpha_1(\theta_1 - 20)\big)\left(\frac{l}{\pi} \cdot r_1^2\right), \tag{8}$$

where I_1 – the current proceeding on a cable vein, A; l – cable length. The superficial effect in uniform cylindrical conductors is described as:

$$\begin{gathered} (R_\omega/R_0) = 1 + (k^4/3) \quad if\ k < 1; \\ (R_\omega/R_0) = 0{,}997 \cdot k + 0{,}277 \quad if\ 1 < k < 10; \\ (R_\omega/R_0) = k + 1/4 + (3/64 \cdot k) \quad if\ k > 10; \\ k = (r_1 \cdot \sqrt{2 \cdot \mu \cdot \mu_0 \cdot \gamma \cdot \omega})/4, \end{gathered} \tag{9}$$

where r_l – the radius of the conductor, especially cable vein radius; μ – relative magnetic permeability of material of a vein; γ – specific conductivity of material of a vein.

Dielectric losses of P_2 and P_4 are calculated by the formulas $P_i = 2 \cdot \pi \cdot U_i^2 \cdot f \cdot C_i \cdot tg\delta_i$,

where i={2;4}; U_i – insulation tension; f – working frequency; C_i – insulation capacity; $tg\delta_i$ — tangent of angle of dielectric losses of insulation.

Value of tension of U_i depends on a way of grounding of the screen of a cable. Generally it is possible to write down $\dot{U}_2 = \dot{U}_\Phi - \dot{U}_3$; $\dot{U}_4 = \dot{U}_3$, where U_3 – complex value of tension on the cable screen concerning the earth. Insulation capacity for cylindrical condensates

$$C_i = (\pi \cdot \varepsilon_i \cdot \varepsilon_0 \cdot l)/(\ln(r_{ext.i}/r_{int.i}), \tag{10}$$

where– insulation radiuses; εi – relative dielectric permeability of material; l – cable length.

Losses of power in the P3 screen depend on a way of grounding of the screen of a cable. If the screen is grounded on the one hand, current through it is approximately equal to zero. Losses of power in the screen pay off on the operating value of current in the screen I_3,

$$P_3 = I_3^2 \cdot \rho 20\big(1 + \alpha_3(\theta_3 - 20)\big)(l/((r_{ext.3})^2 - (r_{int.3})^2)), \tag{11}$$

where ρ20 – the specific resistance of the screen at a temperature of 20 °C; α_3 – screen material TKS; r_{ext3}, r_{int3}– external and internal radiuses of the screen; l – cable length. Having considered a thermal equivalent circuit (Figure 2), we will receive expression for the established value of the temperature of cable veins:

$$\theta_{2.st} = P \cdot (R_2 + R_3) + \theta_4. \tag{12}$$

From comparison of equations (4), (5), (7) and (12) it is understand that values of length of l also don't influence calculations for the set thermal mode.

3. Modeling of thermal processes

Data on the geometrical sizes of elements of the AP PU of - 1x30/25-10 and their help data [3,4]. Diameters of the corresponding uniform bodies of a power cable, are reduced in the Table 1.

Table 1. Cable Settings

Parameters	**Materials**	Copper	XLPE	Lead	Air
Thermal conductivity (W/(m•k))		400	0.38	34	0.024
Density (kg/m3)		8700	2200	11370	1.2
Specific heat (J/(kg•k))		385	1900	3000	1005

Design of a single-conductor power cable as system of thermal uniform bodies, it is presented on the Figure 3.

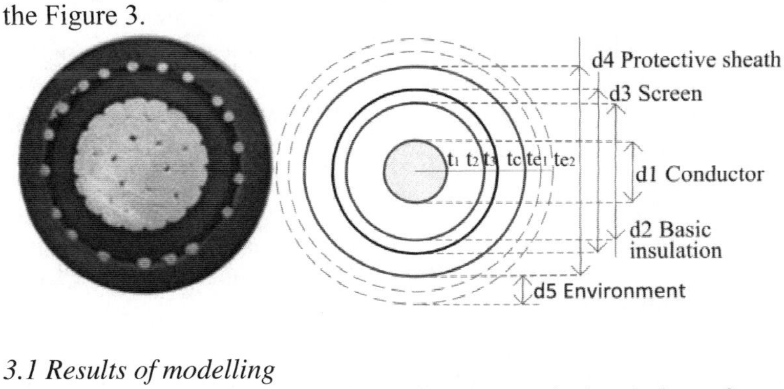

Figure 3. System of uniform thermal bodies.

3.1 Results of modelling

Modelling of non-stationary thermal processes in insulation of a power cable was carried out in the system of Mathcad. On the Figure 3 is taken dependence of process of heating of a cable, on layers. Figure 4 – Settlement chart of process of heating of the cable.

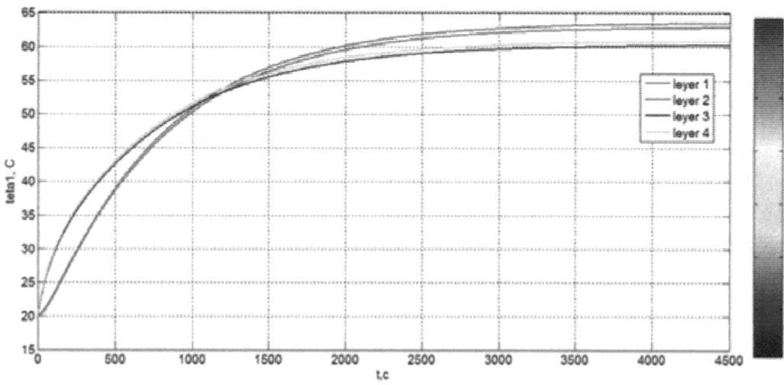

Figure 4. Calculation diagram of the cable heating process.

The considered method of calculation of thermal processes can be applied to cables with any number of veins by increase in quantity of the equations in system (1).

3.2 Research of service life

Research of distribution of temperature and dynamics of its change it is executed in the program COMSOL MULTIPHYSYCS complex on the basis of a method of final elements.

Basic data for calculation are the parameters specified in Table 1. We received the gradient of temperatures on the section of a cable, Figure 5.

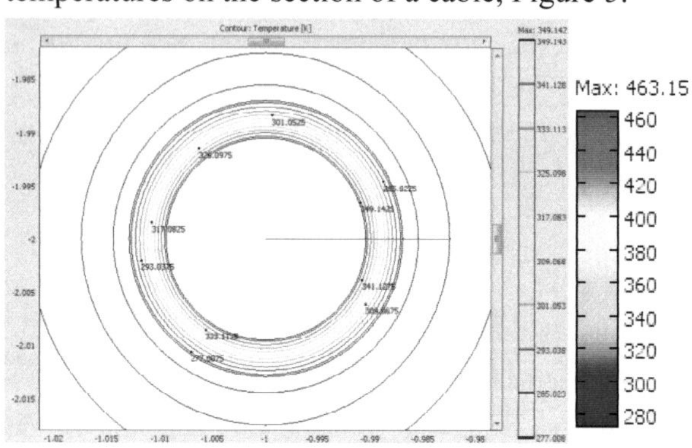

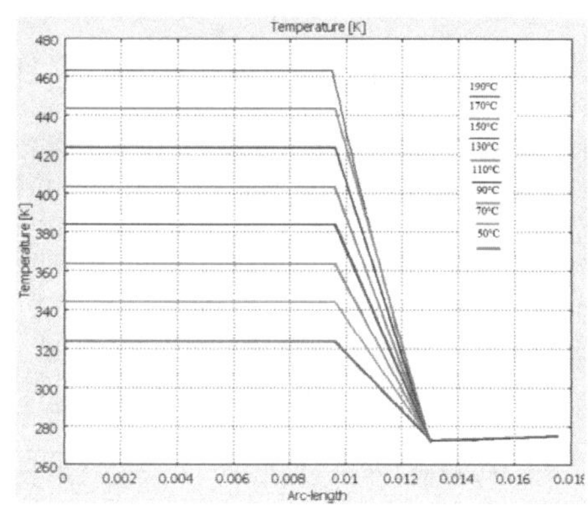

Figure 5. The temperature gradient over the cross section of the cable.

Figure 6. Distribution of temperature in cable section at different temperatures on a vein.

Results of calculation of the set mode (Figure 6) caused by the different current proceeding on a vein distinctly show weak, almost zero, dependence of distribution of temperature of high-heat-conducting elements – a vein and the screen of a cable (horizontal regiments on schedules).

Thus in volume of insulation is observed the big gradient of temperatures that leads to an essential difference of temperatures of the screen of a cable where temperature sensors and the central vein are installed. That is why , service life of the isolating materials of cables can be less than established 25 – 30 years, as is confirmed in practice; the capacity calculated on design stages and actual can significantly differ.

4. Resistance to equipment failure under operating conditions

The operational resource time of life of t of the main and distributive power stations feeling a complex of various influences is described by dependence [5].

$$t = \Theta \cdot b^{a}, \tag{13}$$

where Θ –the constant defining an initial condition of system, b – the basis of an exponential function, a– describes changing character of function in time.

Service life of electric installations it is descry bed by dependence of a look

$$t = \Theta \cdot [1 - b^a], \tag{14}$$

Nature of change of the function described by a formula (11) depends on value a. If to accept the characteristic of resistance of object to refusal in the form of the θ_{ot} parameter, function (11) can be presented in the form

$$\Theta\,(t) = \Theta\,ot[1 - (t\,/t_n)^a], \tag{15}$$

where $\Theta\,(t)$ - resistance to refusal in an initial state, Θot – current time; t— operating time to the full object; t_n– statistical service life in the field of the maximum values:
-by commissioning object when t=0, value Θ_{otis} equal 1,
-by the ending of operation when $t=t_n$, Θ_{ot} value is equal 0.

That is why, Θ(t) is function of time of an operating time to the full and describes resistance of object to refusal, Figure 5, at various values of an indicator a.

Most closer exponential functions of a type of t = A ·(1-b a) (1-ba) at a = 1.3–1.6. Optimum value it is necessary to accept a= 1.5. During the initial period of time of operation, fast wear of the equipment unlike the dependence having an indicator a>1, isn't observed. Materials which service life are described by the dependences having indicators a>1; a<1, can have finally the same time of a resource, t_{pec} if there is no the accelerated aging at emergence of dangerous defects.

If to consider the same object presented by dependences at a=1.5 and a=0.5 (Figure 7), by the time formation of defect at the time of td which leads to identical deterioration of properties in the form of reduction of the Θot parameter by $\Delta\Theta ot_t$ and to various values of the predicted times of refusal of tot1 and t_{ot2}. Thus it is assumed that after the moment of t_d regularities of aging remain.

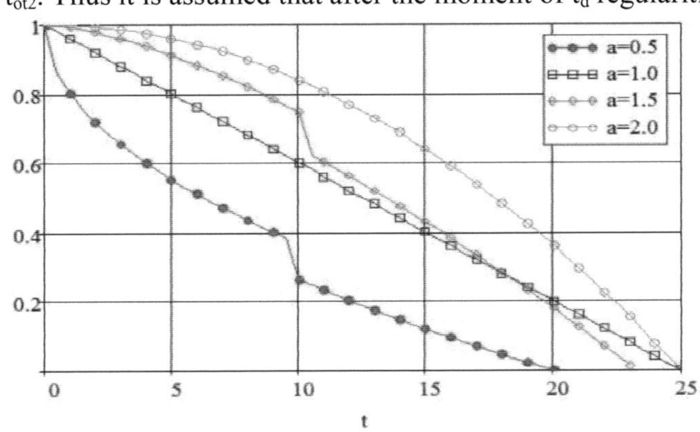

Figure 7. Functions of resistance to refusal at formation of defect and a residual resource respectively.

Function (15) after the moment of formation of defect can be presented in the form

$$\Theta(t) = \Theta ot[1 - (t\,/t_n)^a] - \Delta\Theta ot, \tag{16}$$

At Θ_{ot}=0 the current time will be refusal time $t=t_{otk}$.

$$\Theta ot(t) = \Theta otn[1 - (t\,/t_n)^a] - \Delta\Theta ot = 0, \tag{17}$$

Value t_{otk} will be equal

$$t_{otk} = t_n\big[1 - (1 - \Delta\Theta ot)^{1/a}\big], \tag{18}$$

$$t_{ost} = t_{otk} - t_d = t_n \cdot \big[(1 - \Delta\Theta ot)^{1/a}\big] - t_d \tag{19}$$

Formation of defect in a time point of t_1 *and* t_2 (fig. 8) leads to reduction of resistance to refusal that will be expressed in decrease in *Θot* value on *ΔΘot*. Elimination of part of defects in the period of t_3 leads to increase in relative resistance to refusal. Formation of defects at the time of t_4 leads to reduction of resistance to refusal and at an invariance of regularity of aging – to definition of the predicted refusal time. For the definition of t_{otk} and t_{ost} requires knowledge of the *ΔΘot* values which are function of the measured and rated values of diagnostic parameters. *ΔΘot*, indicator characterizing value of reduction of resistance to refusal is described in a look.

$$\Delta\Theta ot = 0.1 \cdot [X_i/X_H]^m, \tag{20}$$

where X_i – value of the measured diagnostic parameter,X_H– rated value of diagnostic parameter, m – the indicator determined by a type of physical process in controlled object.

By the formation of several defects which values of diagnostic parameters are equal or exceed rated level, the indicator of $\Delta\Theta ot$ is defined as the sum of indicators $\Delta\Theta_{oti}$of each defect:

$$\Delta\Theta ot = \Delta\Theta_{ot1} + \Delta\Theta_{ot2} + \cdots + \Delta\Theta_{otk} = \sum \Delta\Theta_{oti}. \tag{21}$$

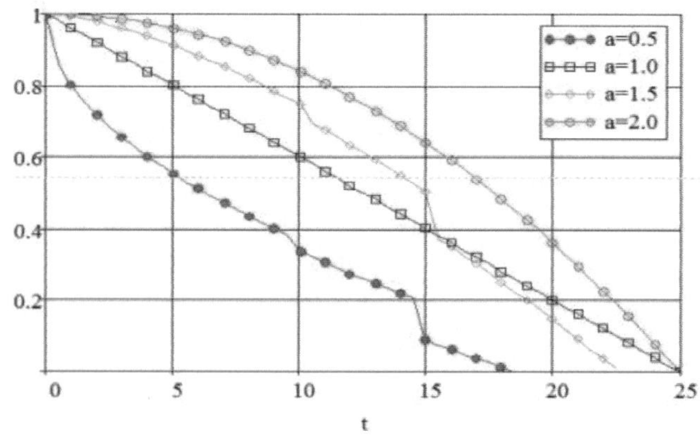

Figure 8. Functions of resistance to refusal by formation of defects and their partial elimination.

5. Conclusion

1. The considered mathematical model can be used as basic at calculation of thermal processes of the isolating materials in problems of nondestructive diagnostics of the main and distributive power stations in real time.
2. The mathematical model taking into account service conditions can be used in devices and systems of continuous diagnosing of power stations in thermal parameters.

Acknowledgements

This work was supported by Russia's Southern Federal University grant: «Development of theoretical foundations and methods for constructing intelligent multiply connected systems for managing the processes of production, transportation, distribution and consumption of energy, No. Grant-07/2017-15».

Reference

[1] Zalizny D I 2009 Mathematical modeling of thermal processes in power cables with plastic isolation *P.Sukhoi Gomel State Technical University* Vol 3(38) (Republic of Belarus) p 65-75
[2] Shevchenko V A, Pshikhopov V K and Medvedev M Y 2014 Multilevel control system of the power grid Advanced Materials Research Vol 1049-1050 pp 736-742
[3] Poluyanovich N K and Dubyago M N 2014 Mathematical description of interaction of the elements in the power network system *Applied Mechanics and Materials* Vols. 448-453 (Trans Tech Publications, Switzerland) pp 2455-2460
[4] Dubyago M N and Poluyanovich N K (2015) Analysis of insulation materials of cable systems by method of partial discharges *Advances in Materials Science and Applications* Vol 4(1) p.23
[5] Sitnikov V F and Skopintsev V A 2007 Using a Probabilistic-Statistical Approach for Assessing the Service Life of Electric Network Equipment During its Operation Vol 11 National Research University Moscow Power Engineering Institute (Russia) pp 9-16

Mechanical Properties and Interfacial Structure of Ti/Al Clad Plates Generated by Differential Temperature Rolling

Zichen Qi[1], Hong Xiao[1, a], Na Li[2], Chao Yu[1] and Zhongkai Ren[1]

[1] National Engineering Research Center for Equipment and Technology of Cold Strip Rolling, Yanshan University, Qinhuangdao, Hebei, China
[2] College of Mechanical Engineering, Yanshan University, Qinhuangdao, Hebei,China.

[a] xhh@ysu.edu.cn

Abstract. In this study, a Ti/Al clad plate was prepared by differential temperature rolling, where only the titanium layer is heated. The effects of the rolling reduction and Ti layer heating temperature on the shear strength and interface of Ti/Al clad plates were investigated. The results indicate that when the titanium layer was heated to 800℃ and the rolling reduction was 50%, the shear strength of the titanium/aluminum clad plates interface reached 107.5 MPa, which is close to the shear strength of aluminum matrix, and the fracture surface presented ductile fracture characteristics. During the heating process, an oxide layer was produced on the surface of the titanium plates. However, the oxide layer was completely broken with a large rolling reduction, and then, aluminum metal extruded into the cracks and made contact with fresh titanium metal. Under the action of high pressure and high temperature, Ti and Al atoms mutually diffuse so that the Ti/Al clad plates achieved a strong metallurgical bond.

1. Introduction

Preparation methods of the titanium/aluminum clad plates mainly include explosion compound, explosive welding, diffusion welding, and hot rolling bonding. Boronski [1] prepared the Al/Ti laminated composite plates by the method of explosive welding and studied the fracture toughness of the composite plates under low temperature. Lazurenko [2] prepared multilayer Ti/Al composites by the method of explosive welding and investigated the structure and transformation during heat treatment. The technology of explosive welding for preparing titanium/aluminum composite plates is very mature. However, the explosion welding process will produce seismic waves, noise and toxic gases, which are not conducive to large-scale production of clad plates. Obielodan [3] used the ultrasonic consolidation method to prepare titanium/aluminum clad plates and found that the resulting clad plates had a relatively low bonding strength. Compared to other methods, the method of hot rolling compound has the advantages of stable product quality, simple equipment, easy automation and mass production.

Currently, the preparation of Ti/Al composite plates by hot rolling has become popular. However, the metal properties (deformation resistance, plasticity, thermal conductivity, melting point, etc.) of titanium and aluminum are quite different, and two problems are found in the preparation process; the first is that the deformation of titanium/aluminum after hot rolling is extremely uncoordinated, and the second is that the bonding strength of the prepared composite plate is low. Chen [4] found that the deformation of the titanium and aluminum layers is uncoordinated in the hot rolling process of Ti /Al composite plates. Ma [5] prepared Al/Ti composite laminates by hot rolling and found that the

Content from this work may be used under the terms of the Creative Commons Attribution 3.0 licence. Any further distribution of this work must maintain attribution to the author(s) and the title of the work, journal citation and DOI.
Published under licence by IOP Publishing Ltd

deformation of titanium and aluminum became increasingly uncoordinated as the rolling temperature increased. When adopting the method of hot rolling directly, due to the low melting point of aluminum, the deformation resistance of titanium is much greater than that of aluminum when the titanium/aluminum composite plates are rolled at a temperature below 600 ℃, and the deformation of the titanium layer after rolling is much smaller than that of the aluminum layer, which leads to extremely uncoordinated deformation. Yu [6] who studied the effect of annealing on the microstructural and mechanical properties of Al/Ti/Al laminate sheets found that the interface contained $TiAl_3$ and other compounds, affecting the bonding properties of composite panels in the processing of hot-rolled Ti/Al clad plates.

To improve the deformation coordination and the bonding strength of titanium/aluminum clad plates, a new method for the preparation of titanium /aluminum composite plates was established in this study. Ti/Al clad plates were prepared by rolling at differential temperatures, where only the titanium layer is heated. The titanium plate was heated to a high temperature to allow for optimal plasticity and to greatly reduce the deformation resistance of titanium, which made the deformation resistance of the titanium plate become close to the deformation resistance of the aluminum plate at room temperature, and then, the effects of the rolling reduction and Ti layer heating temperature on the shear strength and interface of the Ti/Al clad plates were investigated.

The new method used in this study can also provide reference for the composite rolling of two or three kinds of metals, such as Ti/Cu, Ti/Mg and Al/Ti/Mg [7], which have greatly different physical and chemical properties.

2. Experimental

The materials used in the experiments were commercial-purity titanium TA1 sheets and aluminum alloy AA6061 sheets, both with a thickness of 2 mm. The as-received materials were cut in the transverse direction to produce 100 mm×60 mm pieces. To eliminate the internal stress and work-hardening in sheet metal, the TA1 and AA6061 were annealed. AA6061 aluminum alloy sheets were annealed at 350℃ for 2 h and then air-cooled. TA1 sheets were annealed at 780℃ for 2 h and then air-cooled.

The surfaces of the annealed TA1 plates and AA6061 plates were treated to remove grease, dirt and oxide to facilitate the combination of dissimilar metals. In this experiment, grease, dirt and oxides on the surfaces of the metals were removed by use of a grinding machine, and then, the surfaces were repeatedly cleaned with acetone and alcohol and then immediately dried. To be rolled into the rolling mill at the same time, the heated titanium plates needed to be combined with the aluminum plate, which was at room temperature. However, due to the thinness of the titanium plates, a delay of a few seconds would lead to a large temperature drop, which affects the objectivity of the experimental results. Therefore, in the rolling process, a piece of iron was first put into the resistance furnace and heated to the rolling temperature of the titanium plates, and then, the titanium plate was put into the resistance furnace on the iron plate for 5 minutes. Five minutes later, the iron and titanium plates were removed to the entrance of the rolling mill. Then, the iron plate played a role in the heat preservation of the titanium plates. Finally, the heated titanium plate and the aluminum plate were pushed at room temperature into the rolling mill with the guide plate. The schematic diagram for rolling Ti/Al clad plates at differential temperature is shown in Figure 1. The AA6061 plates were at room temperature. The TA1 plates were heated in a furnace for 5 min at temperatures of 500℃, 600℃, 700℃, 800℃ and 900℃. The reductions were 20%, 25%, 29%, 34% and 50%. The parameters for the two-rolling mills in the experiment were the following: the roll size was ϕ200 mm × 200 mm and rolling speed was 50mm / s.

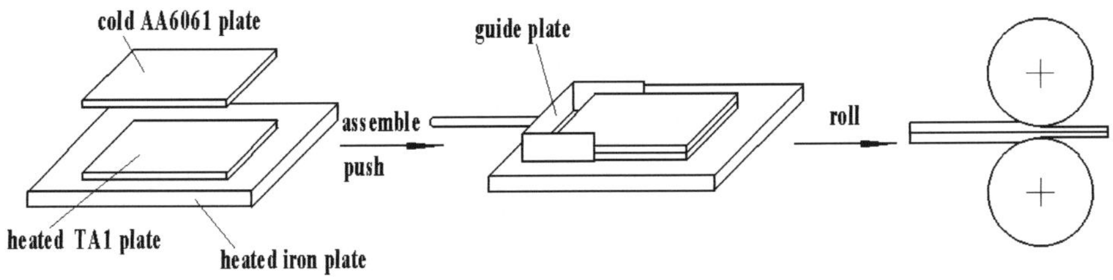

Figure 1. Schematic diagram for rolling Ti/Al clad plates.

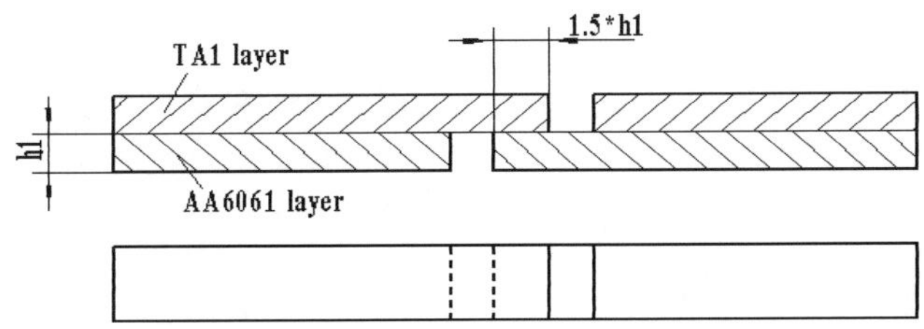

Figure 2. Schematic for the shear test.

The shear strength and the bonding rate of the two metals are the two basic parameters for measuring the properties of the clad plates as described by Gou [8]. Three samples for each plate were cut parallel to the rolling direction and were tested to obtain an average shear strength. The tension-shear test was carried out using an INSPEKT Table 100kN instrument. The stretching speed was 0.5 mm/min. Figure 2 shows the tension-shear test sample.

Specimens for microstructure examination were extracted along the rolling direction. The surfaces of the specimens were ground with emery papers up to No. 4000 and polished with an SiO_2 suspension with a particle size of 0.04 μm, and then, the morphology of the fracture surface along the interface was observed using an optical microscope (OM, ZEISS Scope A1) and scanning electron microscope (SEM, ZEISS Sigma 700). The distribution of elements across the interface was examined by an SEM equipped with an energy dispersive spectrometer (EDS).

3. Results and discussion

3.1. Macroscopic bonding property test of the clad plates

Figure 3 and Figure 4 show the shear strength of Ti / Al clad plates for different Ti layer temperatures and different reductions, respectively. It can be seen from Figure 3 that the shear strength increases with an increase in the titanium layer temperature when the reduction is 29% and the temperature is between 500℃ and 800℃. The shear strength of the composites interface decreases with an increase in temperature above 800℃. The bonding strength of the composite plates is low at temperatures below 700℃. When the temperature is higher than 700℃, the bonding strength is obviously improved, and the shear strength of the composite plates is the largest at 800℃. Therefore, the titanium layer is chosen to be heated to 800℃, and rolling is carried out with different reductions, as shown in Figure 4. The shear strength increases gradually with the increase in total reduction. The growth rate of shear strength is fast initially and then slows. When the reduction is less than 34%, the growth rate of the shear strength is relatively high. The growth rate is relatively low with a reduction above 34%. The shear strength of the titanium/aluminum clad plates interface reaches 107.5 MPa for a 50%reduction at 800℃, which is close to the shear strength of aluminum matrix. The shear strength obtained in this experiment is even higher than that of titanium/aluminum composite plates prepared by explosive compound; for example, Xia [9] prepared Ti/Al clad plates through the explosive compound, and the shear strength was approximately 73.2 MPa.

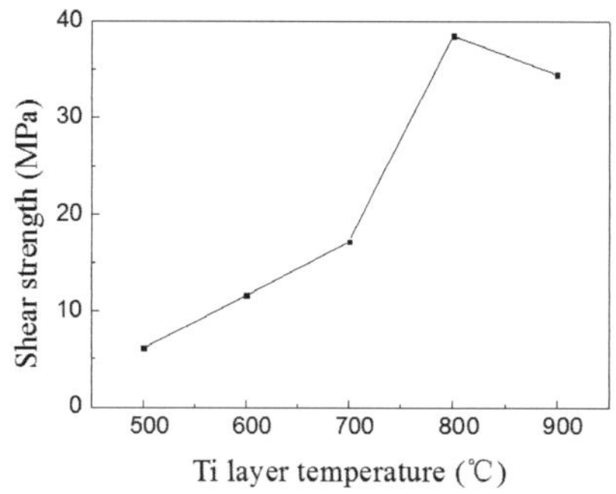

Figure 3. Effect of temperature on the shear strength of the composite plates with a 29% reduction.

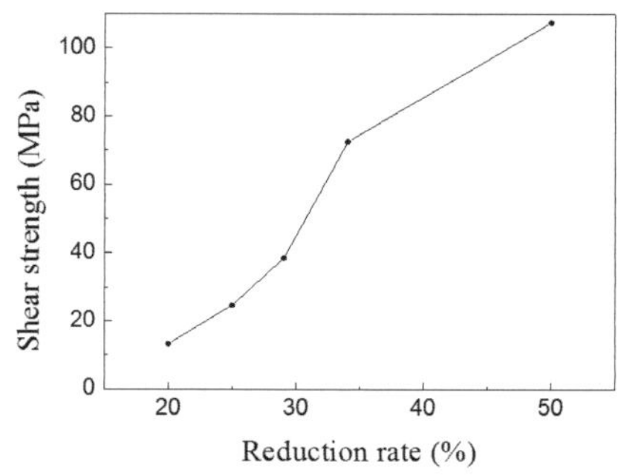

Figure 4. Effect of reduction on the shear strength of the composite plates at 800 ℃.

3.2. Interfacial structure of Ti/Al clad plates

It can be seen from Figure 5a and 5c that some strip-shaped cracks are formed on the titanium side, corresponding to some strip-shaped protrusions formed on the aluminum side (Figure 5b and 5d). Table 1 displays the results for the elemental contents of Ti, Al and O at each point in Figure 5. According to the element point scanning analysis of the fracture surface, the elemental contents at point1 and point5 indicate that there is more than 90% Ti and approximately 5% O inside the strip-shaped cracks on the titanium side. The elemental contents at point2 and point6 shows that O and Ti compose approximately 35% and 60% of the fracture plane, respectively. Based on the above analysis, it can be determined that the fracture surface of the titanium side has a layer of titanium oxide. Fresh titanium metal is also exposed, and a small amount of aluminum metal adheres inside the strip-shaped cracks.

In addition to some strip-shaped protrusions formed on the aluminum side, the surface also contains many lumps (Figure 5b and 5d). The elemental contents at point3and point8 indicate that the strip-shaped protrusions formed on the aluminum side are more than 90% Al and approximately 3% O. The elemental contents at point4 and point7 shows that in the lumps, Al, O and Ti account for approximately 45%, 25% and 25%, respectively. Based on the above analysis, it can be determined that the strip-shaped protrusions are formed by the extrusion of metal aluminum into the titanium cracks and the lumps are the oxide from the surface of the titanium layer adhered to the aluminum plate. From the comparison of Figure 5a and 5c, it can be seen that at a higher temperature for the titanium layer, more and wider cracks formed on the titanium after rolling. As a result, aluminum metal can come into contact with larger amounts of fresh titanium. Therefore, the increase in the temperature of the titanium plate can increase the area of the cracks and increase the bonding area of aluminum and fresh titanium metal, which is beneficial for the combination of the composite plates.

The fracture surface presents ductile fracture features with many dimples, see Figure 5e and 5f. The elemental contents at point9, point10 and point11 shows that the dimple is composed of more than 93% aluminum, and the O and Ti contents are very small. No strip-shaped cracks were found on the titanium side and no striped protrusions on the aluminum side, but there was still a small number of lumps, indicating that the oxide layer of titanium was completely broken with a large 50% rolling reduction at 800℃. The metal aluminum is extruded into the cracks at a large pressure and forms a strong metallurgical bond with a large amount of fresh titanium. In this case, the bonding strength reached 107.5 MPa, basically close to the shear strength of aluminum matrix, and the titanium/aluminum clad plates have very good bonding properties.

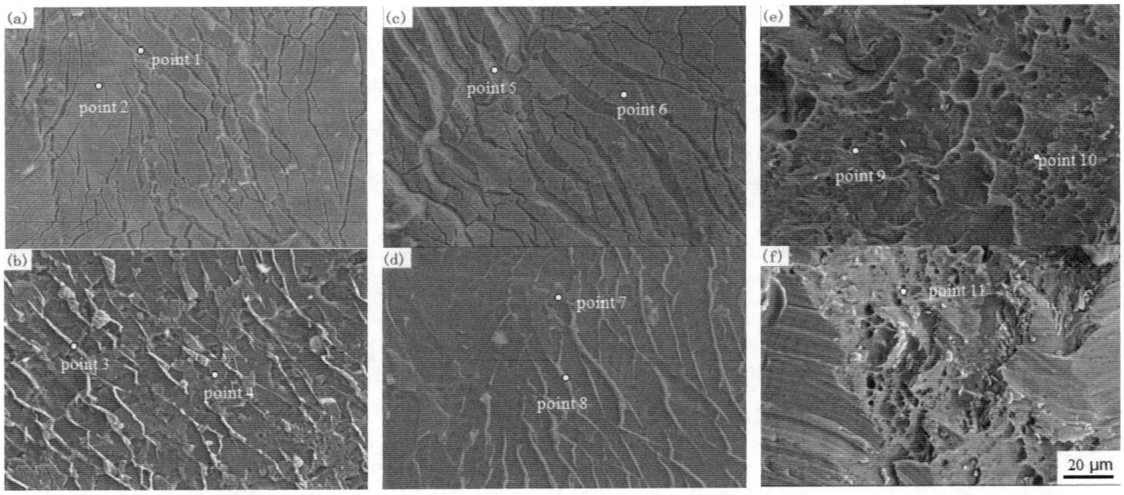

Figure 5. SEM images of the fracture surface and element point scanning under different processes (a) 500℃/29% reduction/Ti side, (b) 500℃/29% reduction/Al side, (c) 800℃/20% reduction/ Ti side, (d) 800℃/20% reduction/ Al side, (e) 800℃/50% reduction/ Ti side, (f) 800℃/50% reduction/ Al side.

Table 1. The elemental contents of Ti, Al and O from element point scanning (wt%).

	point1	point2	point3	point4	point5	point6	point7	point8	point9	point10	point11
Ti	94.06	66.20	0.54	22.61	91.19	60.24	29.10	1.37	0.76	0.87	0.14
Al	0.38	0.44	93.95	47.31	0.42	0.31	39.37	92.18	94.71	93.56	94.60
O	4.54	32.14	3.38	24.35	6.53	37.79	28.15	2.37	0.15	0.42	0.49

Figure 6 shows the SEM image and element line scan of the bonding interface for a 50% reduction at 800℃. It can be seen from Figure 6a that there is a gray stripe in the upper half of the bonding interface and a crack is present, but the lower part does not contain a crack. The upper and lower parts are subjected to element line scanning, and the results are shown in Figure. 6b and 6c. Both the Ti and O distributions exhibit a platform with a width of approximately 2 μ m, indicating that the gray stripe is a layer of titanium oxide. There is no O elemental distribution at the bonding interface of line 2, and the platform of Ti element is not present. It can be determined that there is no titanium oxide in the lower half of the interface. Figure 7c shows that the mutual diffusion depth of Ti and Al is approximately 2 μ m. Based on the above analysis, the oxide layer of titanium at the interface of Ti/Al clad plates is broken with a large 50% rolling reduction at 800℃, and then, metal aluminum is extruded into the cracks and comes into contact with fresh titanium metal. Finally, the Ti/Al clad plates achieve a solid metallurgical bonding with Ti and Al atom mutual diffusion under high pressure and high temperature.

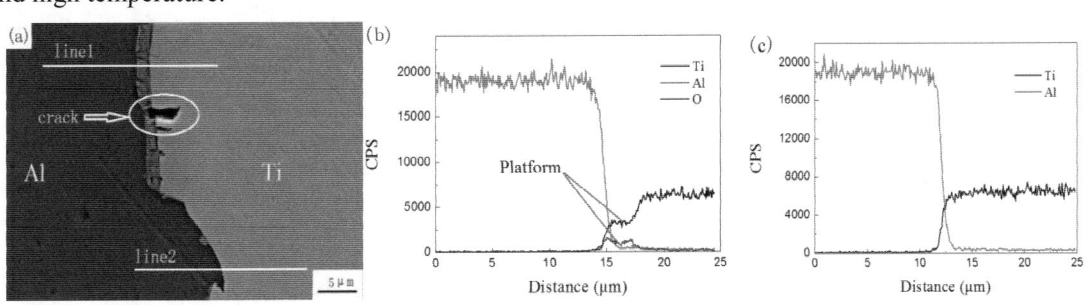

Figure 6. SEM image and element line scan of the bonding interface for a 50% reduction at 800℃ (a) SEM image of the bonding interface, (b) (c) Elemental diffusion on line 1 and line 2, respectively.

4. Conclusions

(1) When the titanium layer was heated to 800℃ and the rolling reduction was 50%, the shear strength of the titanium/aluminum clad plates reached 107.5 MPa, which was close to the shear strength of aluminum matrix, and the fracture surface presented ductile fracture characteristics.

(2) During the heating process, the oxide layer was produced on the surface of the titanium plates. However, in a large rolling reduction, the oxide layer was broken in large area, and then, the metal aluminum extruded into the cracks and came into contact with fresh titanium metal. Finally, the Ti/Al clad plates achieved a solid metallurgical bond with Ti and Al atoms mutually diffusing under high pressure and high temperature.

Acknowledgments

The project was supported by the Natural Science Foundation of HeBei Province (E2015203311) and the National Natural Science Foundation of China (No. 51474190).

References

[1] Boroński D, Kotyk M and Maćkowiak P 2016 Fracture Toughness of Explosively Welded Al/Ti Layered Material in Cryogenic Conditions *Procedia Structural Integrity* **2** 3764-71.

[2] Lazurenko DV, Bataev IA, Mali VI, Bataev AA, Maliutina IN, Lozhkin VS, Esikov MA and Jorge AMJ 2016 Explosively welded multilayer Ti-Al composites: Structure and transformation during heat treatment *Materials & Design* **102** 122-30.

[3] Obielodan JO, Stucker BE, Martinez E, Martinez JL, Hernandez DH, Ramirez DA and Murr LE 2011 Optimization of the shear strengths of ultrasonically consolidated Ti/Al 3003 dual-material structures *Journal of Materials Processing Technology* **211** 988–95.

[4] Chen ZJ, Chen QZ, Huang GJ and Liu XF 2012 Research on Roll Bonding Technology and Microstructure of Al/Ti/Al Three-layer Clad Sheet Fabricated by Hot Rolling *Materials Review* **26** 106-9.

[5] Ma M, Huo P, Liu WC, Wang GJ and Wang DM 2015 Microstructure and mechanical properties of Al/Ti/Al laminated composites prepared by roll bonding *Materials Science & Engineering A* **636** 301-10.

[6] Yu HL, Lu C, Tieu AK, Li HJ, Godbole A and Kong C 2016 Annealing effect on microstructure and mechanical properties of Al/Ti/Al laminate sheets *Materials Science & Engineering A* **660** 195-204.

[7] Parisa DM and Beitllah E 2015 Microstructure and mechanical properties of Tri-metal Al/Ti/Mg laminated composite processed by accumulative roll bonding *Materials Science & Engineering A* **628** 135-42.

[8] Gou JF 1995 Property analysis and application of the clad plate *Petro-Chemical Equipment Technology* **16** 49–52.

[9] Xia HB, Wang SG and Ben HF 2013 Microstructure and Mechanical Properties of Composite Plate between Pure Titanium and Aluminum Alloy with Explosive Welding *Pressure Vessel Technology* **30** 15-20.

IOP Conf. Series: Materials Science and Engineering **229** (2017) 012038 doi:10.1088/1757-899X/229/1/012038

Mechanical Properties of the 2D Re-entrant Honeycomb Made via Direct Metal Printing

Amer Alomarah[12], Jianjun Zhang[1], Dong Ruan[1]*, Syed Masood[1] and Guoxing Lu[1]

[1]Faculty of Science, Engineering and Technology, Swinburne University of Technology, Hawthorn, VIC 3122, Australia
[2]Department of Mechanics, College of Engineering, Wasit University, Al-Kut 52001, Iraq

* druna@swin.edu.au

Abstract. Auxetic structural materials show distinctive properties by exhibiting a negative Poisson's ratio (NPR). When these structures are subjected to uniaxial loading, they expand in tension and contract in compression in both loading and lateral directions. In this paper, two AlSi12 re-entrant honeycomb samples were manufactured using direct metal printing (DMP). Quasi-static uniaxial tension was executed in both X and Y direction. A Digital VIC-2D Image Correlation System was used to record the deformation history. Force and displacement were measured by ZWICK machine. The results show that loading direction has a significant effect on the mechanical properties and auxeticity of the tested structure. Re-entrant honeycomb under X-loading withstand lower force and has smaller magnitude of Poisson's ratio compared with that under Y-loading.

1. Introduction

Honeycombs evoke an extensive exploration because of their attractive mechanical properties represented with lightweight. These structures find dominant applications in different industries including automotive, aerospace, sports, marine and medical industries. Recently, developments in manufacturing technology have boosted the development of novel materials with remarkable performance to meet higher engineering specifications [1]. Unlike conventional materials, auxetic materials present uncommon geometric effect by becoming larger in all directions when stretched and smaller in all directions when compressed [2-4].]. Auxetic property tailors the structures to enhanced mechanical properties: indentation resistance, shear modulus, fracture toughness, variable permeability, and energy absorption [3-6]. Among many auxetic structural designs, re-entrant honeycomb has been studied extensively [7] due its intelligibility, adaptability and ability to produce high NPR which make it a favourable option for many applications [8, 9]. For re-entrant honeycomb, the key point to generate an auxetic behaviour is to select the proper unit cell dimensions and the re-entrant angles in the structure, which result in contraction of the structure in all directions under uniaxial compression, or expansion in all directions under uniaxial tension [10]. Additive Manufacturing (AM) or 3D Printing is becoming a more attractive technology for manufacturing of such structures because of its capability of not only providing geometric freedom but also offering precise control of micro-architecture and dimensions, thus turning any imaginable design into reality. Moreover, AM being a layer by layer building process, introduces its own influence on the performance and properties of the

Content from this work may be used under the terms of the Creative Commons Attribution 3.0 licence. Any further distribution of this work must maintain attribution to the author(s) and the title of the work, journal citation and DOI.
Published under licence by IOP Publishing Ltd

fabricated structures due to anisotropy, build orientation and thermo-mechanical material behaviour during the building process. It is necessary to understand the behaviour of such additive manufactured metallic auxetic structures, especially if they are to be employed in NPR related applications. While there have been some studies on auxetic behaviour of titanium structures made by direct metal 3D printing processes such as electron beam melting (EBM) [11] and selective laser melting (SLM) [12], very little attention has been paid on auxetic behaviour of aluminium structures made by direct metal printing. The aluminium alloy AlSi12 is a favourable powder used in SLM metal printing process due to its good convergence between melting and solidification temperatures. The AlSi12 alloy is widely used in the applications that required superior thermal properties and lightweight such as automotive, aerospace and aviation. In this paper, powder of aluminium alloy (AlSi12) is used to manufacture two re-entrant honeycomb samples using SLM Direct Metal Printer technique. The large deformation behaviour of re-entrant honeycombs subjected to quasi-static tensile load is experimentally investigated. The mechanical properties and the effect of loading direction on the re-entrant honeycomb samples are reported.

2. Design and manufacturing of re-entrant honeycomb samples

In the current study, a unit cell of re-entrant honeycomb is described by five geometric parameters as shown in Figure 1 (a). H is the length of vertical walls, L is the length of inclined walls, θ is the re-entrant angle (the angle between vertical and inclined walls), t_1 is the cell wall thickness and t_2 (not shown) is the out-of-plane depth. The geometric parameters of the re-entrant unit cell are shown in table1. Figures 1 (b) and (c) present the re-entrant honeycomb samples. Moreover, five dog-bone samples in figure 1 (d) were also printed out using the same approach.

Aluminium Silicon Alloy (AlSi12) powder was used to fabricate two re-entrant honeycomb samples figure 1 (b) and (c). The average size of AlSi12 powder was 40 μm. The AlSi12 re-entrant honeycomb samples were manufactured using SLM Direct Metal Printing process based on the geometry model generated in CAD. The ProX200 direct metal printer was employed to build the re-entrant samples layer-by-layer by melting AlSi12 powders using 1.07 μm wavelength laser beam. The laser power and laser focal plane used were 240 W and 0.6 mm respectively. The travelling speed of laser beam was set to 2500 mm/sec. The total layers for each sample were 180 with an approximate thickness 40 μm for each layer. In order to eliminate the effect of building direction on the mechanical properties, both samples were manufactured in the same orientation the out-of-plane direction (depth). After printing re-entrant samples, wire cutting was used to separate the samples from the printing substrate, through CNC Wire cutting machine (Fanuc Robocut) model Alpha 0 C. Afterwards, both samples were coated with white and black paints to create a stochastic black and white spot pattern as required for the digital image correlation (DIC).

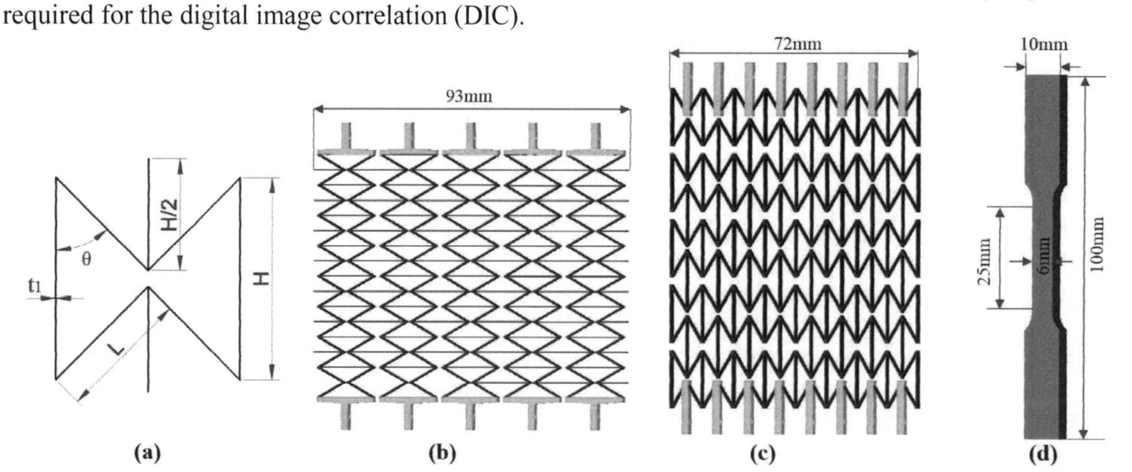

Figure 1. (a) A sketch of a unit cell of re-entrant honeycomb; (b) A re-entrant honeycomb under X-loading, (c) A re-entrant honeycomb under Y-loading, (d) A dog-bone sample used in tensile test to determine the material properties.

Table 1. Geometrical parameters of a unit re-entrant honeycomb cell.

structure	Material	H (mm)	L (mm)	θ (°)	t_1(mm)	t_2 (mm)
Re-entrant samples	AlSi12	16.8	8.4	30	0.36	7.2

3. Experimental setup

3.1 Tensile tests

In order to investigate the properties of the parent material of the re-entrant honeycomb, five tensile tests were firstly conducted on dog-bone samples and the stress-strain curves of these samples is shown in figure 2 (a). The cell wall material made of AlSi12 power is brittle, and the fracture strain is approximately 0.03. To overcome the restriction at both clamped ends during the tests, small stainless steel fixtures were manufactured to hold and grip the test sample at both ends, as in figure 2 (b). The spherical shape ends of these fixtures and lubricant used to reduce friction between contact surfaces. Tensile tests of auxetic samples were conducted using a Zwick Roell machine (Zwick/Z010) with a load cell of 1 kN. One sample was tensioned in the X direction figure 2 (c) and the other sample was tensioned in the Y direction figure 3 (a). The cross-head speed was 3 mm/min for both tests until fracture occurred.

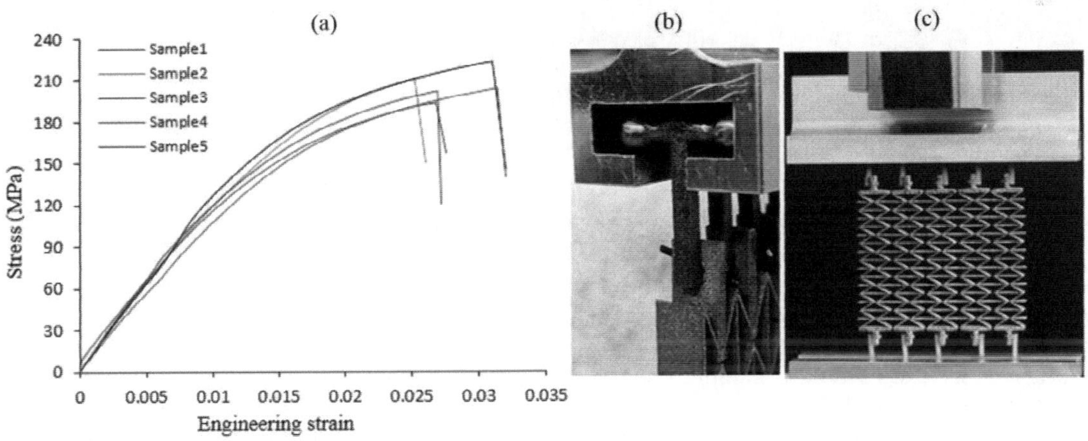

Figure 2. (a) Engineering stress-strain curves of the AlSi12 material, (b) Details of the fixtures holding one end of the sample, (c) Tensile test setup of a re-entrant honeycomb under X-loading.

3.2 Measurement of Poisson's ratio of re-entrant honeycomb and tensile stress

The history of Poisson's ratio of each sample was of primary interest in the current study. For the measurement of Poisson's ratio, a Digital VIC-2D Image Correlation System equipped with a 5-megapixel camera was used to capture the deformation images at one frame per second and calculate the axial and lateral strains of samples. Poisson's ratio, were calculated by the following formula:

$$\nu = -\varepsilon_L/\varepsilon_A \quad (1)$$

where, ν, ε_A and ε_L are Poisson's ratio, axial strain and lateral strain respectively.

4. Results and discussions

4.1 Tensile properties

Figure 3 (a) and (b) show the deformation of re-entrant honeycomb samples. In case of X-loading, the whole structure deforms by bending of incline struts figure 3 (a); while it deforms by bending and stretching of incline and vertical struts in case of Y-loading figure (b). The morphological figures also indicate that the failure of samples are catastrophic resulting from sudden fracture of incline walls. The experimental force-displacement curves of re-entrant honeycomb samples under both loading are shown in figure 3 (c). The force of the sample stretched in the X direction is lower than that loaded in the Y direction. Meanwhile, since the samples made of AlSi12 power does not display great plasticity, as indicated in figure 2(a), the auxetic honeycombs fracture at relatively small displacements (11.4 mm for X-loading and 4 mm for Y-loading). Therefore, the lack of plasticity causes the absence of plateau stage on the force-displacement curves of AlSi12 re-entrant honeycombs under loading in both axes. The comparisons imply that the loading direction plays a critical role in the tensile behaviour of the re-entrant honeycombs.

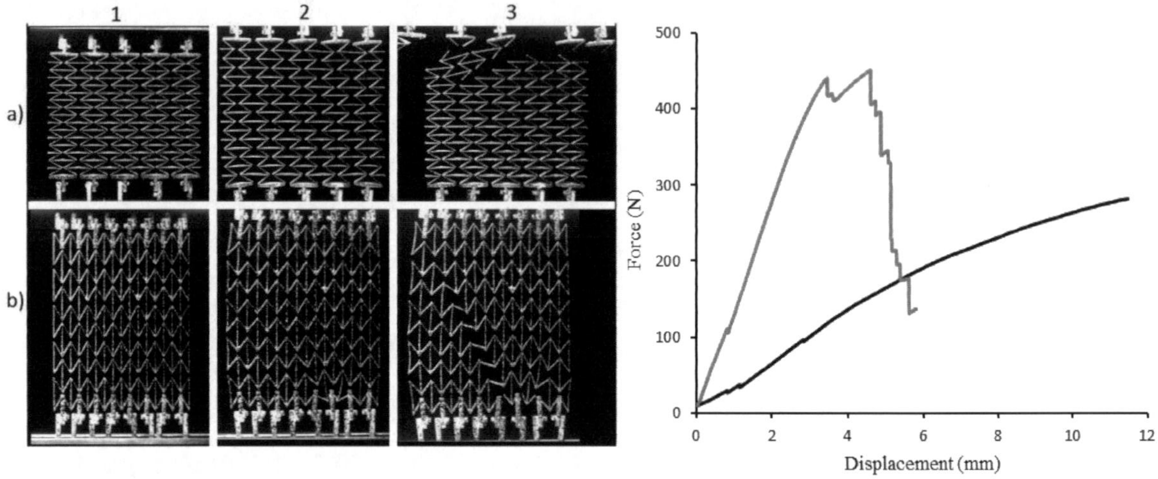

Figure 3. Deformation patterns of re-entrant honeycomb samples under (a) X-loading and (b) Y-loading (1- unreformed samples, 2- maximum displacement, 3- fracture), (c) Experimental force-displacement curves of re-entrant honeycomb samples.

4.2 Poisson's ratios

Digital Image Correlation (DIC) System was used to calculate the lateral and axial strains of the traced points of re-entrant honeycomb samples. Poisson's ratios were then calculated using Eq. (1) and plotted in figure 4. For both loading cases, Poisson's ratio of the re-entrant honeycomb remained negative. In the case of X-loading, the magnitude of Poisson's ratio increased monotonically. Conversely, in the case of Y-loading, the magnitude of Poisson's ratio decreased monotonically. Poisson's ratio of the re-entrant honeycombs is about 10 times as the former case. Under Y-loading, the Poisson's ratio of the re-entrant honeycomb varies from approximately -3.4 to -2.8 at the displacement of 0.7 mm and then remains almost constant till 2.6 mm (Figure 4b). Afterwards, a sudden decrease is observed, might be due to friction. Although lubricant was used, the lateral movement of the small fixtures was not fully free. This is required to be improved in the future tests.

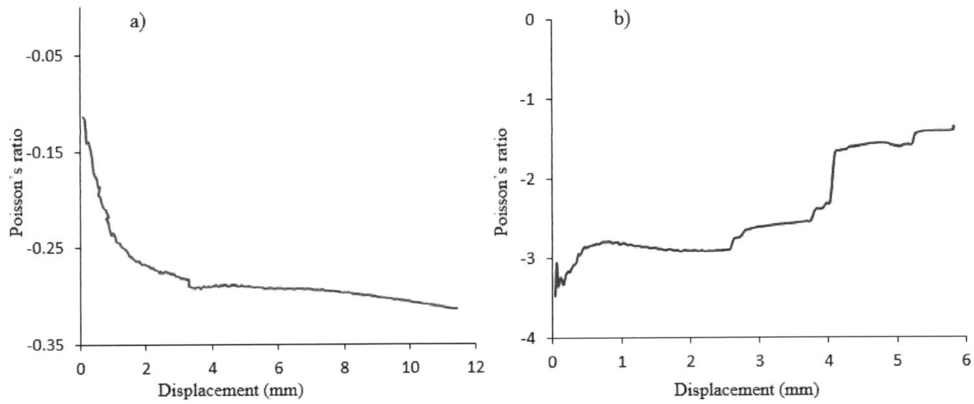

Figure 4. Experimental Poisson's ratios of re-entrant honeycomb under: (a) X-loading, (b) Y-loading.

5. Conclusions

AlSi12 powder was employed to manufacture two re-entrant honeycomb samples with identical geometrical parameters using SLM additive manufacturing technology. Standard tests were performed on dog-bone AlSi12 samples and results indicated the brittleness of this material. Quasi-static tensile tests were carried out on the auxetic honeycombs in the X and Y directions respectively. It was found that the loading direction has a considerable effect on the mechanical and auxetic behaviours of the re-entrant honeycombs. The auxetic samples fractured after a small displacement due to brittle cell wall material. The sample loaded in the Y direction withstood higher tensile force and underwent smaller deformation before fracture than that under X- direction. The magnitude of Poisson's ratios remained negative and increased (X-loading) or decreased (Y-loading) as further tension.

References

[1] Alderson A and Alderson K L 2007 Auxetic Materials *Proceedings of the Institution of* Mechanical Engineers Part G-Journal of Aerospace Engineering. **221(G4)** 565-575

[2] Strek T and Jopek H 2012 Effective mechanical properties of concentric cylindrical composites with auxetic phase *physica status solidi (b)* **249(7)** 1359-65

[3] Javadi A, Faramarzi A and Farmani R 2012 Design and optimization of microstructure of auxetic materials *Engineering Computations*, **29** 260-276.

[4] Lakes R and Elms K 1993 Indentability of Conventional and Negative Poisson's Ratio Foams *Journal of Composite Materials* **27** 1193-1202

[5] Alderson A 1999 A triumph of lateral thought *Chemistry & Industry* 384-391

[6] Wan H, Ohtaki H, Kotosaka S and Hu G 2004 A study of negative Poisson's ratios in auxetic honeycombs based on a large deflection model *European Journal of Mechanics - A/Solids* **23** 95-106

[7] Scarpa F 2008 Auxetic materials for bioprostheses [In the Spotlight] *IEEE Signal Processing Magazine* **25(5)** 128-126

[8] Muslija A and Díaz Lantada A 2014 Deep reactive ion etching of auxetic structures: present capabilities and challenges *Smart Materials and Structures.* **23** 087001

[9] Lakes R 1987 Foam Structures with a Negative Poisson's Ratio Science **235**1038-40

[10] Yang, L 2011 Structural Design Optimization and Application of 3D Re-entrant Auxetic Structures *PhD Thesis, NC State Industrial Engineering Department.* Raleigh

[11] Li Y, Harrysson O, West H, Cormier D, 2012 Compressive properties of Ti-6Al-4V auxetic mesh structures made by electron beam melting, *Acta Mater.* **60 (8)** 3370–3379.

[12] Li S, Hassanin H, Attallah MM, Adkins NJE, Essa K 2016 The development of TiNi based negative Poisson's ratio structure using selective laser melting, *Acta Mater.* **105** 75–83.

IOP Conf. Series: Materials Science and Engineering **229** (2017) 012039 doi:10.1088/1757-899X/229/1/012039

Study on Inverse Method of Soil Elastic Modulus Based on Multi-material ALE Fluid-structure Interaction

Wang Yang[1,2], Xinkai Dang[1] , Jian Yang[1,2,a] and Jianlin Mo[3]

[1]College of Mechanical Engineering, Guangxi University, Nanning, 530004, China
[2]Collaborative Innovation Center for Sugar Industry of Guangxi, Nanning, 530004, China
[3]Guangxi Research Institute of Agricultural Machinery, Nanning 530001, China

[a]gxuyangjian@163.com

Abstract. Because sampling the surface soil appears loose and fracture of soil easily, the elastic modulus of the soil in the tillage layer was difficulty in measuring directly by physical tests. The dynamic simulation model of the cone-air-soil system was established. The multi-material ALE (Arbitrary Lagrangian-Eulerian) fluid-structure interaction method was used in the model. Based on the model, the elastic modulus of the soil was reversed. Meanwhile, the reverse elastic modulus was physically verified by the triaxial test. The results show that the inverse method is feasible. The error is 0.65% compared with the soil elastic modulus obtained by triaxial test. The numerical simulation model of cone-air-soil system has higher precision.

1. Introduction

Soil elastic modulus is not only the important physical and mechanical parameter of soil, but also the basis material parameter of numerical simulation of soil system. The domestic and foreign scholars have studied the elastic modulus of soil material parameter. Yang Guang et al. measured accurately the elastic modulus of clay when they modified side pressure gauge on the soil elastic modulus test and in strict control of the loading speed and rebound times and many other factors[1]. Wang Jianfeng et al. [2] calculated the elastic modulus of clay when applying DCP test method and by measuring the value of cone penetration (DCPI). Mao Xingjun et al. tested the elastic modulus of artificial frozen soil by uniaxial test, and proposed a gray theoretical prediction method for uniaxial elastic modulus of artificial frozen soil[3]. Tian Jia et al. who defined the soil as the ideal elastoplastic material obtained the elastic modulus of soil by triaxial compression test when the soil cohesion and friction angle were obtained by direct shear test[4]. At present, the research mainly focuses on the study of the elastic modulus of soil with deeper soil by physical experiment. However, there are some differences in the nature of the soil at different depths of the tillage layer, and the nature of the surface soil is complicated. At the same time, the soil of the surface layer of the tillage layer is easy to loose and broken, resulting in difficulties in sampling, so that the soil elastic modulus of the soil layer of the tillage layer is difficult to be measured directly by physical test. The method of parameter reversal is an effective method to solve the problem that the physical parameters of materials are difficult to obtain, and there are more applied research[5,6].

In order to avoid the difficulty of sampling in the soil layer of the tillage layer, the indoor remolded soil was chosen as the test object. Firstly, physical experiments were used to determine the basic

Content from this work may be used under the terms of the Creative Commons Attribution 3.0 licence. Any further distribution of this work must maintain attribution to the author(s) and the title of the work, journal citation and DOI.
Published under licence by IOP Publishing Ltd

IOP Conf. Series: Materials Science and Engineering **229** (2017) 012039 doi:10.1088/1757-899X/229/1/012039

physical and mechanical parameters of soil. Then the kinetic numerical simulation model of cone-air-soil system was constructed to reverse the soil elastic modulus.

2. Reverse Method of Soil Elastic Modulus

The TYD-1 soil hardness tester was tested by the WDW3100 microcomputer-controlled electronic universal machine before the soil elastic modulus was reversed, as shown in Figure 1. Because the test machine can show the displacement of the soil hardness tester during the test and the value of the force received. In other words, it can show the penetration depth(h) of the soil hardness tester and the value of the resistance(F). Moreover, in the course of the test, the soil hardness meter indicates that the hardness value(D) can be displayed. Therefore, soil hardness value(D), soil penetration tester penetration depth(h) and penetration resistance(F) will be related. With continuous compression test 20 times, through the regression analysis of the test data, the mathematical model of the depth(h) and the hardness value(D) and the cone pressure(F) of the taper cone inserted into the soil was established respectively[7], as shown in equation (1), equation (2).

$$D = 2103.9e^{-0.14h} \tag{1}$$

$$F = -3.21h + 162.96 \tag{2}$$

Where, D is called soil hardness; h is called soil penetration tester penetration depth; F is called penetration resistance.

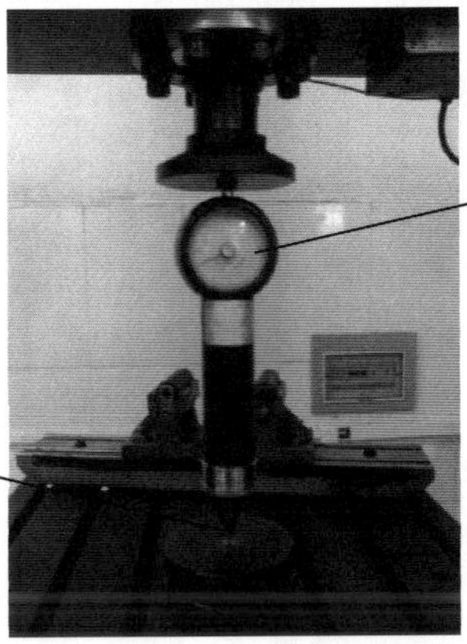

Figure 1. Compression test of hardness tester

The soil hardness(D) was obtained by using the TYD-1 target soil hardness tester. Based on the measured values and the regression results of the above hardness test, the actual penetration depth(h) and the penetration resistance(F) of the hardness tester were obtained. The soil parameters obtained from the physical test include the moisture content(n), density(ρ), cohesion(C), internal friction angle(φ), and the shear modulus(G) and bulk modulus(K) obtained from the initial set modulus of elastic modulus(E) as the material parameters. Then, based on the multi-material ALE fluid-structure interaction method, the dynamic simulation model of the cone-air-soil system was established. By controlling the single variable and adjusting constantly the initial modulus of elasticity, the elastic modulus is the inverse elastic modulus when the penetration depth and penetration resistance of the cone in the simulation test are close to the actual value and the error is within ± 2%. According to reference [8], the relationship between soil bulk modulus, shear modulus and elastic modulus is shown in equation (3) and (4).

$$K = \frac{E}{3(1-2v)} \tag{3}$$

$$G = \frac{E}{2(1+v)} \tag{4}$$

Where, K is bulk modulus; G is shear modulus; E is elastic modulus; v is Poisson's ratio. The Poisson's ratio of the soil is generally 0.25-0.37, where v is 0.3.

3. Measurement of basic parameters of soil

3.1. Determination methods and equipment

In order to compare the elastic modulus obtained by the reverse method and the triaxial test respectively and to avoid the sampling difficulty of the soil in the tillage layer, the remodeled soil was selected and a total of 6 soils were tested. According to the physical test method in the geotechnical test procedure[9], the basic parameters of 6 soils were measured, including hardness(D), water content(n), density(ρ), cohesion(C) and internal friction angle(φ). The equipment used for the determination of these parameters are electrothermal thermostatic blast oven, electronic scale, aluminum box, standard ring knife (volume V = 60cm^3), strain controlled direct shear apparatus and sclerometer.

3.2. Determination results

Through the measurement and analysis, the hardness(D), water content(n), density(ρ), cohesion(C) and internal friction angle(φ) of 6 soils were obtained, and the average value was obtained, as shown in Table 1.

Table 1. Soil material parameter

	Soil hardness (N·cm^{-2})	Water content (%)	Density (g·cm^{-3})	Cohesion (KPa)	Internal friction angle (°)
mean	26	25.57	1.9473	13.843	3.434

The soil hardness(D) is substituted into the equation (1) and (2), and the actual penetration depth (h) and penetration resistance(F) of the soil hardness tester are obtained, as shown in Table 2.

Table 2. Results of soil hardness conversion

Soil hardness (N·cm^{-2})	Actual penetration depth (mm)	Actual penetration resistance (N)
26	31.38	62.23

4. Simulation model of cone-air-soil system dynamics

That the multi-material ALE (Arbitrary Lagrangian-Eulerian) fluid-structure interaction method is selected in the establishment of cone-air-soil system dynamics simulation model because the real soil is prone to large deformation.

The ALE method encompasses the advantages of the Lagrangian method and the Euler method. It not only utilizes the Lagrangian method to effectively track the movement of the boundary of the material structure, but also uses the Euler method to separate the internal grid from the material, so that the grid in the ALE method can be adjusted as needed during the calculation to avoid the serious distortion of the grid.

The simulation steps are as follows: First, a rectangular model is established. The model needs to be divided into two parts. The upper part is the air layer and the lower part is the soil layer. Second, a cone-head model was established above the model to construct a simulation model of ALE fluid-structure interaction for cone-air-soil system. Third, since the entire system simulation model is symmetric, so a quarter of the model has been established in order to reduce the amount of computation. Besides, the grid has been built on the model. Then, add necessary constraint at the

IOP Conf. Series: Materials Science and Engineering **229** (2017) 012039 doi:10.1088/1757-899X/229/1/012039

bottom of the soil and the symmetry of the soil. Finally, applying a downward velocity with the actual value on the cone head to simulate the process of inserting the cone into the soil.

Therefore, the geometric dimensions of the selected model should be based on the above requirements and the actual situation. Among them, the cone bottom radius is 0.01m, its height is 0.05m. The size of the soil model is 0.08m × 0.1m × 0.08m, where the air layer thickness is 0.002 m. At the time of modeling, the unit type of the cone is Thin Shell 163 housing finite element, density is 7830kg/m^3, elastic modulus is 2×10^{11}Pa, Poisson's ratio is 0.3[7]. The element type of soil and air layer is Solid 164 solid finite element element. Since the solid model is spatially symmetric, the solid model is separated by 1/4, and the boundary is treated according to reference [10]. There is a point contact between the cone and the soil.The static friction factor is 0.7, the dynamic friction factor is 0.6, and the cone insertion speed is 0.03m/s[11]. The cone-air-soil system ALE fluid-structure interaction model is shown in Figure 2.The cone-air-soil system ALE fluid-structure interaction simulation model was meshed, as shown in Figure 3. Boundary conditions of the model are also shown in Figure 4. The soil-penetrometer interaction is shown in Figure 5.

Figure 2. Dynamics simulation model of cone-air-soil system

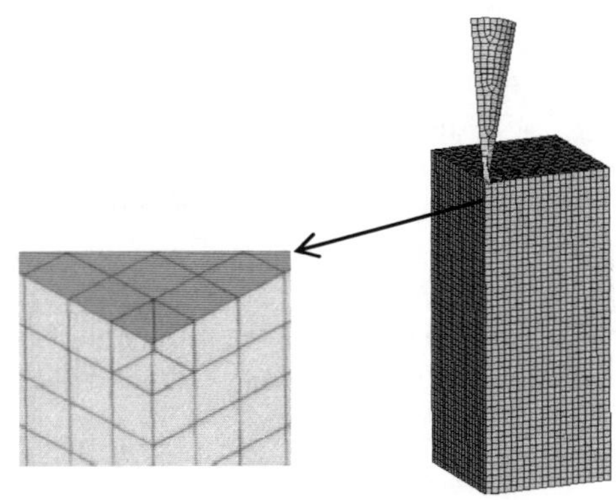

Figure 3. Finite element mesh of 1/4 cone-air-soil model

Figure 4. Soil boundary conditions

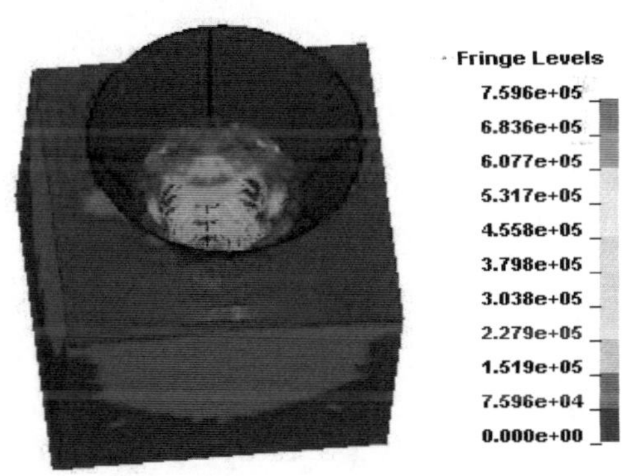

Figure 5. The soil- penetrometer interaction

5. The results and analysis of the reverse

In the cone-air-soil system dynamics simulation model, the same penetration rate is 0.03 m/s as the actual test, and the same depth is 31.38 mm as the actual test. The penetration resistance of the cone with the penetration time changes in the curve can be get when the elastic modulus E=0.46MPa after controlling the elastic modulus of a single variable and several tests, as shown in Figure 6. As can be seen from Figure 6, the penetration depth is 31.38 mm when the penetration time t=1.044s, which is the same as the actual penetration depth. And the penetration resistance of the cone is $F'_{1/4}$=15.8N, as shown in point (1.04,-15.8). In the simulation test, the penetration of the entire cone received resistance F'=63.2N because the simulation model was selected for a quarter of the model before. At the same time, the actual resistance of the penetration resistance F is 62.23N, the error within ± 2%, so at this time, the elastic modulus in the simulation model is the reverse elastic modulus.

The soil elastic modulus(E), the corresponding bulk modulus(K) and shear modulus(G) obtained by the inverse method are shown in Table 3.

Table 3. Reverse results of elasticity modulus

Elastic modulus (Mpa)	Bulk modulus (Mpa)	Shear modulus (Mpa)	relative error between $F'$and F{%}
0.46	0.3833	0.1769	1.5

6. Verification and Analysis of Soil Elastic Modulus

In this paper, the soil elastic modulus of the above method is verified by indoor triaxial test. The triaxial test is a shear test method for determining the shear strength parameters of the soil under three-way pressure conditions. With reference to the soil of the triaxial test method in the geotechnical test procedure [9], the rubber film will be smoothly wrapped in the soil sample into the pressure chamber. Under the confining pressure of 50 MPa, the axial pressure was applied to shear the cylindrical soil samples until the soil samples were destroyed. Then, according to the relationship between soil stress and strain, the elastic modulus of soil samples is obtained.

According to the mechanics of materials, the relationship between soil stress and strain is shown in equation (5). Therefore, the elastic modulus of soil is obtained because the relationship between soil stress and strain is shown in Figure 7.The elastic modulus of soil E is 0.457MPa.

$$E = \frac{\Delta\sigma}{\Delta\varepsilon} \tag{5}$$

Where, E is elastic modulus of soil; $\Delta\sigma$ is soil stress; $\Delta\varepsilon$ is soil strain.

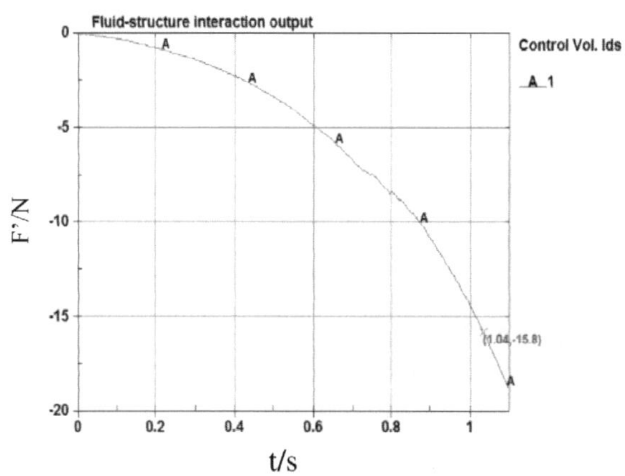

Figure 6. Relationship between penetration resistance and penetration time

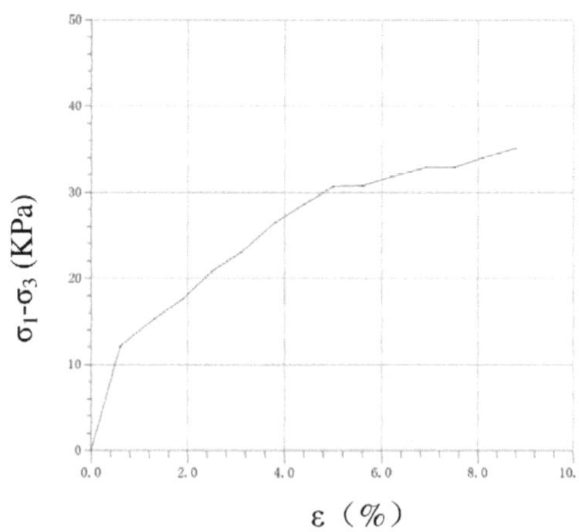

Figure 7. Stress strain relationship of soil samp

The error of elastic modulus of the soil obtained by the triaxial test is 0.65% (within ± 2%) compared with the elastic modulus obtained by the inverse method. It is proved that the soil elastic modulus obtained by the inverse method is correct and feasible.

7. Summary

The dynamic simulation model of the cone-air-soil system was established. The multi-material ALE (Arbitrary Lagrangian-Eulerian) fluid-structure interaction method was used in the model. Based on the model, the elastic modulus of the soil was reversed. Meanwhile, the reverse elastic modulus was physically verified by the triaxial test. The results show that the inverse method is feasible. The error is 0.65% compared with the soil elastic modulus obtained by triaxial test. The numerical simulation model of cone-air-soil system has higher precision.

Acknowledgments
This work was supported by a grant from the National Natural Science Foundation of China (Grant No.51565003), Guangxi Natural Science Foundation (Grant No.2014GXNSFBA118279), College Scientific Research Foundation of Guangxi Education Department (Grant No.YB2014002), Young Doctoral Research Foundation of Guangxi University (Grant No.XBZ160111).

References

[1] Yang G. and Tian K.L. (2006). Study of Soil Elastic Modular Test Method.*Urban Roads Bridges & Flood Control*, vol **5**, pp. 145-147+208.
[2] Wang J.F., Wang Z., and Deng W.D.(2007). Simple Method to Evaluate Resilient Modulus of Subgrade Cohesive Soils. *Journal of Wuhan University of Technology(Transportation Science& Engineering)*, vol **5**, pp. 800-803.
[3] Mao X.J.(2014). Grey prediction of uniaxial elastic modulus of artificial frozen soil. *Sichuan Building Materials*, vol **3**, pp. 95-97.
[4] Tian J., Cao B. and Ji J.N. (2015). Numerical simulation and validation test of direct shear test for root-soil composite of Hedysarum scoparium using finite element method. *Transactions of the Chinese Society of Agricultural Engineering*, vol **16**, pp. 152-158.
[5] Cooreman S, Lecompte D and Sol H,et al(2007). Elasto-plastic material parameter identification by inverse methods: Calculation of the sensitivity matrix. *International Journal of Solids and Structures*, vol **44**, pp. 4329-4341.
[6] Molimard J, Riche R L and Vautrin A, et al(2005). Identification of the four orthotropic plate stiffnesses using a single open-hole tensile test. *Experimental Mechanics*, vol **45**, pp. 404-411.
[7] Yang W., Zhang S. and Chen K.Y.(2016). Dynamics Simulation Model of Cassava Tuber Lifting System Based on Soil Layering. *Research on Agricultural Mechanization*, vol **8**, pp. 51-55.
[8] Zhang S. *Study on dynamic simulation model of cassava root pulling system based on soil layer*, Nanning: Guangxi University，2014.
[9] *Ministry of water resources and electric power of People's Republic of China, Geotechnical test code SD128-84 First fascicle*. Beijing:Water conservancy and electric power press,1987.(Book)
[10] Yang Wang,Li Juanjuan and Yang Jian,et al(2015). Numerical simulation of an experienced farmer lifting tubers of cassava for designing a bionic harvester. *CMES: Computer Modeling in Engineering Sciences*, vol **104**, pp. 471-491.
[11] Li Y.J.,Lin J.H. and Xu Y.(2011). Simulation of penetration test of cone index instrument by discrete element method. *Journal of agricultural machinery*, vol **11**, pp. 44-48.

Institute of Physics
Dirac House, Temple Back
Bristol BS1 6BE UK

ISSN: 1757-8981
ISBN 978-1-5108-4868-9
9 781510 848689

CURRAN ASSOCIATES INC.
proceedings.com

Bicycle Transportation 2018

Transportation Research Record: Journal of the Transportation Research Board Volume 2672, Issue 36

SAGE Publishing